INVITATION TO DISCRETE MATHEMATICS

"Only mathematicians could appreciate this work . . ."

Illustration by G. Roux from the Czech edition of *Sans dessus dessous* by Jules Verne, published by J. R. Vilímek, Prague, 1931 (English title: *The purchase of the North Pole*).

Invitation to
Discrete Mathematics

JIŘÍ MATOUŠEK

and

JAROSLAV NEŠETŘIL

Department of Applied Mathematics
Charles University, Prague

CLARENDON PRESS · OXFORD
1998

Oxford University Press, Great Clarendon Street, Oxford OX2 6DP

Oxford New York

Athens Auckland Bangkok Bogota Buenos Aires Calcutta
Cape Town Chennai Dar es Salaam Delhi Florence Hong Kong Istanbul
Karachi Kuala Lumpur Madrid Melbourne Mexico City Mumbai
Nairobi Paris São Paolo Singapore Taipei Tokyo Toronto Warsaw

and associated companies in
Berlin Ibadan

Oxford is a registered trade mark of Oxford University Press

Published in the United States by
Oxford University Press Inc., New York

© Jiří Matoušek and Jaroslav Nešetřil, 1998

A catalogue record for this book is available from the British Library

Library of Congress Cataloging in Publication Data
(Data available)

ISBN 0–19–850208 7 (Hbk)
0–19–850207 9 (Pbk)

Typeset by the authors

Printed in Great Britain by
Bookcraft Ltd Midsomer Norton, Avon

Preface

Why should an introductory textbook on discrete mathematics have such a long preface, and what do we want to say in it? There are many ways of presenting discrete mathematics, and first we list some of the guidelines we tried to follow in our writing; the reader may judge later how we succeeded. Then we add some more technical remarks concerning a possible course based on the book, the exercises, the existing literature, and so on.

So, here are some features which may perhaps distinguish this book from some others with a similar title and subject:

- *Developing mathematical thinking.* Our primary aim, besides teaching some factual knowledge, and perhaps more importantly than that, is to lead the student to understand and appreciate mathematical notions, definitions, and proofs, to solve problems requiring more than just standard recipes, and to express mathematical thoughts precisely and rigorously. Mathematical habits may give great advantages in many human activities, say in programming or in designing complicated systems.[1] It seems that many private (and well-paying) companies are aware of this. They are not really interested in whether you know mathematical induction by heart, but they may be interested in whether you have been trained to think about and absorb complicated concepts quickly—and mathematical theorems seem to provide a good workout for such a training. The choice of specific material for this preparation is probably not essential—if you're enchanted by algebra, we certainly won't try to convert you to combinatorics! But we believe that discrete mathematics is especially suitable for such a first immersion into mathematics, since the initial problems and notions are more elementary than in analysis, for instance, which starts with quite deep ideas at the outset.

---

[1] On the other hand, one should keep in mind that in many other human activities, mathematical habits should better be suppressed.

- *Methods, techniques, principles.* In contemporary university curricula, discrete mathematics usually means the mathematics of finite sets, often including diverse topics like logic, finite automata, linear programming, or computer architecture. Our text has a narrower scope; the book is essentially an introduction to combinatorics and graph theory, and even there, everyone familiar with these areas may miss some favorite topics (such as matchings and network flows, Ramsey theory, or NP-completeness, to name just a few). We concentrate on relatively few basic methods and principles, aiming to display the rich variety of mathematical techniques even at this basic level, and the choice of material is subordinated to this.

- *Joy.* The book is written for a reader who, every now and then, enjoys mathematics, and our boldest hope is that our text might help some readers to develop some positive feelings towards mathematics that might have remained latent so far. In our opinion, this is a key prerequisite: an aesthetic pleasure from an elegant mathematical idea, sometimes mixed with a triumphant feeling when the idea was difficult to understand or to discover. Not all people seem to have this gift, just as not everyone can enjoy music, but without it, we imagine, studying mathematics could be a most boring thing.

- *All cards on the table.* We try to present arguments in full and to be mathematically honest with the reader. When we say that something is easy to see, we really mean it, and if the reader can't see it then something is probably wrong—we may have misjudged the situation, but it may also indicate a reader's problem in following and understanding the preceding text. Whenever possible, we make everything self-contained (sometimes we indicate proofs of auxiliary results in exercises with hints), and if a proof of some result cannot be presented rigorously and in full (as is the case for some results about planar graphs, say), we emphasize this and indicate the steps that aren't fully justified.

- *CS.* A large number of discrete mathematics students nowadays are those specializing in computer science. Still, we believe that even people who know nothing about computers and computing, or find these subjects repulsive, should have free access to discrete mathematics knowledge, so we have intentionally avoided overburdening the text with computer science terminology and examples. However, we have not forgotten computer scientists and have included several passages on efficient algorithms and their analysis plus a number of exercises concerning algorithms (see below).

- *Other voices, other rooms.* In the material covered, there are several opportunities to demonstrate concepts from other branches of mathematics in action, and while we intentionally restrict the factual scope of the book, we want to emphasize these connections. Our experience tells us that students like such applications, provided that they are done thoroughly enough and not just by hand-waving.

Prerequisites and readership. In most of the book, we do not assume much previous mathematical knowledge beyond a standard high-school course. Several more abstract notions that are very common in all mathematics but go beyond the usual high-school level are explained in the first chapter. In several places, we need some concepts from undergraduate-level algebra, and these are summarized in an appendix. There are also a few excursions into calculus (encountering notions such as limit, derivative, continuity, and so on), but we believe that a basic calculus knowledge should be generally available to almost any student taking a course related to our book.

The readership can include early undergraduate students of mathematics or computer science with a standard mathematical preparation from high school (as is usual in most of Europe, say), and more senior undergraduate or early graduate students (in the United States, for instance). Also nonspecialist graduates, such as biologists or chemists, might find the text a useful source. For mathematically more advanced readers, the book could serve as a fast introduction to combinatorics.

Teaching it. This book is based on an undergraduate course we have been teaching for a long time to students of mathematics and computer science at the Charles University in Prague. The second author also taught parts of it at the University of Chicago, at the University of Bonn, and at Simon Fraser University in Vancouver. Our one-semester course in Prague (13 weeks, with one 90-minute lecture and one 90-minute tutorial per week) typically included material from Chapters 1–8, with many sections covered only partially and some others omitted (such as 2.6, 3.5, 3.6, 4.5, 7.3–7.5, 8.2).

While the book sometimes proves one result in several ways, we only presented one proof in a lecture, and alternative proofs were occasionally explained in the tutorials. Sometimes we inserted two lectures on generating functions (10.1–10.3) or a lecture on the cycle space of a graph (11.4).

To our basic course outline, we have added a lot of additional (and sometimes more advanced) material in the book, hoping that the reader might also read a few other things besides the sections that are necessary for an exam. Some chapters, too, can serve as introductions to more specialized courses (on the probabilistic method or on the linear algebra method). The last three chapters and the last section in the chapter on counting spanning trees are of this type.

> This type of smaller print is used for "second-level" material, namely things which we consider interesting enough to include but less essential. These are additional clarifications, comments, and examples, sometimes on a more advanced level than the basic text. The main text should mostly make sense even if this smaller-sized text is skipped.
>
> We also tried to sneak a lot of further related information into the exercises. So even those who don't intend to solve the exercises may want to read them.

On exercises. At the end of most of the sections, the reader will find a smaller or larger collection of exercises. Some of them are only loosely related to the theme covered and are included for fun and for general mathematical education. Solving at least some exercises is an essential part of studying this book, although we know that the pace of modern life and human nature hardly allow the reader to invest the time and effort to solve the majority of the 451 exercises offered (although this might ultimately be the fastest way to master the material covered).

Mostly we haven't included completely routine exercises requiring only an application of some given "recipe", such as "Apply the algorithm just explained to this specific graph". We believe that most readers can check their understanding by themselves.

We classify the exercises into three groups of difficulty (no star, one star, and two stars). We imagine that a good student who has understood the material of a given section should be able to solve most of the no-star exercises, although not necessarily effortlessly. One-star exercises usually need some clever idea or some slightly more advanced mathematical knowledge (from calculus, say), and finally two-star exercises probably require quite a bright idea. Almost all the exercises have short solutions; as far as we know, long and tedious computations can always be avoided. Our classification of difficulty is subjective, and an exercise which looks easy to some may be insurmountable for others. So if you can't solve some no-star exercises don't get desperate.

Some of the exercises are also marked by ^{CS}, a shorthand for computer science. These are usually problems in the design of efficient algorithms, sometimes requiring an elementary knowledge of data structures. The designed algorithms can also be programmed and tested, thus providing material for an advanced programming course. Some of the ^{CS} exercises with stars may serve (and have served) as project suggestions, since they usually require a combination of a certain mathematical ingenuity, algorithmic tricks, and programming skills.

Hints to many of the exercises are given in a separate chapter of the book. They are really hints, not complete solutions, and although looking up a hint spoils the pleasure of solving a problem, writing down a detailed and complete solution might still be quite challenging for many students.

On the literature. In the citations, we do not refer to all sources of the ideas and results collected in this book. Here we would like to emphasize, and recommend, one of the sources, namely a large collection of solved combinatorial problems by Lovász [7]. This book is excellent for an advanced study of combinatorics, and also as an encyclopedia of many results and methods. It seems impossible to ignore when writing a new book on combinatorics, and, for example, a significant number of our more difficult exercises are selected from, or inspired by, Lovász' (less advanced) problems. Biggs [1] is a nice introductory textbook with a somewhat different scope to ours. Slightly more advanced ones (suitable as a continuation of our text, say) are by Van Lint and Wilson [6] and Cameron [3]. The beautiful introductory text in graph theory by Bollobás [2] was probably written with somewhat similar goals as our own book, but it proceeds at a less leisurely pace and covers much more on graphs. A very recent textbook on graph theory at graduate level is by Diestel [4]. The art of combinatorial counting and asymptotic analysis is wonderfully explained in a popular book by Graham, Knuth, and Patashnik [5] (and also in Knuth's monograph [38]). If you're looking for something specific in combinatorics and don't know where to start, we suggest the *Handbook of Combinatorics* [35]. Other recommendations to the literature are scattered throughout the text. The number of textbooks in discrete mathematics is vast, and we only mention some of our favorite titles.

On the index. For most of the mathematical terms, especially those of general significance (such as relation, graph), the index only refers to their definition. Mathematical symbols composed of Latin letters (such as C_n) are placed at the beginning of the appropriate letter's section. Notation including special symbols (such as $X \setminus Y$, $G \cong H$) and Greek letters are listed at the beginning of the index.

Acknowledgments. A preliminary Czech version of this book was developed gradually during our teaching in Prague. We thank our colleagues in the Department of Applied Mathematics of the Charles University, our teaching assistants, and our students for a stimulating environment and helpful comments on the text and exercises. In particular, Pavel Socha, Eva Matoušková, Tomáš Holan, and Robert Babilon discovered a number of errors in the Czech version. Martin Klazar and Jiří Otta compiled a list of a few dozen problems and exercises; this list was a starting point of our collection of exercises. Our colleague Jan Kratochvíl provided invaluable remarks based on his experience in teaching the same course. We thank Tomáš Kaiser for substantial help in translating one chapter into English. Adam Dingle and Tim Childers helped us with some comments on the English at early stages of the translation. Jan Nekovář was so kind as to leave the peaks of number theory for a moment and provide pointers to a suitable proof of Fact 10.7.1.

Several people read parts of the English version at various stages and provided insights that would probably never have occurred to us. Special thanks go to Jeff Stopple for visiting us in Prague, carefully reading the whole manuscript, and sharing some of his teaching wisdom with us. We are much indebted to Mari Inaba and Helena Nešetřilová for comments that were very useful and different from those made by most of other people. Also opinions in several reports obtained by Oxford University Press from anonymous referees were truly helpful. Most of the finishing and polishing work on the book was done by the first author during a visit to the ETH Zürich. Emo Welzl and the members of his group provided a very pleasant and friendly environment, even after they were each asked to read through a chapter, and so the help of Hans-Martin Will, Beat Trachsler, Bernhard von Stengel, Lutz Kettner, Joachim Giesen, Bernd Gärtner, Johannes Blömer, and Artur Andrzejak is gratefully acknowledged. We also thank Hee-Kap Ahn for reading a chapter.

Next, we would like to thank Karel Horák for several expert suggestions helping the first author in his struggle with the layout of the book (unfortunately, the times when books used to be typeset by professional typographers seem to be over), and Jana Chlebíková for a long list of minor typographic corrections.

Almost all the figures were drawn by the first author using the graphic editor Ipe 5.0. In the name of humankind, we thank Otfried Cheong (formerly Schwarzkopf) for its creation.

Finally, we should not forget to mention that Sönke Adlung has been extremely nice to us and very helpful during the editorial process, and that it was a pleasure to work with Julia Tompson in the final stages of the book preparation.

A final appeal. A long mathematical text usually contains a substantial number of mistakes. Compared to various preliminary versions of this book, we have corrected a large number of them, but certainly some still remain. So we plead with readers who discover errors, bad formulations, wrong hints to exercises, etc., to let us know about them.[2]

Prague
February 1998

J. M.
J. N.

---

[2]Please send us e-mails concerning this book to `idm@kam.ms.mff.cuni.cz`. Hopefully, readers will understand that we may not be able to answer all correspondence in this matter. If there is substantial feedback from readers, we will set up an Internet home page concerning the book under `http://www.ms.mff.cuni.cz/acad/kam/idm`.

Contents

1 Introduction and basic concepts   1
   1.1 An assortment of problems   2
   1.2 Numbers and sets: notation   7
   1.3 Mathematical induction and other proofs   16
   1.4 Functions   25
   1.5 Relations   32
   1.6 Equivalences   36
   1.7 Ordered sets   40

2 Combinatorial counting   47
   2.1 Functions and subsets   47
   2.2 Permutations and factorials   52
   2.3 Binomial coefficients   55
   2.4 Estimates: an introduction   66
   2.5 Estimates: the factorial function   73
   2.6 Estimates: binomial coefficients   81
   2.7 Inclusion–exclusion principle   86
   2.8 The hatcheck lady & co.   91

3 Graphs: an introduction   97
   3.1 The notion of a graph; isomorphism   97
   3.2 Subgraphs, components, adjacency matrix   106
   3.3 Graph score   112
   3.4 Eulerian graphs   117
   3.5 An algorithm for an Eulerian tour   123
   3.6 Eulerian directed graphs   127
   3.7 2-connectivity   132

4 Trees   138
   4.1 Definition and characterizations of trees   138
   4.2 Isomorphism of trees   144
   4.3 Spanning trees of a graph   151
   4.4 The minimum spanning tree problem   155
   4.5 Jarník's algorithm and Borůvka's algorithm   161

5 Drawing graphs in the plane   167
  5.1 Drawing in the plane and on other surfaces   167
  5.2 Cycles in planar graphs   174
  5.3 Euler's formula   181
  5.4 Coloring maps: the four-color problem   191

6 Double-counting   202
  6.1 Parity arguments   202
  6.2 Sperner's theorem on independent systems   211
  6.3 A result in extremal graph theory   218

7 The number of spanning trees   223
  7.1 The result   223
  7.2 A proof via score   224
  7.3 A proof with vertebrates   226
  7.4 A proof using the Prüfer code   229
  7.5 A proof working with determinants   231

8 Finite projective planes   240
  8.1 Definition and basic properties   240
  8.2 Existence of finite projective planes   250
  8.3 Orthogonal Latin squares   255
  8.4 Combinatorial applications   258

9 Probability and probabilistic proofs   262
  9.1 Proofs by counting   262
  9.2 Finite probability spaces   269
  9.3 Random variables and their expectation   279
  9.4 Several applications   285

10 Generating functions   294
  10.1 Combinatorial applications of polynomials   294
  10.2 Calculation with power series   298
  10.3 Fibonacci numbers and the golden section   309
  10.4 Binary trees   317
  10.5 On rolling the dice   322
  10.6 Random walk   323
  10.7 Integer partitions   326

11 Applications of linear algebra   333
  11.1 Block designs   333
  11.2 Fisher's inequality   338

11.3 Covering by complete bipartite graphs   342

11.4 Cycle space of a graph   345

11.5 Circulations and cuts: cycle space revisited   349

11.6 Probabilistic checking   353

Appendix: Prerequisites from algebra   363

Bibliography   371

Hints to selected exercises   377

Index   399

1
Introduction and basic concepts

In this introductory chapter, we first give a sample of the problems and questions to be treated in the book. Then we explain some basic notions and techniques, mostly fundamental and simple ones common to most branches of mathematics. We assume that the reader is already familiar with many of them or has at least heard of them. Thus, we will mostly review the notions, give precise formal definitions, and point out various ways of capturing the meaning of these concepts by diagrams and pictures. A reader preferring a more detailed and thorough introduction to these concepts may refer to the book by Stewart and Tall [8], for instance.

Section 1.1 presents several problems to be studied later on in the book and some thoughts on the importance of mathematical problems and similar things.

Section 1.2 is a review of notation. It introduces some common symbols for operations with sets and numbers, such as $\cup$ for set union or $\sum$ for summation of a sequence of numbers. Most of the symbols are standard, and the reader should be able to go through this section fairly quickly, relying on the index to refresh memory later on.

In Section 1.3, we discuss mathematical induction, an important method for proving statements in discrete mathematics. Here it is sufficient to understand the basic principle; there will be many opportunities to see and practice various applications of induction in subsequent chapters. We will also say few words about mathematical proofs in general.

Section 1.4 recalls the notion of a function and defines special types of functions: injective functions, surjective functions, and bijections. These terms will be used quite frequently in the text.

Sections 1.5 through 1.7 deal with relations and with special types of relations, namely equivalences and orderings. These again belong to the truly essential phrases in the vocabulary of mathematics. However, since they are simple general concepts which we have not yet fleshed out by many interesting particular examples, some readers may find them "too abstract"—a polite phrase for "boring"—on the first reading. Such readers may want to skim through these sections and return to them later. For instance, ordered sets (Section 1.7) are only needed for a full understanding of Section 6.2 and for some exercises in this book, but they certainly should be a part of any deeper mathematical education. (When learning a new language, say, it is not very thrilling to memorize the grammatical forms of the verb "to be", but after some time you may find it difficult to speak fluently knowing only "I am" and "he is". Well, this is what we have to do in this chapter: we must review some of the *language* of mathematics.)

1.1 An assortment of problems

Let us look at some of the problems we are going to consider in this book. Here we are going to present them in a popular form, so you may well know some of them as puzzles in recreational mathematics.

A well-known problem concerns three houses and three wells. Once upon a time, three fair white houses stood in a meadow in a distant kingdom, and there were three wells nearby, their water clean and fresh. All was well, until one day a seed of hatred was sown, fights started among the three households and would not cease, and no reconciliation was in sight. The people in each house insisted that they have three pathways leading from their gate to each well, three pathways which should not cross any of their neighbors' paths. Can they ever find paths that will satisfy everyone and let peace set in?

A solution would be possible if there were only two wells:

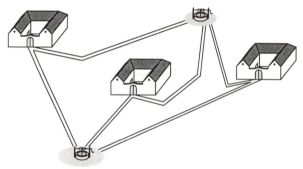

But with three wells, there is no hope (unless these proud men and women would be willing to use tunnels or bridges, which sounds quite unlikely). Can you state this as a mathematical problem and prove that it has no solution?

Essentially, this is a problem about drawing in the plane. Many other problems to be studied in this book can also be formulated in terms of drawing. Can one draw the following picture without lifting the pencil from the paper, drawing each line only once?

And what about this one?

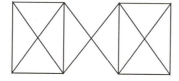

If not, why not? Is there a simple way to distinguish pictures that can be drawn in this way from those that cannot? (And, can you find nice accompanying stories to this problem and the ones below?)

For the subsequent set of problems, draw 8 dots in the plane in such a way that no 3 of them lie on a common line. (The number 8 is quite arbitrary; in general we could consider n such dots.) Connect some pairs of these points by segments, obtaining a picture like the following:

What is the maximum number of segments that can be drawn so that no triangle with vertices at the dots arises? The following picture has 13 segments:

Can you draw more segments for 8 dots with no triangle? Probably you can. But can you prove your result is already the best possible?

Next, suppose that we want to draw some segments so that any two dots can be connected by a path consisting of the drawn segments. The path is not allowed to make turns at the crossings of the segments, only at the dots, so the left picture below gives a valid solution while the right one doesn't:

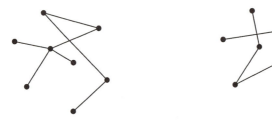

What is the minimum number of segments we must draw? How many different solutions with this minimum number of segments are there? And how can we find a solution for which the total length of all the drawn segments is the smallest possible?

All these problems are popular versions of simple basic questions in *graph theory*, which is one of main subjects of this book (treated in Chapters 3, 4, and 5). For the above problems with 8 dots in the plane, it is easily seen that the way of drawing the dots is immaterial; all that matters is which pairs of dots are connected by a segment and which are not. Most branches of graph theory deal with problems which can be pictured geometrically but in which geometry doesn't really play a role. On the other hand, the problem about wells and houses belongs to a "truly" geometric part of graph theory. It is important that the paths should be built in the plane. If the houses and wells were on a tiny planet shaped like a tire-tube then the required paths would exist:

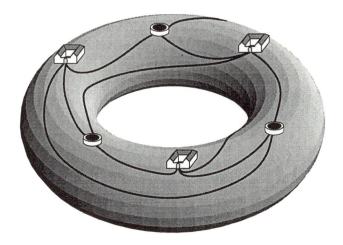

Another important theme of this book is *combinatorial counting*, treated in Chapters 2 and 10. The problems there usually begin with "How many ways are there..." or something similar. One question of this type was mentioned in our "8 dots" series (and it is a nice question—the whole of Chapter 7 is devoted to it). The reader has probably seen lots of such problems; let us add one more. How many ways are there to divide n identical coins into groups? For instance, 4 coins can be divided in 4 ways: $1 + 1 + 1 + 1$ (4 groups of 1 coin each), $1 + 1 + 2$, $1 + 3$, and 4 (all in one group, which is not really a "division" in the sense most people understand it, but what do you expect from mathematicians!). For this problem, we will not be able to give an exact formula; such a formula does exist but its derivation is far beyond the scope of this book. Nonetheless, we will at least derive estimates for the number in question. This number is a function of n, and the estimates will allow us to say "how fast" this function grows, compared to simple and well-known functions like n^2 or 2^n. Such a comparison of complicated functions to simple ones is the subject of the so-called *asymptotic analysis*, which will also be touched on below and which is important in many areas, for instance for comparing several algorithms which solve the same problem.

Although the problems presented may look like puzzles, each of them can be regarded as the starting point of a theory with numerous applications, both in mathematics and in practice.

In fact, distinguishing a good mathematical problem from a bad one is one of the most difficult things in mathematics, and the "quality" of a problem can often be judged only in hindsight, after the problem has been solved and the consequences of its solution mapped. What is a good problem? It is one whose solution will lead to new insights, methods, or even a whole new fruitful theory. Many problems in recreational mathematics are not good in this sense, although their solution may require considerable skill or ingenuity.

A pragmatically minded reader might also object that the problems shown above are useless from a practical point of view. Why take a whole course about them, a skeptic might say, when I have to learn so many practically important things to prepare for my future career? Objections of this sort are quite frequent and cannot be simply dismissed, if only because the people controlling the funding are often pragmatically minded.

One possible answer is that for each of these puzzle-like problems, we can exhibit an eminently practical problem that is its cousin. For instance, the postal delivery service in a district must deliver mail to all houses, which means passing through each street at least once. What is the shortest route to take? Can it be found in a reasonable time using a supercomputer? Or with a personal computer? In order to understand this postal delivery problem, one should be familiar with simple results about drawing pictures without lifting a pencil from the paper.

Or, given some placement of components of a circuit on a board, is it possible to interconnect them in such a way that the connections go along the surface of the board and do not cross each other? What is the most economical placement of components and connections (using the smallest area of the board, say)? Such questions are typical of VLSI design (designing computer chips and similar things). Having learned about the three-wells problem and its relatives (or, scientifically speaking, about planar graphs) it is much easier to grasp ways of designing the layout of integrated circuits.

These "practical" problems also belong to graph theory, or to a mixture of graph theory and the design of efficient algorithms. This book doesn't provide a solution to them, but in order to comprehend a solution in some other book, or even to come up with a new good solution, one should master the basic concepts first.

We would also like to stress that the most valuable mathematical research was very seldom directly motivated by practical goals. Some great mathematical ideas of the past have only found applications quite recently. Mathematics does have impressive applications (it might be easier to list those human activities where it is not applied than those where it is), but anyone trying to restrict mathematical research to the directly applicable parts would be left with a lifeless fragment with most of the creative power gone.

Exercises are unnecessary in this section. Can you solve some of the problems sketched here, or perhaps all of them? Even if you try and get only partial results or fail completely, it will still be of great help in reading further.

So what *is* this discrete mathematics they're talking about, the reader may (rightfully) ask? The adjective "discrete" here is an opposite of "continuous". Roughly speaking, objects in discrete mathematics, such as the natural numbers, are clearly separated and distinguishable from each other and we can perceive them individually (like trees in a forest which surrounds us). In contrast, for a typical "continuous" object, such as the set of all points on a line segment, the points are indiscernible (like the trees in a forest seen from a high-flying airplane). We can focus our attention on some individual points of the segment and see them clearly, but there are always many more points nearby that remain indistinguishable and form the totality of the segment.

According to this explanation, such parts of mathematics as algebra or set theory might also be considered "discrete". But in the common usage of the term, discrete mathematics is most often understood as mathematics dealing with finite sets. In many current university curricula, a course on discrete mathematics has quite a wide range, including some combinatorics, counting, graph theory, but also elements of mathematical logic, some set theory, basics from the theory of computing (finite automata, formal languages, elements of computer architecture), and other things. We prefer a more narrowly focussed scope, so perhaps a more descriptive title for this book would be "Invitation to combinatorics and graph theory", covering most of the contents. But the name of the course we have been teaching happened to be "Discrete mathematics" and we decided to stick to it.

1.2 Numbers and sets: notation

Number domains. For the set of all natural numbers, i.e. the set $\{1, 2, 3, \ldots\}$, we reserve the symbol $\mathbf{N}$. The letters n, m, k, i, j, p and possibly some others usually represent natural numbers.

Using the natural numbers, we may construct other well-known number domains: the integers, the rationals, and the reals (and also the complex numbers, but we will seldom hear about them here).

The *integer numbers* or simply *integers* arise from the natural numbers by adding the negative integer numbers and 0. The set of all integers is denoted by $\mathbf{Z}$.

The *rational numbers* are fractions with integer numerator and denominator. This set is usually denoted by $\mathbf{Q}$ but we need not introduce any symbol for it in this book. The construction of the set $\mathbf{R}$ of all *real numbers* is more complicated, and it is treated in introductory courses of mathematical analysis. Famous examples of real numbers which are not rational are numbers such as $\sqrt{2}$, some important constants like π, and generally numbers whose decimal notation has an infinite and aperiodic sequence of digits following the decimal point, such as $0.12112111211112\ldots$.

The *closed interval* from a to b on the real axis is denoted by $[a, b]$, and the *open interval* with the same endpoints is written as (a, b).

Operations with numbers. Most symbols for operations with numbers, such as $+$ for addition, $\sqrt{}$ for square root, and so on, are generally well known. We write *division* either as a fraction, or sometimes with a slash, i.e. either in the form $\frac{a}{b}$ or as a/b.

We introduce two less common functions. For a real number x, the symbol $\lfloor x \rfloor$ is called[1] the *lower integer part of x* (or the *floor function* of x), and its value is the largest integer smaller than or equal to x. Similarly $\lceil x \rceil$, the *upper integer part of x* (or the *ceiling function*), denotes the smallest integer greater than or equal to x. For instance, $\lfloor 0.999 \rfloor = 0$, $\lfloor -0.1 \rfloor = -1$, $\lceil 0.01 \rceil = 1$, $\lceil \frac{17}{3} \rceil = 6$, $\lfloor \sqrt{2} \rfloor = 1$.

Later on, we will introduce some more operations and functions for numbers, which have an important combinatorial meaning and which we will investigate in more detail. Examples are $n!$ and $\binom{n}{k}$.

Sums and products. If $a_1, a_2, \ldots, a_n$ are real numbers, their sum $a_1 + a_2 + \cdots + a_n$ can also be written using the *summation sign* $\sum$, in the form

$$\sum_{i=1}^{n} a_i.$$

This notation somewhat resembles the FOR loop in many programming languages. Here are a few more examples:

$$\sum_{j=2}^{5} \frac{1}{2j} = \frac{1}{4} + \frac{1}{6} + \frac{1}{8} + \frac{1}{10}$$

$$\sum_{i=2}^{5} \frac{1}{2j} = \frac{1}{2j} + \frac{1}{2j} + \frac{1}{2j} + \frac{1}{2j} = \frac{2}{j}$$

$$\sum_{i=1}^{n} \sum_{j=1}^{n} (i+j) = \sum_{i=1}^{n} \left((i+1) + (i+2) + \cdots + (i+n) \right)$$

$$= \sum_{i=1}^{n} \left(ni + (1 + 2 + \cdots + n) \right)$$

---

[1] In the older literature, one often finds $[x]$ used for the same function.

$$= n\left(\sum_{i=1}^{n} i\right) + n(1 + 2 + \cdots + n)$$

$$= 2n(1 + 2 + \cdots + n).$$

Similarly as sums are written using $\sum$ (which is the capital Greek letter "sigma", from the word *sum*), products may be expressed using the sign $\prod$ (capital Greek "pi"). For example,

$$\prod_{i=1}^{n} \frac{i+1}{i} = \frac{2}{1} \cdot \frac{3}{2} \cdot \ldots \cdot \frac{n+1}{n} = n+1.$$

Sets. Another basic notion we will use is that of a set. Most likely you have already encountered sets in high school (and, thanks to the permanent modernization of the school system, maybe even in elementary school). Sets are usually denoted by capital letters:

$$A, B, \ldots, X, Y, \ldots, M, N, \ldots$$

and so on, and the elements of sets are mostly denoted by lowercase letters: $a, b, \ldots, x, y, \ldots, m, n, \ldots$.

The fact that a set X contains an element x is traditionally written using the symbol $\in$, which is a somewhat stylized Greek letter ε—"epsilon". The notation $x \in X$ is read "x is an element of X", "x belongs to X", "x is in X", and so on.

Let us remark that the concept of a set and the symbol $\in$ are so-called primitive notions. This means that we do not define them using other "simpler" notions (unlike the rational numbers, say, which are defined in terms of the integers). To understand the concept of a set, we rely on intuition (supported by numerous examples) in this book. It turned out at the beginning of the 20th century that if such an intuitive notion of a set is used completely freely, various strange situations, the so-called paradoxes, may arise.[2] In order to exclude such paradoxes, the theory of sets has been rebuilt on a formalized basis, where all properties of sets are derived formally from several precisely formulated basic assumptions (axioms). For the sets used in this text, which are mostly finite, we need not be afraid of any paradoxes, and so we can keep relying on the intuitive concept of a set.

---

[2]The most famous one is probably Russell's paradox. One possible formulation is about an army barber. An army barber is supposed to shave all soldiers who do not shave themselves—should he, as one of the soldiers, shave himself or not? This paradox can be translated into a rigorous mathematical language and it implies the inconsistency of notions like "the set of all sets".

The set with elements 1, 37, and 55 is written as $\{1, 37, 55\}$. This, and also the notations $\{37, 1, 55\}$ and $\{1, 37, 1, 55, 55, 1\}$, express the same thing. Thus, a multiple occurrence of the same element is ignored: the same element cannot be contained twice in the same set! Three dots (an ellipsis) in $\{2, 4, 6, 8, \ldots\}$ mean "and further similarly, using the same pattern", i.e. this notation means the set of all even natural numbers. The appropriate pattern should be apparent at first sight. For instance, $\{2^1, 2^2, 2^3, \ldots\}$ is easily understandable as the set of all powers of 2, while $\{2, 4, 8, \ldots\}$ may be less clear.

Ordered and unordered pairs.   The symbol $\{x, y\}$ denotes the set containing exactly the elements x and y, as we already know. In this particular case, the set $\{x, y\}$ is sometimes called the *unordered pair* of x and y. Let us recall that $\{x, y\}$ is the same as $\{y, x\}$, and if $x = y$, then $\{x, y\}$ is a 1-element set.

We also introduce the notation (x, y) for the *ordered pair* of x and y. For this construct, the order of the elements x and y is important. We thus assume the following:

$$(x, y) = (z, t) \text{ if and only if } x = z \text{ and } y = t. \qquad (1.1)$$

Interestingly, the ordered pair can be defined using the notion of unordered pair, as follows:

$$(x, y) = \{\{x\}, \{x, y\}\}.$$

Verify that ordered pairs defined in this way satisfy the condition (1.1). However, in this text it will be simpler for us to consider (x, y) as another primitive notion.

Similarly, we write $(x_1, x_2, \ldots, x_n)$ for the *ordered n-tuple* consisting of elements $x_1, x_2, \ldots, x_n$. A particular case of this convention is writing a point in the plane with coordinates x and y as (x, y), and similarly for points or vectors in higher-dimensional spaces.

Defining sets.   More complicated and interesting sets are usually created from known sets using some rule. The sets of all squares of natural numbers can be written

$$\{i^2 \colon i \in \mathbf{N}\}$$

or also

$$\{n \in \mathbf{N} \colon \text{ there exists } k \in \mathbf{N} \text{ such that } k^2 = n\}$$

or using the symbol ∃ for "there exists":

$$\{n \in \mathbf{N}\colon \exists k \in \mathbf{N}\, (k^2 = n)\}.$$

Another example is a formal definition of the open interval (a, b) introduced earlier:

$$(a, b) = \{x \in \mathbf{R}\colon a < x < b\}.$$

Note that the symbol (a, b) may mean either the open interval, or also the ordered pair consisting of a and b. These two meanings must (and usually can) be distinguished by the context. This is not at all uncommon in mathematics: many symbols, like parentheses in this case, are used in several different ways. For instance, (a, b) also frequently denotes the greatest common divisor of natural numbers a and b (but we avoid this meaning in this book).

With modern typesetting systems, it is no problem to use any kind of alphabets and symbols including hieroglyphs, so one might think of changing the notation in such cases. But mathematics tends to be rather conservative and the existing literature is vast, and so such notational inventions are usually short-lived.

The empty set. An important set is the one containing no element at all. There is just one such set, and it is customarily denoted by $\emptyset$ and called the *empty set*. Let us remark that the empty set can be an element of another set. For example, $\{\emptyset\}$ is the set containing the empty set as an element, and so it is not the same set as $\emptyset$!

Set systems. In mathematics, we often deal with sets whose elements are other sets. For instance, we can define the set

$$M = \{\{1, 2\}, \{1, 2, 3\}, \{2, 3, 4\}, \{4\}\},$$

whose elements are 4 sets of natural numbers, more exactly 4 subsets of the set $\{1, 2, 3, 4\}$. One meets such sets in discrete mathematics quite frequently. To avoid saying a "set of sets", we use the notions *set system* or *family of sets*. We could thus say that M is a system of sets on the set $\{1, 2, 3, 4\}$. Such set systems are sometimes denoted by calligraphic capital letters, such as $\mathcal{M}$.

However, it is clear that such a distinction using various types of letters cannot always be quite consistent—what do we do if we encounter a set of sets of sets?

The system consisting of all possible subsets of some set X is denoted by the symbol[3] 2^X and called the *power set* of X. Another notation for the power set common in the literature is $\mathcal{P}(X)$.

Set size. A large part of this book is devoted to counting various kinds of objects. Hence a very important notation for us is that for the number of elements of a finite set X. We write it using the same symbol as for the absolute value of a number: $|X|$.

A more general notation for sums and products. Sometimes it is advantageous to use a more general way to write down a sum than using the pattern $\sum_{i=1}^n a_i$. For instance,

$$\sum_{i \in \{1,3,5,7\}} i^2$$

means the sum $1^2 + 3^2 + 5^2 + 7^2$. Under the summation sign, we first write the summation variable and then we write out the set of values over which the summation is to be performed. We have a lot of freedom in denoting this set of values. Sometimes it can in part be described by words, as in the following:

$$\sum_{\substack{i:\,1\leq i\leq 10 \\ i\text{ a prime}}} i = 2 + 3 + 5 + 7.$$

Should the set of values for the summation be empty, we define the value of the sum as 0, no matter what appears after the summation sign. For example:

$$\sum_{i=1}^{0}(i + 10) = 0, \qquad \sum_{\substack{i \in \{2,4,6,8\} \\ i\text{ odd}}} i^4 = 0.$$

A similar "set notation" can also be employed for products. An empty product, such as $\prod_{j:\,2\leq j<1} 2^j$, is always defined as 1 (*not* 0 as for an empty sum).

Operations with sets. Using the primitive notion of set membership, $\in$, we can define further relations among sets and operations with sets. For example, two sets X and Y are considered identical (equal) if they have the same elements. In this case we write $X = Y$.

---

[3]This notation may look strange, but it is traditional and has its reasons. For instance, it helps to remember that an n-element set has 2^n subsets; see Proposition 2.1.2.

Other relations among sets can be defined similarly. If X, Y are sets, $X \subseteq Y$ (in words: "X is a subset of Y") means that each element of X also belongs to Y.

The notation $X \subset Y$ sometimes denotes that X is a subset of Y but X is not equal to Y. This distinction between $\subseteq$ and $\subset$ is not quite unified in the literature, and some authors may use $\subset$ synonymously with our $\subseteq$.

The notations $X \cup Y$ (the union of X and Y) and $X \cap Y$ (the intersection of X and Y) can be defined as follows:

$$X \cup Y = \{z : z \in X \text{ or } z \in Y\}, \quad X \cap Y = \{z : z \in X \text{ and } z \in Y\}.$$

If we want to express that the sets X and Y in the considered union are disjoint, we write the union as $X \dot{\cup} Y$. The expression $X \setminus Y$ is the *difference* of the sets X and Y, i.e. the set of all elements belonging to X but not to Y.

Enlarged symbols $\cup$ and $\cap$ may be used in the same way as the symbols $\sum$ and $\prod$. So, if $X_1, X_2, \ldots, X_n$ are sets, their union can be written

$$\bigcup_{i=1}^{n} X_i \tag{1.2}$$

and similarly for intersection.

Note that this notation is possible (or correct) only because the operations of union and intersection are *associative*; that is, we have

$$X \cap (Y \cap Z) = (X \cap Y) \cap Z$$

and

$$X \cup (Y \cup Z) = (X \cup Y) \cup Z$$

for any triple X, Y, Z of sets. As a consequence, the way of "parenthesizing" the union of any 3, and generally of any n, sets is immaterial, and the common value can be denoted as in (1.2). The operations $\cup$ and $\cap$ are also *commutative*, in other words they satisfy the relations

$$X \cap Y = Y \cap X,$$

$$X \cup Y = Y \cup X.$$

The commutativity and the associativity of the operations $\cup$ and $\cap$ are complemented by their *distributivity*. For any sets X, Y, Z we have

$$X \cap (Y \cup Z) = (X \cap Y) \cup (X \cap Z),$$

$$X \cup (Y \cap Z) = (X \cup Y) \cap (X \cup Z).$$

The validity of these relations can be checked by proving that any element belongs to the left-hand side if and only if it belongs to the right-hand side. The relations can be generalized for an arbitrary number of sets as well. For instance,

$$A \cap \left(\bigcup_{i=1}^{n} X_i \right) = \bigcup_{i=1}^{n} (A \cap X_i);$$

$$A \cup \left(\bigcap_{i=1}^{n} X_i \right) = \bigcap_{i=1}^{n} (A \cup X_i).$$

Such relations can be proved by induction; see Section 1.3 below. Other popular relations for sets are

$$X \setminus (A \cup B) = (X \setminus A) \cap (X \setminus B) \quad \text{and} \quad X \setminus (A \cap B) = (X \setminus A) \cup (X \setminus B)$$

(the so-called *de Morgan laws*), and their generalizations

$$X \setminus \left(\bigcup_{i=1}^{n} A_i \right) = \bigcap_{i=1}^{n} (X \setminus A_i)$$

$$X \setminus \left(\bigcap_{i=1}^{n} A_i \right) = \bigcup_{i=1}^{n} (X \setminus A_i).$$

The last operation to be introduced here is the *Cartesian product*, denoted by $X \times Y$, of two sets X and Y. The Cartesian product of X and Y is the set of all ordered pairs of the form (x, y), where $x \in X$ and $y \in Y$. Written formally,

$$X \times Y = \{(x, y) \colon x \in X, \, y \in Y\}.$$

Note that generally $X \times Y$ is not the same as $Y \times X$, i.e. the operation is not commutative.

The name "Cartesian product" comes from a geometric interpretation. If, for instance, $X = Y = \mathbf{R}$, then $X \times Y$ can be interpreted as all points of the plane, since a point in the plane is uniquely described by an ordered pair of real numbers, namely its Cartesian coordinates[4]—the x-coordinate and the y-coordinate (Fig. 1.1(a)). This geometric view can also be useful for Cartesian products of sets whose elements are not numbers (Fig. 1.1(b)).

---

[4]These are named after their inventor, René Descartes.

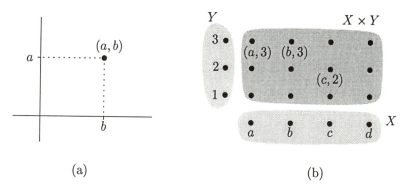

Fig. 1.1 Illustrating the Cartesian product: (a) $\mathbf{R} \times \mathbf{R}$; (b) $X \times Y$ for finite sets X, Y.

The Cartesian product of a set X with itself, i.e. $X \times X$, may also be denoted by X^2.

Exercises

1. Which of the following formulas are correct?
   (a) $\lfloor \frac{(n+1)^2}{2} \rfloor = \lfloor \frac{n^2}{2} \rfloor + n$,
   (b) $\lfloor \frac{n+k}{2} \rfloor = \lfloor \frac{n}{2} \rfloor + \lfloor \frac{k}{2} \rfloor$,
   (c) $\lceil (\lfloor x \rfloor) \rceil = \lceil x \rceil$ (for a real number x),
   (d) $\lceil (\lfloor x \rfloor + \lfloor y \rfloor) \rceil = \lfloor x \rfloor + \lfloor y \rfloor$.

2. *Prove that the equality $\lfloor \sqrt{x} \rfloor = \lfloor \sqrt{\lfloor x \rfloor} \rfloor$ holds for any positive real number x.

3. (a) Define a "parenthesizing" of a union of n sets $\bigcup_{i=1}^{n} X_i$. Similarly, define a "parenthesizing" of a sum of n numbers $\sum_{i=1}^{n} a_i$.

   (b) Prove that any two parenthesizings of the intersection $\bigcap_{i=1}^{n} X_i$ yield the same result.

   (c) How many ways are there to parenthesize the union of 4 sets $A \cup B \cup C \cup D$?

   (d) **Try to derive a formula or some other way to count the number of ways to parenthesize the union of n sets $\bigcup_{i=1}^{n} X_i$.

4. True or false? If $2^X = 2^Y$ holds for two sets X and Y, then $X = Y$.

5. Is a "cancellation" possible for the Cartesian product? That is, if $X \times Y = X \times Z$ holds for some sets X, Y, Z, does it necessarily follow that $Y = Z$?

6. Prove that for any two sets A, B we have
   $$(A \setminus B) \cup (B \setminus A) = (A \cup B) \setminus (A \cap B).$$

7. *Consider the numbers $1, 2, \ldots, 1000$. Show that among any 501 of them, two numbers exist such that one divides the other one.

8. In this problem, you can test your ability to discover simple but "hidden" solutions. Divide the following figure into 7 parts, all of them congruent (they only differ by translation, rotation, and possibly by a mirror reflection). All the bounding segments in the figure have length 1, and the angles are 90, 120, and 150 degrees.

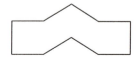

1.3   Mathematical induction and other proofs

Let us imagine that we want to calculate, say, the sum $1 + 2 + 2^2 + 2^3 + \cdots + 2^n = \sum_{i=0}^{n} 2^i$ (and that we can't remember a formula for the sum of a geometric progression). We suspect that one can express this sum by a nice general formula valid for all the n. By calculating numerical values for several small values of n, we can guess that the desired formula will most likely be $2^{n+1} - 1$. But even if we verify this for a million specific values of n with a computer, this is still no proof. The million-and-first number might, in principle, be a counterexample. The correctness of the guessed formula for all n can be proved by so-called *mathematical induction*. In our case, we can proceed as follows:

1. The formula $\sum_{i=0}^{n} 2^i = 2^{n+1} - 1$ holds for $n = 1$, as one can check directly.

2. Let us suppose that the formula holds for some value $n = n_0$. We prove that it also holds for $n = n_0 + 1$. Indeed, we have

$$\sum_{i=0}^{n_0+1} 2^i = \left(\sum_{i=0}^{n_0} 2^i \right) + 2^{n_0+1}.$$

The sum in parentheses equals $2^{n_0+1} - 1$ by our assumption (the validity for $n = n_0$). Hence

$$\sum_{i=0}^{n_0+1} 2^i = 2^{n_0+1} - 1 + 2^{n_0+1} = 2 \cdot 2^{n_0+1} - 1 = 2^{n_0+2} - 1.$$

This is the required formula for $n = n_0 + 1$.

This establishes the validity of the formula for an arbitrary n: by step 1, the formula is true for $n = 1$, by step 2 we may thus infer it is also true for $n = 2$ (using step 2 with $n_0 = 1$), then, again by step 2, the formula holds for $n = 3 \ldots$, and in this way we can reach any natural number. Note that this argument only works because the value of n_0 in step 2 was quite arbitrary. We have made the step from n_0 to $n_0 + 1$, where any natural number could equally well appear as n_0.

Step 2 in this type of proof is called the *inductive step*. The assumption that the statement being proved is already valid for some value $n = n_0$ is called the *inductive hypothesis*.

One possible general formulation of the principle of mathematical induction is the following:

1.3.1 Proposition. *Let X be a set of natural numbers with the following properties:*

(i) The number 1 belongs to X.

(ii) If some natural number n is an element of X, then the number $n + 1$ belongs to X as well.

Then X is the set of all natural numbers $(X = \mathbf{N})$.

In applications of this scheme, X would be the set of all numbers n such that the statement being proved, $S(n)$, is valid for n.

The scheme of a proof by mathematical induction has many variations. For instance, if we need to prove some statement for all $n \geq 2$, the first step of the proof will be to check the validity of the statement for $n = 2$. As an inductive hypothesis, we can sometimes use the validity of the statement being proved not only for $n = n_0$, but for all $n \leq n_0$, and so on; these modifications are best mastered by examples.

Mathematical induction can either be regarded as a basic property of natural numbers (an axiom, i.e. something we take for granted without a proof), or be derived from the following other basic property (axiom): *Any nonempty subset of natural numbers possesses a smallest element.* This is expressed by saying that the usual ordering of natural numbers by magnitude is a *well-ordering*. In fact, the principle of mathematical induction and the well-ordering property are equivalent to each other,[5] and either one can be taken as a basic axiom for building the theory of natural numbers.

---

[5] Assuming that each natural number $n > 1$ has a unique predecessor $n - 1$.

Proof of Proposition 1.3.1 from the well-ordering property.
For contradiction, let us assume that a set X satisfies both (i) and (ii),
but it doesn't contain all natural numbers. Among all natural numbers
n not lying in X, let us choose the smallest one and denote it by n_0.
By condition (i) we know that $n_0 > 1$, and since n_0 was the smallest
possible, the number $n_0 - 1$ is an element of X. However, using (ii) we
get that n_0 is an element of X, which is a contradiction. □

Let us remark that this type of argument (saying "Let n_0 be the
smallest number violating the statement we want to prove" and deriving a contradiction, namely that a yet smaller violating number must
exist) sometimes replaces mathematical induction. Both ways, this one
and induction, essentially do the same thing, and it depends on the
circumstances or personal preferences which one is actually used.

We will use mathematical induction quite often. It is one of our
basic proof methods, and the reader can thus find many examples
and exercises on induction in subsequent chapters.

Mathematical proofs and not-quite proofs. Mathematical
proof is an amazing invention. It allows one to establish the truth of
a statement beyond any reasonable doubt, even when the statement
deals with a situation so complicated that its truth is inaccessible to
direct evidence. Hardly anyone can see directly that no two natural
numbers m, n exist such that $\frac{m}{n} = \sqrt{2}$ and yet we can trust this
fact completely, because it can be proved by a chain of simple logical
steps.

Students often don't like proofs, even students of mathematics.
One reason might be that they have never experienced satisfaction
from understanding an elegant and clever proof or from making a
nice proof by themselves. One of our main goals is to help the reader
to acquire the skill of rigorously proving simple mathematical statements.

A possible objection is that most students will never need such
proofs in their future jobs. We believe that learning how to prove mathematical theorems helps to develop useful habits in thinking, such as
working with clear and precise notions, exactly formulating thoughts
and statements, and not overlooking less obvious possibilities. For instance, such habits are invaluable for writing software that doesn't crash
every time the circumstances become slightly non-standard.

The art of finding and writing proofs is mostly taught by examples,[6] by showing many (hopefully) correct and "good" proofs to the

---

[6]We will not even try to say what a proof is and how to do one!

student and by pointing out errors in the student's own proofs. The latter "negative" examples are very important, and since a book is a one-way communication device, we decided to include also a few negative examples in this book, i.e. students' attempts at proofs with mistakes which are, according to our experience, typical. These intentionally wrong proofs are presented in a special font like this. In the rest of this section, we discuss some common sources of errors. (We hasten to add that types of errors in proofs are as numerous as grains of sand, and by no means do we want to attempt any classification.)

One quite frequent situation is where the student doesn't understand the problem correctly. There may be subtleties in the problem's formulation which are easy to overlook, and sometimes a misunderstanding isn't the student's fault at all, since the author of the problem might very well have failed to see some double meaning. The only defense against this kind of misunderstanding is to pay the utmost attention to reading and understanding a problem before trying to solve it. Do a preliminary check: does the problem make sense in the way you understand it? Does it have a suspiciously trivial solution? Could there be another meaning?

With the current abundance of calculators and computers, errors are sometimes caused by the uncritical use of such equipment. Asked how many zeros does the decimal notation of the number $50! = 50 \cdot 49 \cdot 48 \cdots \cdot 1$ end with, a student answered 60, because a pocket calculator with an 8-digit display shows that $50! = 3.04140 \cdot 10^{64}$. Well, a more sophisticated calculator or computer programmed to calculate with integers with arbitrarily many digits would solve this problem correctly and calculate that

$$50! = 30414093201713378043612608166064768844377641568960512000000000000$$

with 12 trailing zeros. Several software systems can even routinely solve such problems as finding a formula for the sum $1^2 \cdot 2^1 + 2^2 \cdot 2^2 + 3^2 \cdot 2^3 + \cdots + n^2 2^n$, or for the number of binary trees on n vertices (see Section 10.4). But even programmers of such systems can make mistakes and so it's better to double-check such results. Moreover, the capabilities of these systems are very limited; artificial intelligence researchers will have to make enormous progress before they can produce computers that can discover and prove a formula for the number of trailing zeros of $n!$, or solve a significant proportion of the exercises in this book, say.

Next, we consider the situation where a proof has been written down but it has a flaw, although its author believes it to be satisfactory.

In principle, proofs can be written down in such detail and in such a formal manner that they can be checked automatically by a computer. If such a completely detailed and formalized proof is wrong, some step has to be clearly false, but the catch is that formalizing proofs completely is very laborious and impractical. All textbook proofs and problem solutions are presented somewhat informally.

While some informality may be necessary for a reasonable presentation of a proof, it may also help to hide errors. Nevertheless, a good rule for writing and checking proofs is that *every statement in a correct proof should be literally true*. Errors can often be detected by isolating a specific false statement in the proof, a mistake in calculation, or a statement that makes no sense ("Let ℓ_1, ℓ_2 be two arbitrary lines in the 3-dimensional space, and let ρ be a plane containing both of them..." etc.). Once detected and brought out into the light, such errors become obvious to (almost) everyone. Still, they are frequent. If, while trying to come up with a proof, one discovers an idea seemingly leading to a solution and shouts "This must be IT!", caution is usually swept aside and one is willing to write down the most blatant untruths. (Unfortunately, the first idea that comes to mind is often nonsense, rather than "it", at least as far as the authors' own experience with problem solving goes.)

A particularly frequent mistake, common perhaps to all mathematicians of the world, is a *case omission*. The proof works for some objects it should deal with, but it fails in some cases the author overlooked. Such a case analysis is mostly problem specific, but one keeps encountering variations on favorite themes. Dividing an equation by $x - y$ is only allowed for $x \neq y$, and the $x = y$ case must be treated separately. An intersection of two lines in the plane can only be used in a proof if the lines are not parallel. Deducing $a^2 > b^2$ from $a > b$ may be invalid if we know nothing about the sign of a and b, and so on and so on.

Many proofs created by beginners are wrong because of a *confused application of theorems*. Something seems to follow from a theorem presented in class or in a textbook, say, but in reality the theorem says something slightly different, or some of its assumptions don't hold. Since we have covered no theorems worth mentioning so far, let us give an artificial geometric example: "Since ABC is an isosceles triangle with the sides adjacent to A having equal length, we have $|AB|^2 + |AC|^2 = |BC|^2$ by the theorem of Pythagoras." Well, wasn't there something about a right angle in Pythagoras' theorem?

A rich source of errors and misunderstandings is *relying on unproved statements*.

Many proofs, including correct and even textbook ones, contain unproved statements intentionally, marked by clauses like "obviously...". In an honest proof, the meaning of such clauses should ideally be "I, the author of this proof, can see how to prove this rigorously, and since I consider this simple enough, I trust that you, my reader, can also fill in all the details without too much effort". Of course, in many mathematical papers, the reader's impression about the author's thinking is more in the spirit of "I can see it somehow since I've been working on this problem for years, and if you can't it's your problem". Hence omitting parts of proofs that are "clear" is a highly delicate social task, and one should always be very careful with it. Also, students shouldn't be surprised if their teacher insists that such an "obvious" part be proved in detail. After all, what would be a better hiding place for errors in a proof than in the parts that are missing?

A more serious problem concerns parts of a proof that are omitted unconsciously. Most often, the statement whose proof is missing is not even formulated explicitly.[7] For a teacher, it may be a very challenging task to convince the proof's author that something is wrong with the proof, especially when the unproved statement is actually true.

One particular type of incomplete proof, fairly typical of students' proofs in discrete mathematics, could be labeled as *mistaking the particular for the general*. To give an example, let us consider the following Mathematical Olympiad problem:

1.3.2 Problem. Let $n > 1$ be an integer. Let M be a set of closed intervals. Suppose that the endpoints u, v of each interval $[u, v] \in M$ are natural numbers satisfying $1 \le u < v \le n$, and, moreover, for any two distinct intervals $I, I' \in M$, one of the following possibilities occurs: $I \cap I' = \emptyset$, or $I \subset I'$, or $I' \subset I$ (i.e. two intervals must not partially overlap). Prove that $|M| \le n - 1$.

An insufficient proof attempt. In order to construct an M as large as possible, we first insert as many unit-length intervals as possible, as in

---

[7]Even proofs by the greatest mathematicians of the past suffer from such incompleteness, partly because the notion of a proof has been developing over the ages (towards more rigor, that is).

the following figure:

1   2   . . .   13

These $\lfloor n/2 \rfloor$ intervals are all disjoint. Now any other interval in M must contain at least two of these unit intervals (or, for n odd, possibly the last unit interval plus the point that remains). Hence, to get the maximum number of intervals, we put in the next "layer" of shortest possible intervals, as illustrated below:

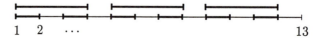

1   2   . . .   13

We continue in this manner, adding one layer after another, until we finally add the last layer consisting of the whole interval $[1, n]$:

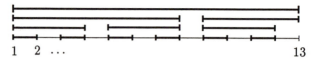

1   2 . . .   13

It remains to show that the set M created in this way has at most $n - 1$ intervals. We note that every interval I in the kth layer contains a point of the form $i + \frac{1}{2}$, $1 \leq i \leq n - 1$, that was not contained in any interval of the previous layers, because the space between the two intervals in the previous layer was not covered before adding the kth layer. Therefore, $|M| \leq n - 1$ as claimed.   ☐

    This "proof" looks quite clever (after all, the way of counting the intervals in the particular M constructed in the proof is quite elegant). So what's wrong with it? Well, we have shown that *one particular M* satisfies $|M| \leq n - 1$. The argument tries to make the impression of showing that this particular M is the worst possible case, i.e. that no other M may have more intervals, but in reality it doesn't prove anything like that! For instance, the first step seems to argue that an M with the maximum possible number of intervals should contain $\lfloor n/2 \rfloor$ unit-length intervals. But this is not true, as is witnessed by $M = \{[1, 2], [1, 3], [1, 4], \ldots, [1, n]\}$. Saving the "proof" above by justifying its various steps seems more difficult than finding another, correct, proof. Although the demonstrated "proof" contains some useful hints (the counting idea at the end of the proof can in fact be made to work for any M), it's still quite far from a valid solution.

    The basic scheme of this "proof", apparently a very tempting one, says "this object X must be the worst one", and then proves that this particular X is OK. But the claim that nothing can be worse than X is not substantiated (although it usually looks plausible that by constructing this X, we "do the worst possible thing" concerning the statement being proved).

Another variation of "mistaking the particular for the general" often appears in proofs by induction, and is shown in several examples in Sections 4.1 and 5.3.

Exercises

1. Prove the following formulas by mathematical induction:

    (a) $1 + 2 + 3 + \cdots + n = n(n+1)/2$

    (b) $\sum_{i=1}^{n} i \cdot 2^i = (n-1)2^{n+1} + 2$.

2. The numbers $F_0, F_1, F_2, \ldots$ are defined as follows (this is a definition by mathematical induction, by the way): $F_0 = 0$, $F_1 = 1$, $F_{n+2} = F_{n+1} + F_n$ for $n = 0, 1, 2, \ldots$. Prove that for any $n \geq 0$ we have $F_n \leq ((1 + \sqrt{5})/2)^{n-1}$ (see also Section 10.3).

3. (a) Let us draw n lines in the plane in such a way that no two are parallel and no three intersect in a common point. Prove that the plane is divided into exactly $n(n+1)/2 + 1$ parts by the lines.

    (b) *Similarly, consider n planes in the 3-dimensional space in general position (no two are parallel, any three have exactly one point in common, and no four have a common point). What is the number of regions into which these planes partition the space?

4. Prove *de Moivre's theorem* by induction: $(\cos \alpha + i \sin \alpha)^n = \cos(n\alpha) + i \sin(n\alpha)$. Here i is the imaginary unit.

5. In ancient Egypt, fractions were written as sums of fractions with numerator 1. For instance, $\frac{3}{5} = \frac{1}{2} + \frac{1}{10}$. Consider the following algorithm for writing a fraction $\frac{m}{n}$ in this form ($1 \leq m < n$): write the fraction $\frac{1}{\lceil n/m \rceil}$, calculate the fraction $\frac{m}{n} - \frac{1}{\lceil n/m \rceil}$, and if it is nonzero repeat the same step. Prove that this algorithm always finishes in a finite number of steps.

6. *Consider a $2^n \times 2^n$ chessboard with one (arbitrarily chosen) square removed, as in the following picture (for $n = 3$):

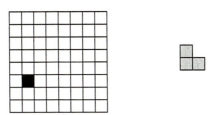

Prove that any such chessboard can be tiled without gaps or overlaps by L-shapes consisting of 3 squares each.

7. Let $n \geq 2$ be a natural number. We consider the following game. Two players write a sequence of 0s and 1s. They start with an empty line and alternate their moves. In each move, a player writes 0 or 1 to the end of the current sequence. A player loses if his digit completes a block of n consecutive digits that repeats itself in the sequence for the second time (the two occurrences of the block may overlap). For instance, for $n = 4$, a sequence produced by such a game may look as follows: 00100001101011110011 (the second player lost by the last move because 0011 is repeated).

(a) Prove that the game always finishes after finitely many steps.

(b) *Suppose that n is odd. Prove that the second player (the one who makes the second move) has a winning strategy.

(c) *Show that for $n = 4$, the first player has a winning strategy. Unsolved question: Can you determine who has a winning strategy for some even $n > 4$?

8. *On an infinite sheet of white graph paper (a paper with a square grid), n squares are colored black. At moments $t = 1, 2, \ldots$, squares are recolored according to the following rule: each square gets the color occurring at least twice in the triple formed by this square, its top neighbor, and its right neighbor. Prove that after the moment $t = n$, all squares are white.

9. At time 0, a particle resides at the point 0 on the real line. Within 1 second, it divides into 2 particles that fly in opposite directions and stop at distance 1 from the original particle. Within the next second, each of these particles again divides into 2 particles flying in opposite directions and stopping at distance 1 from the point of division, and so on. Whenever particles meet they annihilate (leaving nothing behind). How many particles will there be at time $2^{11} + 1$?

10. *Let $M \subseteq \mathbf{R}$ be a set of real numbers, such that any nonempty subset of M has a smallest number and also a largest number. Prove that M is necessarily finite.

11. We will prove the following statement by mathematical induction: *Let $\ell_1, \ell_2, \ldots, \ell_n$ be $n \geq 2$ distinct lines in the plane, no two of which are parallel. Then all these lines have a point in common.*

   1. For $n = 2$ the statement is true, since any 2 nonparallel lines intersect.

   2. Let the statement hold for $n = n_0$, and let us have $n = n_0 + 1$ lines $\ell_1, \ldots, \ell_n$ as in the statement. By the inductive hypothesis, all these lines but the last one (i.e. the lines $\ell_1, \ell_2, \ldots, \ell_{n-1}$) have some point in common; let us denote this point by x. Similarly the $n - 1$ lines $\ell_1, \ell_2, \ldots, \ell_{n-2}, \ell_n$ have a point in common; let us denote it by y. The line ℓ_1 lies in both groups, so it contains both x and y. The same is true for the line ℓ_{n-2}. Now ℓ_1 and ℓ_{n-2} intersect at a single point only, and

so we must have $x = y$. Therefore all the lines $\ell_1, \ldots, \ell_n$ have a point in common, namely the point x.

Something must be wrong. What is it?

12. Let $n_1, n_2, \ldots, n_k$ be natural numbers, each of them at least 1, and let $n_1 + n_2 + \cdots + n_k = n$. Prove that $n_1^2 + n_2^2 + \cdots + n_k^2 \le (n-k+1)^2 + k - 1$.

    "Solution": In order to make $\sum_{i=1}^{k} n_i^2$ as large as possible, we must set all the n_i but one to 1. The remaining one is therefore $n - k + 1$, and in this case the sum of squares is $(n - k + 1)^2 + k - 1$.

    Why isn't this a valid proof? *Give a correct proof.

13. *Give a correct proof for Problem 1.3.2.

14. *Let $n > 1$ and k be given natural numbers. Let $I_1, I_2, \ldots, I_m$ be closed intervals (not necessarily all distinct), such that for each interval $I_j = [u_j, v_j]$, u_j and v_j are natural numbers with $1 \le u_j < v_j \le n$, and, moreover, no number is contained in more than k of the intervals $I_1, \ldots, I_m$. What is the largest possible value of m?

1.4 Functions

The notion of a function is a basic one in mathematics. It took a long time for today's view of functions to emerge. For instance, around the time when differential calculus was invented, only real or complex functions were considered, and an "honest" function had to be expressed by some formula, such as $f(x) = x^2 + 4$, $f(x) = \sqrt{\sin(x/\pi)}$, $f(x) = \int_0^x (\sin t/t)\,dt$, $f(x) = \sum_{n=0}^{\infty} (x^n/n!)$, and so on. By today's standards, a real function may assign to each real number an arbitrary real number without any restrictions whatsoever, but this is a relatively recent invention.

Let X and Y be some quite arbitrary sets. Intuitively, a function f is "something" assigning a unique element of Y to each element of X. To depict a function, we can draw the sets X and Y, and draw an arrow from each element $x \in X$ to the element $y \in Y$ assigned to it:

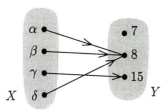

Note that each element of X must have exactly one outgoing arrow, while the elements of Y may have none, one, or several ingoing arrows.

Instead of saying that a function is "something", it is better to define it using objects we already know, namely sets and ordered pairs.

1.4.1 Definition. *A function f from a set X into a set Y is a set of ordered pairs (x, y) with $x \in X$ and $y \in Y$ (in other words, a subset of the Cartesian product $X \times Y$), such that for any $x \in X$, f contains exactly one pair with first component x.*

Of course, an ordered pair (x, y) being in f means just that the element x is assigned the element y. We then write $y = f(x)$, and we also say that f *maps x to y* or that y is the *image* of x.

For instance, the function depicted in the above figure consists of the ordered pairs $(\alpha, 8)$, $(\beta, 8)$, $(\gamma, 15)$ and $(\delta, 8)$.

A function, as a subset of the Cartesian product $X \times Y$, is also drawn using a *graph*. We depict the Cartesian product as in Fig. 1.1, and then we mark the ordered pairs belonging to the function. This is perhaps the most usual way used in high school or in calculus. The following figure shows a graph of the function $f : \mathbf{R} \to \mathbf{R}$ given by $f(x) = x^3 - x + 1$:

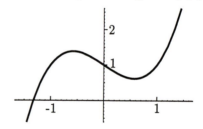

The fact that f is a function from a set X into a set Y is written as follows:

$$f : X \to Y .$$

And the fact that the function f assigns some element y to an element x can also be written

$$f : x \mapsto y.$$

We could simply write $y = f(x)$ instead. So why this new notation? The symbol $\mapsto$ is advantageous when we want to speak about some function without naming it. (Those who have programmed in LISP, Mathematica, or a few other programming languages might recall the existence of unnamed functions in these languages.) For instance, it is not really correct to say "consider the function x^2", since we do not say what the variable is. In this particular case, one can be reasonably sure that we mean the function assigning x^2 to each real number x, but if we say "consider the function $zy^2 + 5z^3 y$", it is not clear whether we mean the dependence on y, on z, or on both. By writing $y \mapsto zy^2 + 5z^3 y$, we indicate that we want to study the dependence on y, treating z as some parameter.

Instead of "function", the words "mapping" or "map" are used with the same meaning.[8]

Sometimes we also write $f(X)$ for the set $\{f(x): x \in X\}$ (the set of those elements of Y that are images of something). Also other terms are usually introduced for functions. For example, X is called the *domain* and Y is the *range*, etc., but here we try to keep the terminology and formalism to a minimum.

We definitely need to mention that functions can be composed.

1.4.2 Definition (Composition of functions). *If $f: X \to Y$ and $g: Y \to Z$ are functions, we can define a new function $h: X \to Z$ by*

$$h(x) = g(f(x))$$

for each $x \in X$. In words, to find the value of $h(x)$, we first apply f to x and then we apply g to the result.

The function h *(check that h is indeed a function) is called the composition of the functions g and f and it is denoted by $g \circ f$. We thus have*

$$\left(g \circ f \right)(x) = g(f(x))$$

for each $x \in X$.

The composition of functions is associative but not commutative. For example, if $g \circ f$ is well defined, $f \circ g$ need not be. In order that two functions can be composed, the "middle set" must be the same.

Composing functions can get quite exciting. For example, consider the mapping $f: \mathbf{R}^2 \to \mathbf{R}^2$ (i.e. mapping the plane into itself) given by

$$f: (x, y) \mapsto \left(\sin(ax) + b\sin(ay), \sin(cx) + d\sin(cy) \right)$$

with $a = 2.879879$, $b = 0.765145$, $c = -0.966918$, $d = 0.744728$. Except for the rather hairy constants, this doesn't look like a very complicated function. But if one takes the initial point $p = (0.1, 0.1)$ and plots

---

[8]In some branches of mathematics, the word "function" is reserved for function into the set of real or complex numbers, and the word mapping is used for functions into arbitrary sets. For us, the words "function" and "mapping" will be synonymous.

Fig. 1.2 The "King's Dream" fractal (formula taken from the book by C. Pickover: *Chaos in Wonderland*, St Martin's Press, New York 1994).

the first several hundred thousand or million points of the sequence p, $f(p)$, $f(f(p))$, $f(f(f(p)))$,..., a picture like Fig. 1.2 emerges.[9] This is one of the innumerable species of the so-called *fractals*. There seems to be no universally accepted mathematical definition of a fractal, but fractals are generally understood as complicated point sets defined by iterations of relatively simple mappings. The reader can find colorful and more sophisticated pictures of various fractals in many books on the subject or download them from the Internet. Fractals can be not only pleasant to the eye (and suitable for killing an unlimited amount of time by playing with them on a personal computer) but also important for describing various phenomena in nature.

After this detour, let us return to the basic definitions concerning functions.

1.4.3 Definition (Important special types of functions). *A function $f: X \to Y$ is called*

- *a one-to-one function if $x \neq y$ implies $f(x) \neq f(y)$,*
- *a function onto if for every $y \in Y$ there exists $x \in X$ satisfying $f(x) = y$, and*

---

[9]To be quite honest, the way such pictures are generated by a computer is actually by iterating an *approximation* to the mapping given by the formula, because of limited numerical precision.

- a bijective function, *or* bijection, *if f is one-to-one and onto.*

A one-to-one function is also called an *injective function* or an *injection*, and a function onto is also called a *surjective function* or a *surjection*.

In a pictorial representation of functions by arrows, these types of functions can be recognized as follows:

- for a one-to-one function, each point $y \in Y$ has *at most one* ingoing arrow,
- for a function onto, each point $y \in Y$ has *at least one* ingoing arrow, and
- for a *bijection*, each point $y \in Y$ has *exactly one* ingoing arrow.

The fact that a function $f: X \to Y$ is one-to-one is sometimes expressed by the notation

$$f: X \hookrightarrow Y.$$

The $\hookrightarrow$ symbol is a combination of the inclusion sign $\subset$ with the mapping arrow $\to$. Why? If $f: X \hookrightarrow Y$ is an injective mapping, then the set $Z = f(X)$ can be regarded as a "copy" of the set X within Y (since f considered as a map $X \to Z$ is a bijection), and so an injective mapping $f: X \hookrightarrow Y$ can be thought of as a "generalized inclusion" of X in Y. This point can probably be best appreciated in more abstract and more advanced parts of mathematics like topology or algebra.

There are also symbols for functions onto and for bijections, but these are still much less standard in the literature than the symbol for an injective function, so we do not introduce them.

Since we will be interested in counting objects, bijections will be especially significant for us, for the following reason: if X and Y are sets and there exists a bijection $f: X \to Y$, then X and Y have the same number of elements. Let us give a simple example of using a bijection for counting (more sophisticated ones come later).

1.4.4 Example. How many 8-digit sequences consisting of digits 0 through 9 are there? How many of them contain an even number of odd digits?

Solution. The answer to the first question is 10^8. One easy way of seeing this is to note that each 8-digit sequence can be read as the decimal notation of an integer number between 0 and $10^8 - 1$, and conversely, each such integer can be written in decimal notation and, if necessary, padded with zeros on the left to produce an 8-digit sequence. This defines a bijection between the set $\{0, 1, \ldots, 10^8 - 1\}$ and the set of all 8-digit sequences.

Well, this bijection was perhaps too simple (or, rather, too customary) to impress anyone. What about the 8-digit sequences with an even number of odd digits? Let E be the set of all these sequences (E for "even"), and let O be the remaining ones, i.e. those with an odd number of odd digits. Consider any sequence $s \in E$, and define another sequence, $f(s)$, by changing the first digit of s: 0 is changed to 1, 1 to 2,..., 8 to 9, and 9 to 0. It is easy to check that the modified sequence $f(s)$ has an odd number of odd digits and hence f is a mapping from E to O. From two different sequences $s, s' \in E$, we cannot get the same sequence by the described modification, so f is one-to-one. And any sequence $t \in O$ is obtained as $f(s)$ for some $s \in E$, i.e. s arises from t by changing the first digit "back", by replacing 1 by 0, 2 by 1,..., 9 by 8, and 0 by 9. Therefore, f is a bijection and $|E| = |O|$. Since $|E| + |O| = 10^8$, we finally have $|E| = 5 \cdot 10^7$.   □

In the following proposition, we prove some simple properties of functions.

Proposition. *Let $f: X \to Y$ and $g: Y \to Z$ be functions. Then*
 (i) *If f, g are one-to-one, then $g \circ f$ is also a one-to-one function.*
 (ii) *If f, g are functions onto, then $g \circ f$ is also a function onto.*
 (iii) *If f, g are bijective functions, then $g \circ f$ is a bijection as well.*
 (iv) *For any function $f: X \to Y$ there exist a set Z, a one-to-one function $h: Z \hookrightarrow Y$, and a function onto $g: X \to Z$, such that $f = h \circ g$. (So any function can be written as a composition of a one-to-one function and a function onto.)*

Proof. Parts (i), (ii), (iii) are obtained by direct verification from the definition. As an example, let us prove (ii).

We choose $z \in Z$, and we are looking for an $x \in X$ satisfying $(g \circ f)(x) = z$. Since g is a function onto, there exists a $y \in Y$ such that $g(y) = z$. And since f is a function onto, there exists an $x \in X$ with $f(x) = y$. Such an x is the desired element satisfying $(g \circ f)(x) = z$.

The most interesting part is (iv). Let $Z = f(X)$ (so $Z \subseteq Y$). We define mappings $g: X \to Z$ and $h: Z \to Y$ as follows:

$$g(x) = f(x) \quad \text{for } x \in X$$
$$h(z) = z \quad\quad \text{for } z \in Z.$$

Clearly g is a function onto, h is one-to-one, and $f = h \circ g$.   □

Finishing the remaining parts of the proof may be a good exercise for understanding the notions covered in this section.

Inverse function. If $f: X \to Y$ is a bijection, we can define a function $g: Y \to X$ by setting $g(y) = x$ if x is the unique element of X with $y = f(x)$. This g is called the *inverse function* of f, and it is commonly denoted by f^{-1}. Pictorially, the inverse function is obtained by reversing all the arrows. Another equivalent definition of the inverse function is given in Exercise 4. It may look more complicated, but from a certain "higher" mathematical point of view it is better.

Exercises

1. Show that if X is a finite set, then a function $f: X \to X$ is one-to-one if and only if it is onto.

2. Find an example of:

   (a) A one-to-one function $f: \mathbf{N} \hookrightarrow \mathbf{N}$ which is not onto.

   (b) A function $f: \mathbf{N} \to \mathbf{N}$ which is onto but not one-to-one.

3. Decide which of the following functions $\mathbf{Z} \to \mathbf{Z}$ are injective and which are surjective: $x \mapsto 1 + x$, $x \mapsto 1 + x^2$, $x \mapsto 1 + x^3$, $x \mapsto 1 + x^2 + x^3$. Does anything in the answer change if we consider them as functions $\mathbf{R} \to \mathbf{R}$? (You may want to sketch their graphs and/or use some elementary calculus methods.)

4. For a set X, let $\mathrm{id}_X: X \to X$ denote the function defined by $\mathrm{id}_X(x) = x$ for all $x \in X$ (the *identity function*). Let $f: X \to Y$ be some function. Prove:

   (a) A function $g: Y \to X$ such that $g \circ f = \mathrm{id}_X$ exists if and only if f is one-to-one.

   (b) A function $g: Y \to X$ such that $f \circ g = \mathrm{id}_Y$ exists if and only if f is onto.

   (c) A function $g: Y \to X$ such that both $f \circ g = \mathrm{id}_Y$ and $g \circ f = \mathrm{id}_X$ exist if and only if f is a bijection.

   (d) If $f: X \to Y$ is a bijection, then the following three conditions are equivalent for a function $g: Y \to X$:

   (i) $g = f^{-1}$,
   (ii) $g \circ f = \mathrm{id}_X$, and
   (iii) $f \circ g = \mathrm{id}_Y$.

5. (a) If $g \circ f$ is an onto function, does g have to be onto? Does f have to be onto?

   (b) If $g \circ f$ is a one-to-one function, does g have to be one-to-one? Does f have to be one-to-one?

6. Prove that the following two statements about a function $f: X \to Y$ are equivalent (X and Y are some arbitrary sets):

   (i) f is one-to-one.

(ii) For any set Z and any two distinct functions $g_1: Z \to X$ and $g_2: Z \to X$, the composed functions $f \circ g_1$ and $f \circ g_2$ are also distinct. (First, make sure you understand what it means that two functions are equal and what it means that they are distinct.)

7. In everyday mathematics, the number of elements of a set is understood in an intuitive sense and no definition is usually given. In the logical foundations of mathematics, however, the number of elements is defined via bijections: $|X| = n$ means that there exists a bijection from X to the set $\{1, 2, \ldots, n\}$. (Also other, alternative definitions of set size exist but we will consider only this one here.)

(a) Prove that if X and Y have the same size according to this definition, then there exists a bijection from X to Y.

(b) Prove that if X has size n according to this definition, and there exists a bijection from X to Y, then Y has size n too.

(c) *Prove that a set cannot have two different sizes m and n, $m \neq n$, according to this definition. Be careful not to use the intuitive notion of "size" but only the definition via bijections. Proceed by induction.

1.5 Relations

It is remarkable how many mathematical notions can be expressed using sets and various set-theoretic constructions. It is not only remarkable but also surprising, since set theory, and even the notion of a set itself, are notions which appeared in mathematics relatively recently, and some 100 years ago, set theory was rejected even by some prominent mathematicians. Today, set theory has entered the mathematical vocabulary and it has become the language of all mathematics (and mathematicians), a language which helps to understand mathematics, with all its diversity, as a whole with common foundations.

We will show how more complicated mathematical notions can be built using the simplest set-theoretical tools. The key notion of a relation,[10] which we now introduce, is a common generalization of such diverse notions as equivalence, function, and ordering.

1.5.1 Definition. *A* relation *is a set of ordered pairs.*[11] *If X and Y are sets, any subset of the Cartesian product $X \times Y$ is called a* relation between X and Y. *The most important case is $X = Y$; then we speak of a* relation on X, *which is thus an arbitrary subset $R \subseteq X \times X$.*

---

[10] As a mathematical object; you know "relation" as a word in common language.

[11] In more detail, we could say a *binary relation* (since pairs of elements are being related). Sometimes also *n*-ary relations are considered for $n \neq 2$.

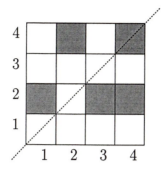

Fig. 1.3 A graphic presentation of the relation $R = \{(1,2), (2,4), (3,2),$ $(4,2), (4,4)\}$ on the set $\{1,2,3,4\}$.

If an ordered pair (x, y) belongs to a relation R, i.e. $(x, y) \in R$, we say that x and y *are related by* R, and we also write xRy.

We have already seen an object which was a subset of a Cartesian product, namely a function. Indeed, a function is a special type of relation, where we require that any $x \in X$ is related to precisely one $y \in Y$. In a general relation, an $x \in X$ can be related to several elements of Y, or also to none.

Many symbols well known to the reader can be interpreted as relations in this sense. For instance, $=$ (equality) and $\geq$ (non-strict inequality) are both relations on the set **N** of all natural numbers. The first one consists of the pairs $(1,1), (2,2), (3,3), \ldots$, the second one of the pairs $(1,1)$, $(2,1)$, $(2,2)$, $(3,1)$, $(3,2)$, $(3,3)$, $(4,1), \ldots$. We could thus also write $(5,2) \in \geq$ instead of the usual $5 \geq 2$, which we usually don't do, however. Note that we had to specify the set on which the relation $\geq$, say, is considered: as a relation on **R** it would be a quite different set of ordered pairs.

Many interesting "real life" examples of relations come from various kinds of relationships among people, e.g. "to be the mother of", "to be the father of", "to be a cousin of" are relations on the set of all people, usually well defined although not always easy to determine.

A relation R on a set X can be captured pictorially in (at least) two quite different ways. The first way is illustrated in Fig. 1.3. The little squares correspond to ordered pairs in the Cartesian product, and for pairs belonging to the relation we have shaded the corresponding squares. This kind of picture emphasizes the definition of a relation on X and it captures its "overall shape".

This figure is also very close in spirit to an alternative way of describing a relation on a set X using the notion of a matrix.[12] If R is a relation on some n-element set $X = \{x_1, x_2, \ldots, x_n\}$ then R is completely described by an $n \times n$ matrix $A = (a_{ij})$, where

$$a_{ij} = 1 \quad \text{if } (x_i, x_j) \in R$$
$$a_{ij} = 0 \quad \text{if } (x_i, x_j) \notin R.$$

The matrix A is called the *adjacency matrix* of the relation R. For instance, for the relation in Fig. 1.3, the corresponding adjacency matrix would be

$$\begin{pmatrix} 0 & 1 & 0 & 0 \\ 0 & 0 & 0 & 1 \\ 0 & 1 & 0 & 0 \\ 0 & 1 & 0 & 1 \end{pmatrix}.$$

Note that this matrix is turned by 90 degrees compared to Fig. 1.3. This is because, for a matrix element, the first index is the number of a row and the second index is the number of a column, while for Cartesian coordinates it is customary for the first coordinate to determine the horizontal position and the second coordinate the vertical position.

The adjacency matrix is one possible computer representation of a relation on a finite set.

Another picture of the same relation as in Fig. 1.3 is shown below:

Here the dots correspond to elements of the set X. The fact that a given ordered pair (x, y) belongs to the relation R is marked by drawing an arrow from x to y:

and, in the case $x = y$, by a loop:

A relation between X and Y can be depicted in a similar way:

---

[12] An $n \times m$ matrix is a rectangular table of numbers with n rows and m columns. Any reader who hasn't met matrices yet can consult the Appendix for the definitions and basic facts, or, preferably, take a course of linear algebra or refer to a good textbook.

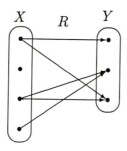

This way was suggested for drawing functions in Section 1.4.

Composition of relations. Let X, Y, Z be sets, let $R \subseteq X \times Y$ be a relation between X and Y, and let $S \subseteq Y \times Z$ be a relation between Y and Z. The *composition of the relations R and S* is the relation $T \subseteq X \times Z$ defined as follows: for given $x \in X$ and $z \in Z$, xTz holds if and only if there exists some $y \in Y$ such that xRy and ySz. The composition of relations R and S is usually denoted by $R \circ S$.

The composition of relations can be nicely illustrated using a drawing with arrows. In the following figure,

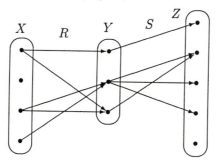

a pair (x, z) is in the relation $R \circ S$ whenever one can get from x to z along the arrows (i.e. via some $y \in Y$).

Have you noticed? Relations are composed in the same way as functions, but the notation is unfortunately different! For *relations* it is customary to write the composition "from left to right", and for functions it is usually written "from right to left". So if $f : X \to Y$ and $g : Y \to Z$ are functions, their composition is written $g \circ f$, but if we understood them as relations, we would write $f \circ g$ for the same thing! Both ways of notation have their reasons, such a notation has been established historically, and probably there is no point in trying to change it. In this text, we will talk almost exclusively about composing functions.

Similarly as for functions, the composition is not defined for arbitrary two relations. In order to compose relations, they must have the "middle" set in common (which was denoted by Y in the definition). In particular, it may happen that $R \circ S$ is defined while $S \circ R$ makes no sense! However, if both R and S are relations on the same set X, their composition is always well defined. But also in this case the result of composing relations depends on the order, and $R \circ S$ is in general different from $S \circ R$—see Exercise 2.

Exercises

1. Describe the relation $R \circ R$, if R stands for

   (a) the equality relation "$=$" on the set $\mathbf{N}$ of all natural numbers,

   (b) the relation "less than or equal" ("$\leq$") on $\mathbf{N}$,

   (c) the relation "strictly less" ("$<$") on $\mathbf{N}$,

   (d) the relation "strictly less" ("$<$") on the set $\mathbf{R}$ of all real numbers.

2. Find relations R, S on some set X such that $R \circ S \neq S \circ R$.

3. For a relation R on a set X we define the symbol R^n by induction: $R^1 = R$, $R^{n+1} = R \circ R^n$.

   (a) Prove that if X is finite and R is a relation on it, then there exist $r, s \in \mathbf{N}$, $r < s$, such that $R^r = R^s$.

   (b) Find a relation R on a finite set such that $R^n \neq R^{n+1}$ for every $n \in \mathbf{N}$.

   (c) Show that if X is infinite, the claim (a) need not hold (i.e. a relation R may exist such that all the relations R^n, $n \in \mathbf{N}$, are distinct).

4. (a) Let $X = \{x_1, x_2, \ldots, x_n\}$ and $Y = \{y_1, y_2, \ldots, y_m\}$ be finite sets, and let $R \subseteq X \times Y$ be a relation. Generalize the definition of the adjacency matrix of a relation to this case.

   (b) *Let X, Y, Z be finite sets, let $R \subseteq X \times Y$ and $S \subseteq Y \times Z$ be relations, and let A_R and A_S be their adjacency matrices, respectively. If you have defined the adjacency matrix in (a) properly, the matrix product $A_R A_S$ should be well defined. Discover and describe the connection of the composed relation $R \circ S$ to the matrix product $A_R A_S$.

5. Prove the associativity of composing relations: if R, S, T are relations such that $(R \circ S) \circ T$ is well defined, then also $R \circ (S \circ T)$ is well defined and equals $(R \circ S) \circ T$.

1.6   Equivalences

Besides the functions, equivalences are another important special type of relations. Informally, an equivalence on a set X is a relation describing which pairs of elements of X are "of the same type" in some sense. For instance, let X be the set of all triangles in the plane. By saying that two triangles are related if and only if they are congruent (i.e. one can be transformed into the other by translation and rotation), we have defined one equivalence on X. Another equivalence is defined by relating all pairs of similar triangles (two triangles are similar if one can be obtained from the other one by translation, rotation, and scaling; in other words, if their corresponding angles are the same). And a third equivalence arises by saying that each triangle is only related to itself.

These are three particular examples of equivalence relations, and the reader may look forward to many more examples later on. In general, in order to be called an equivalence, a relation must satisfy three conditions. These conditions are so useful that they each deserve a name.

1.6.1 Definition. *We say that a relation R on a set X is*

- reflexive *if xRx for every $x \in X$,*
- symmetric *if xRy implies yRx, for all $x, y \in X$,*
- transitive *if xRy and yRz imply xRz, for all $x, y, z \in X$.*

In a drawing like that in Fig. 1.3, a reflexive relation is one containing all squares on the diagonal (drawn by a dotted line). In drawing using arrows, a reflexive relation has loops at all points.

For a symmetric relation, a picture of the type in Fig. 1.3 has the diagonal as an axis of symmetry. In a picture using arrows, the arrows between two points always go in both directions:

The condition of transitivity can be well explained using arrows. If there are arrows $x \to y$ and $y \to z$, then the $x \to z$ arrow is present as well:

1.6.2 Definition. *We say that a relation R on X is an* equivalence *on X if it is reflexive, symmetric, and transitive.*

(You may want to contemplate for a while why these properties are natural for a relation that should express something like "being of the same type".) The notion of equivalence is a common generalization of notions expressing identity, isomorphism, similarity, etc. Relations of equivalence are often denoted by symbols like $=$, $\equiv$, $\simeq$, $\approx$, $\cong$, and so on.

Although an equivalence R on a set X is a special type of relation and we can thus depict it by either of the above methods, more often a picture similar to the one below is used:

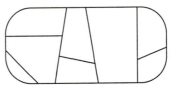

The key to this type of drawing is the following notion of *equivalence class*. Let R be an equivalence on a set X and let x be an element of X. By the symbol $R[x]$, we denote the set of all elements $y \in X$ that are equivalent to x; in symbols, $R[x] = \{y \in X : xRy\}$. $R[x]$ is called the *equivalence class of R determined by x*.

1.6.3 Proposition. *For any equivalence R on X, we have*

(i) *$R[x]$ is nonempty for every $x \in X$.*

(ii) *For any two elements $x, y \in X$, either $R[x] = R[y]$ or $R[x] \cap R[y] = \emptyset$.*

(iii) *The equivalence classes determine the relation R uniquely.*

Before we start proving this, we should explain the meaning of (iii). It means the following: if R and S are two equivalences on X and if the equality $R[x] = S[x]$ holds for every element $x \in X$, then $R = S$.

Proof. The proof is simple using the three requirements in the definition of equivalence.

(i) The set $R[x]$ always contains x since R is a reflexive relation.

(ii) Let x, y be two elements. We distinguish two cases:

    (a) If xRy, then we prove $R[x] \subseteq R[y]$ first. Indeed, if $z \in R[x]$, then we also know that zRx (by symmetry of R) and therefore zRy (by transitivity of R). Thus also $z \in R[y]$. By using symmetry again, we get that xRy implies $R[x] = R[y]$.

    (b) Suppose that xRy doesn't hold. We show that $R[x] \cap R[y] = \emptyset$. We proceed by contradiction. Suppose that there exists $z \in R[x] \cap R[y]$. Then xRz and zRy (by symmetry of R), and so xRy (by transitivity of R), which is a contradiction.

(iii) This part is obvious, since the equivalence classes determine R as follows:

$$xRy \text{ if and only if } \{x, y\} \subseteq R[x].\qquad\Box$$

The above proposition explains the preceding picture. It guarantees that the equivalence classes form a *partition* of the set X; that is, they are pairwise disjoint subsets of X whose union is the whole X. Conversely, any partition of X determines exactly one equivalence on X. That is, there exists a bijective mapping of the set of all equivalences on X onto the set of all partitions of X.

Exercises

1. Formulate the conditions for reflexivity of a relation, for symmetry of a relation, and for its transitivity using the adjacency matrix of the relation.

2. Prove that a relation R is transitive if and only if $R \circ R \subseteq R$.

3. (a) Prove that for any relation R, the relation $T = R \cup R \circ R \cup R \circ R \circ R \cup \ldots$ (the union of all multiple compositions of R) is transitive.

   (b) Prove that any transitive relation containing R as a subset also contains T.

   (c) Prove that if $|X| = n$, then $T = R \cup R \circ R \cup \cdots \cup \underbrace{R \circ R \circ \cdots \circ R}_{(n-1)\times}$.

   Remark. The relation T as in (a), (b) is the smallest transitive relation containing R, and it is called the *transitive closure* of R.

4. Let R and S be arbitrary equivalences on a set X. Decide which of the following relations are necessarily also equivalences (if yes, prove; if not, give a counterexample).

   (a) $R \cap S$

   (b) $R \cup S$

   (c) $R \setminus S$

   (d) $R \circ S$.

5. (a) Suppose that R is a transitive relation on the set $\mathbf{Z}$ of all integers, and we know that for any two integers $a, b \in \mathbf{Z}$, if $|a - b| = 2$ then aRb. Is *every* R satisfying these conditions necessarily an equivalence? (Note that a pair of elements can perhaps be in R even if it is not enforced by the given conditions!)

   (b) Suppose that R is a transitive relation on $\mathbf{Z}$, and we know that for any two integers $a, b \in \mathbf{Z}$, if $|a - b| \in \{3, 4\}$ then aRb. Is R necessarily an equivalence?

6. Call an equivalence $\sim$ on the set $\mathbf{Z}$ (the integers) a *congruence* if the following condition holds for all $a, x, y \in \mathbf{Z}$: if $x \sim y$ then also $a + x \sim a + y$.

(a) Let q be a nonzero integer. Define a relation $\equiv_q$ on $\mathbf{Z}$ by letting $x \equiv_q y$ if and only if q divides $x - y$. Check that $\equiv_q$ is a congruence according to the above definition.

(b) *Prove that any congruence on $\mathbf{Z}$ is either of the form $\equiv_q$ for some q or the *diagonal relation* $\{(x, x): x \in \mathbf{Z}\}$.

(c) Suppose we replaced the condition "$a + x \sim a + y$" in the definition of a congruence by "$ax \sim ay$". Would the claim in (a) remain true for this kind of "multiplicative congruence"? *And how about the claim in (b)?

1.7 Ordered sets

The reader will certainly be familiar with the ordering of natural numbers and of other number domains by magnitude (the "usual" ordering of numbers). In mathematics, such an ordering is considered as a special type of a relation, i.e. a set of pairs of numbers. In the case just mentioned, this relation is usually denoted by the symbol "$\leq$" ("less than or equal"). Various orderings can be defined on other sets too, such as the set of all words in some language, and one set can be ordered in many different and perhaps exotic ways.

Before introducing the general notion of an ordered set, we define an auxiliary notion. A relation R on a set X is called *antisymmetric*[13] if xRy and yRx imply that $x = y$, for all $x, y \in X$. When depicting a relation by arrows, the following situation is forbidden in an antisymmetric relation:

1.7.1 Definition. *An* ordering *on a set X is any relation on X that is reflexive, antisymmetric, and transitive. An* ordered set *is a pair (X, R), where X is a set and R is an ordering on X.*

Note the formal similarity of this definition to the definition of an equivalence. The definitions are almost the same, "only" the symmetry

---

[13]Sometimes this is called *weakly antisymmetric*, while for a *strongly antisymmetric* relation xRy and yRx never happen at the same time, i.e. xRx is also excluded.

has been replaced by antisymmetry. Yet equivalences and orderings are very different concepts.

For orderings, the symbols $\preceq$ or $\leq$ are commonly used. The first of them is useful, e.g. when we want to speak of some other ordering of the set of natural numbers than the usual ordering by magnitude, or if we consider some arbitrary ordering on a general set.

If we have some ordering $\preceq$, we define a derived relation of "strict inequality", $\prec$, as follows: $a \prec b$ if and only if $a \preceq b$ and $a \neq b$. Further we can introduce the "reverse inequality" $\succeq$, by letting $a \succeq b$ if and only if $b \preceq a$.

Examples. We have already mentioned several examples of ordered sets—these were $(\mathbf{N}, \leq)$, $(\mathbf{R}, \leq)$, and similar ones, where $\leq$ of course denotes the usual ordering, formally understood as a relation.

As is easy to check, if R is an ordering on a set X, and $Y \subseteq X$ is some subset of X, the relation $R \cap Y^2$ (the restriction of R on Y) is an ordering on Y. Intuitively, we order the elements of Y in the same way as before but we forget the others. This yields further examples of ordered sets, namely various subsets of real numbers with the usual ordering.

Linear orderings. The examples discussed so far have a significant feature in common: any two elements of the underlying set can be compared; in other words, for any two distinct elements x and y either $x \leq y$ or $y \leq x$ holds. This property is *not* a part of the definition of an ordering, and orderings having it are called *linear* orderings (sometimes the term *total ordering* is used with the same meaning).

Other examples of orderings. What do orderings which are not linear look like? For example, on any set X, we may define a relation Δ in which each element x is in relation with itself only, i.e. $\Delta = \{(x, x): x \in X\}$. It is easily checked that this relation satisfies the definition of an ordering, but this is a rather dull example. Before giving more examples, let us insert a remark about terminology.

In order to emphasize that we speak of an ordering which is not necessarily linear, we sometimes use the longer term *partial ordering*. A partial ordering thus means exactly the same as ordering (without further adjectives), so a partial ordering may also happen to be linear. Similarly, instead of an ordered set, one sometimes speaks of a *partially ordered set*. To abbreviate this long term, the artificial word *poset* is frequently used.

Let us describe more interesting examples of partially ordered sets.

1.7.2 Example. Let us imagine we intend to buy a refrigerator, say. We simplify the complicated real situation by a mathematical abstraction, and we suppose that we only look at three numerical parameters of refrigerators: their cost, electricity consumption, and the volume of the inner space. If we consider two types of refrigerators, and if the first type is more expensive, consumes more power, and a smaller amount of food fits into it, then the second type can be considered a better one—a large majority of buyers of refrigerators would agree with that. On the other hand, someone may prefer a smaller and cheaper refrigerator, another may prefer a larger refrigerator even if it costs more, and an environmentally concerned customer may even buy an expensive refrigerator if it saves power.

The relation "to be clearly worse" (denote it by $\preceq$) in this sense is thus a partial ordering on refrigerators or, mathematically reformulated, on the set of triples (c, p, v) of real numbers (c stands for cost, p for power consumption, and v for volume), defined as follows:

$$(c_1, p_1, v_1) \preceq (c_2, p_2, v_2) \text{ if and only if} \tag{1.3}$$
$$c_1 \geq c_2, \; p_1 \geq p_2, \text{ and } v_1 \leq v_2.$$

1.7.3 Example. For natural numbers a, b, the symbol $a|b$ means "a divides b". In other words, there exists a natural number c such that $b = ac$. The relation "$|$" is a partial ordering on $\mathbf{N}$. We leave the verification of this to the reader.

1.7.4 Example. Let X be a set. Recall that the symbol 2^X denotes the system of all subsets of the set X. The relation "$\subseteq$" (to be a subset) defines a partial ordering on 2^X.

Drawing partially ordered sets. Finite orderings can be drawn using arrows, as with any other relations. Typically, such drawings will contains lots of arrows. For instance, for a 10-element linearly ordered set we would have to draw $10 + 9 + \cdots + 1 = 55$ arrows and loops. A number of arrows can be reconstructed from transitivity, however: if we know that $x \preceq y$ and $y \preceq z$, then also $x \preceq z$, so we may leave out the arrow from x to z. Similarly, we need not draw the loops, since we know they are always there. For finite ordered sets, all the information is captured by the relation of "immediate predecessor", which we are now going to define.

Let $(X, \preceq)$ be an ordered set. We say that an element $x \in X$ is an *immediate predecessor of* an element $y \in X$ if

- $x \prec y$, and
- there is no $t \in X$ such that $x \prec t \prec y$.

Let us denote the just-defined relation of immediate predecessor by $\lhd$.

The claim that the ordering $\preceq$ can be reconstructed from the relation $\lhd$ may be formulated precisely as follows:

1.7.5 Proposition. *Let $(X, \preceq)$ be a finite ordered set, and let $\lhd$ be the corresponding immediate predecessor relation. Then for any two elements $x, y \in X$, $x \prec y$ holds if and only if there exist elements $x_1, x_2, \ldots, x_k \in X$ such that $x \lhd x_1 \lhd \cdots \lhd x_k \lhd y$ (possibly with $k = 0$, i.e. we may also have $x \lhd y$).*

Proof. One implication is easy to see: if we have $x \lhd x_1 \lhd \cdots \lhd x_k \lhd y$, then also $x \preceq x_1 \preceq \cdots \preceq x_k \preceq y$ (since the immediate predecessor relation is contained in the ordering relation), and by the transitivity of $\preceq$, we have $x \preceq y$.

The reverse implication is not difficult either, and we prove it by induction. We prove the following statement:

Lemma. *Let $x, y \in X$, $x \prec y$, be two elements such that there exist at most n elements $t \in X$ satisfying $x \prec t \prec y$ (i.e. "between" x and y). Then there exist $x_1, x_2, \ldots, x_k \in X$ such that $x \lhd x_1 \lhd \cdots \lhd x_k \lhd y$.*

For $n = 0$, the assumption of this lemma asserts that there exists no t with $x \prec t \prec y$, and hence $x \lhd y$, which means that the statement holds (we choose $k = 0$).

Let the lemma hold for all n up to some n_0, and let us have $x \prec y$ such that the set $M_{xy} = \{t \in X : x \prec t \prec y\}$ has $n = n_0 + 1$ elements. Let us choose an element $u \in M_{xy}$, and consider the sets $M_{xu} = \{t \in X : x \prec t \prec u\}$ and M_{uy} defined similarly. By the transitivity of $\prec$ it follows that $M_{xu} \subset M_{xy}$ and $M_{uy} \subset M_{xy}$. Both M_{xu} and M_{uy} have at least one element less than M_{xy} (since $u \notin M_{xu}$, $u \notin M_{uy}$), and by the inductive hypothesis, we find elements $x_1, \ldots, x_k$ and $y_1, \ldots, y_\ell$ in such a way that $x \lhd x_1 \lhd \cdots \lhd x_k \lhd u$ and $u \lhd y_1 \lhd \cdots \lhd y_\ell \lhd y$. By combining these two "chains" we obtain the desired sequence connecting x and y.
□

By the above proposition, it is enough to draw the relation of immediate predecessor by arrows. If we accept the convention that all arrows in the drawing will be directed upwards (this means that if $x \prec y$ then x is drawn higher than y), we need not even draw the direction of the arrows—it is enough to draw segments connecting the points. Such a picture of a partially ordered set is called its *Hasse diagram*. The following figure shows a 7-element linearly ordered set, such as $(\{1, 2, \ldots, 7\}, \leq)$:

The next drawing depicts the set $\{1, 2, \ldots, 10\}$ ordered by the divisibility relation (see Example 1.7.3):

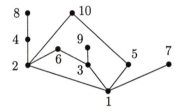

The following figure is a Hasse diagram of the set $\{1, 2, 3\} \times \{1, 2, 3\}$ with ordering $\preceq$ given by the rule $(a_1, b_1) \preceq (a_2, b_2)$ if and only if $a_1 \leq a_2$ and $b_1 \leq b_2$:

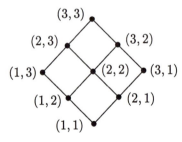

Finally, here is a Hasse diagram of the set of all subsets of $\{1, 2, 3\}$ ordered by inclusion:

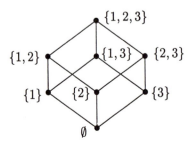

Further examples and notions concerning posets are left to the exercises. The theory of finite posets is an important and flourishing branch of combinatorics. The reader can learn about it in Trotter [28].

Exercises

1. Describe all relations on a set X which are equivalences and (partial) orderings at the same time.

2. Let R and S be arbitrary partial orderings on a set X. Decide which of the following relations are necessarily partial orderings:

   (a) $R \cap S$

   (b) $R \cup S$

   (c) $R \setminus S$

   (d) $R \circ S$.

3. Verify that the relation (1.3) in Example 1.7.2 indeed defines a partial ordering.

4. *Let R be a relation on a set X such that there is no finite sequence of elements $x_1, x_2, \ldots, x_k$ of X satisfying $x_1 R x_2, x_2 R x_3, \ldots, x_{k-1} R x_k$, $x_k R x_1$ (we say that such an R is *acyclic*). Prove that there exists an ordering $\preceq$ on X such that $R \subseteq \preceq$. You may assume that X is finite if this helps.

5. (a) Consider the set $\{1, 2, \ldots, n\}$ ordered by the divisibility relation $\mid$ (see Example 1.7.3). What is the maximum possible number of elements of a set $X \subseteq \{1, 2, \ldots, n\}$ which is ordered linearly by the relation $\mid$ (such a set X is called a *chain*)?

   (b) Solve the same question for the set $2^{\{1,2,\ldots,n\}}$ ordered by the relation $\subseteq$ (see Example 1.7.4).

6. Show that Proposition 1.7.5 does not hold for infinite sets.

7. Let $(X, \preceq)$ be a poset. An element $a \in X$ is called

   - a *largest element* of X if for every $x \in X$, $x \preceq a$ holds, and
   - a *maximal element* of X if there exists no $y \in X$ such that $a \prec y$.

   A *smallest element* and a *minimal element* are defined similarly.

   (a) Show that a largest element is always maximal, and find an example of a poset with a maximal element but no largest element.

   (b) Find a poset having no smallest element and no minimal element either, but possessing a largest element.

8. *Let $\preceq$ be any (partial) ordering on a set X. A *linear extension* of $\preceq$ is any linear ordering $\leq$ on X such that $x \preceq y$ implies $x \leq y$ for all $x, y \in X$. (If it didn't look so strange we could write this condition compactly as $\preceq \subseteq \leq$.) Prove that any partial ordering on a finite set X has at least one linear extension.

9. Let $(X, \leq)$, $(Y, \preceq)$ be ordered sets. We say that they are *isomorphic* (meaning that they "look the same" from the point of view of ordering) if there exists a bijection $f: X \to Y$ such that for every $x, y \in X$, we have $x \leq y$ if and only if $f(x) \preceq f(y)$.

   (a) Draw Hasse diagrams for all nonisomorphic 3-element posets.

   (b) Prove that any two n-element linearly ordered sets are isomorphic.

   (c) Find two nonisomorphic linear orderings of the set of all natural numbers.

   (d) Can you find infinitely many nonisomorphic linear orderings of **N**? *Uncountably many (for readers knowing something about the cardinalities of infinite sets)?

10. *Show that for every finite poset $(X, \preceq)$ there exists a finite set A and a system $\mathcal{M}$ of subsets of A such that the ordered set $(\mathcal{M}, \subseteq)$ is isomorphic to $(X, \preceq)$. (Isomorphism of posets has been defined in Exercise 9.)

11. Let $(X, \preceq)$ be a poset and let $A \subseteq X$ be its subset. An element $s \in X$ is called a *supremum* of the set A if the following holds:

    - $a \preceq s$ for each $a \in A$,
    - if $a \preceq s'$ holds for all $a \in A$, where s' is some element of X, then $s \preceq s'$.

    The *infimum* of a subset $A \subseteq X$ is defined analogously, but with all inequalities going in the opposite direction.

    (a) Check that any subset $A \subseteq X$ has at most one supremum and at most one infimum. (The supremum of A, if it exists, is denoted by $\sup A$. Similarly $\inf A$ denotes the infimum.)

    (b) Which element is the supremum of the empty set (according to the definition just given)?

    (c) Find an example of a poset in which every nonempty subset has an infimum, but there are nonempty subsets having no supremum.

    (d) *Let $(X, \preceq)$ be a poset in which every subset (including the empty one) has a supremum. Show then that every subset has an infimum as well.

12. Consider the poset $(\mathbf{N}, |)$ (ordering by divisibility).

    (a) Decide whether each nonempty subset of **N** has a supremum.

    (b) Decide whether each nonempty finite subset of **N** has a supremum.

    (c) Decide whether each nonempty subset has an infimum.

2

Combinatorial counting

In this chapter, we are going to consider problems concerning the number of various configurations, such as "How many ways are there to send n distinct postcards to n friends?", "How many mappings of an n-element set to an m-element set are there?", and so on. We begin with simple examples that can usually be solved with common sense (plus, maybe, some cleverness) and require no special knowledge. Later on, we will come to somewhat more advanced techniques.

2.1 Functions and subsets

As promised, we begin with a simple problem with postcards.

Problem. Professor X. (no real person meant), having completed a successful short-term visit at the School of Mathematical Contemplation and Machine Cleverness in the city of Y., strolls around one sunny day and decides to send a postcard to each of his friends Alice, Bob, Arthur, Merlin, and HAL-9000. A street vendor nearby sells 26 kinds of postcards with great sights of Y.'s historical center. How many possibilities does Professor X. have for sending postcards to his 5 friends?

Since the postcard for each friend can be picked in 26 ways, and the 5 selections are independent (making some of them doesn't influence the remaining ones), the answer to this problem is 26^5. In a more abstract language, we have thereby counted the number of all mappings of a 5-element set (Prof. X.'s friends) to a 26-element set (the postcard types). Here is another closely related problem:

Problem. How many distinct 5-letter words are there (using the 26-letter English alphabet, and including meaningless words such as $ywizp$[1])?

---

[1] Who can tell which words are meaningless? It might mean something in Tralfamadorian, say.

Since each of the 5 letters can be picked independently in 26 ways, it is not hard to see that the answer is again 26^5. And, indeed, a 5-letter word can be understood as a mapping of the set $\{1, 2, \ldots, 5\}$ to the set $\{a, b, \ldots, z\}$ of letters: for each of the 5 positions in the word, numbered $1, 2, \ldots, 5$, we specify the letter in that position. Finding such simple transformations of counting problems is one of the basic skills of the art of counting.

In the next proposition, we count mappings of an n-element set to an m-element set. The idea is exactly the same as for counting the ways of sending the postcards, but we use this opportunity to practice more rigorous mathematical proofs in a simple situation.

2.1.1 Proposition. *Let N be an n-element set (it may also be empty, i.e. we admit $n = 0, 1, 2, \ldots$) and let M be an m-element set, $m \geq 1$. Then the number of all possible mappings $f : N \to M$ is m^n.*

Proof. We can proceed by induction on n. What does the proposition say for $n = 0$? In this case, we consider all mappings f of the set $N = \emptyset$ to some set M. The definition of a mapping tells us that such an f should be a set of ordered pairs (x, y) with $x \in N = \emptyset$ and $y \in M$. Since the empty set has no elements, f cannot possibly contain any such ordered pairs, and hence the only possibility is that f is the empty set (no ordered pairs). On the other hand, $f = \emptyset$ does indeed satisfy the definition of a mapping in this case: the definition says that for each $x \in N$ something should be true, but there are no $x \in N$. Therefore, exactly 1 mapping $f : \emptyset \to M$ exists. This agrees with the formula, because $m^0 = 1$ for any $m \geq 1$. We have verified the $n = 0$ case as a basis for the induction.

Many would object that a mapping of the empty set makes no sense and so it is useless to consider it, and we could really start the induction with $n = 1$ without any difficulty. But, in mathematical considerations, it often pays to clarify such "borderline" cases, to find out what exactly the general definition says about them. This allows us to avoid various exceptions and special cases later on, or missing cases and mistakes in proofs. It is similar to the usefulness of defining an empty sum (with no addends) as 0 etc.

Next, suppose that the proposition has been proved for all $n \leq n_0$ and for all $m \geq 1$. We set $n = n_0 + 1$ and we consider an n-element set N and an m-element set M. Let us fix an arbitrary element $a \in N$. To specify a mapping $f : N \to M$ is the same as specifying the value

$f(a) \in M$ plus giving a mapping $f': N \setminus \{a\} \to M$. The value $f(a)$ can be chosen in m ways, and for the choice of f' we have m^{n-1} ways by the inductive hypothesis. Each choice of $f(a)$ can be combined with any choice of f', and so the total number of possibilities for f equals $m \cdot m^{n-1} = m^n$. Here is a picture for the more visually oriented reader:

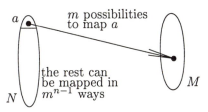

$\square$

2.1.2 Proposition. *Any n-element set X has exactly 2^n subsets* $(n \geq 0)$.

This is another simple and well-known counting result. Let us give two proofs.

First proof (by induction). For $X = \emptyset$, there exists a single subset, namely $\emptyset$, and this agrees with the formula $2^0 = 1$.

Having an $(n+1)$-element set X, let us fix one element $a \in X$, and divide the subsets of X into two classes: those not containing a and those containing it. The first class are exactly all the subsets of the n-element set $X \setminus \{a\}$, and their number is 2^n by the inductive hypothesis. For each subset A of the second class, let us consider the set $A' = A \setminus \{a\}$. This is a subset of $X \setminus \{a\}$. Clearly, each subset $A' \subseteq X \setminus \{a\}$ is obtained from exactly one set A of the second class, namely from $A' \cup \{a\}$. In other words, there is a bijection between all subsets of the first class and all subsets of the second class. Hence the number of subsets of the second class is 2^n as well, and altogether we have $2^n + 2^n = 2^{n+1}$ subsets of the $(n+1)$-element set X as it should be. $\square$

Second proof (reduction to a known result). Consider an arbitrary subset A of a given n-element set X, and define a mapping $f_A: X \to \{0, 1\}$. For an element $x \in X$ we put

$$f_A(x) = \begin{cases} 1 & \text{if } x \in A \\ 0 & \text{if } x \notin A. \end{cases}$$

(This mapping is often encountered in mathematics; it is called the *characteristic function* of the set A.) Schematically,

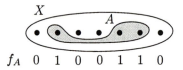

$$f_A \quad 0 \quad 1 \quad 0 \quad 0 \quad 1 \quad 1 \quad 0$$

Distinct sets A have distinct mappings f_A, and conversely, any given mapping $f: X \to \{0,1\}$ determines a subset $A \subseteq X$ with $f = f_A$. Hence the number of subsets of X is the same as the number of all mappings $X \to \{0,1\}$, and this is 2^n by Proposition 2.1.1. $\square$

Now, a somewhat more difficult result:

2.1.3 Proposition. *Let $n \geq 1$. Each n-element set has exactly 2^{n-1} subsets of an odd size and exactly 2^{n-1} subsets of an even size.*

Proof. We make use of Proposition 2.1.2. Let us fix an element $a \in X$. Any subset $A \subseteq X \setminus \{a\}$ can be completed to a subset $A' \subseteq X$ with an odd number of elements, by the following rule: if $|A|$ is odd, we put $A' = A$, and for $|A|$ even, we put $A' = A \cup \{a\}$. It is easy to check that this defines a bijection between the system of all subsets of $X \setminus \{a\}$ and the system of all odd-size subsets of X. Therefore, the number of subsets of X of odd cardinality is 2^{n-1}.

For subsets of an even size, we can proceed similarly, or we can simply say that their number must be 2^n minus the number of odd-size subsets, i.e. $2^n - 2^{n-1} = 2^{n-1}$. $\square$

Later on, we will examine several more proofs.

Injective mappings.

Problem. Professor X., having spent some time contemplating the approximately 12 million possibilities of sending his postcards, returned to the street vendor and wanted to buy his selection. But the vendor had already sold all the postcards and was about to close. After some discussion, he admitted he still had one sample of each of the 26 postcards, and was willing to sell 5 of them to Professor X. for $5 apiece. In this situation, Professor X. has to make a new decision about which postcard is best for whom. How many possibilities does he have now?

As Professor X. (and probably the reader too) recognized, one has to count one-to-one mappings from a 5-element set to a 26-element set. This is the same as counting 5-letter words with all letters distinct.

2.1.4 Proposition. *For given numbers $n, m \geq 0$, there exist exactly*

$$m(m-1)\ldots(m-n+1) = \prod_{i=0}^{n-1}(m-i)$$

one-to-one *mappings of a given n-element set to a given m-element set.*

Proof. We again proceed by induction on n (and more concisely this time). For $n = 0$, the empty mapping is one-to-one, and so exactly 1 one-to-one mapping exists, and this agrees with the fact that the value of an empty product has been defined as 1. So the formula holds for $n = 0$.

Next, we note that no one-to-one mapping exists for $n > m$, and this again agrees with the formula (since one factor equal to 0 appears in the product).

Let us now consider an n-element set N, $n \geq 1$, and an m-element set M, $m \geq n$. Fix an element $a \in N$ and choose the value $f(a) \in M$ arbitrarily, in one of m possible ways. It remains to choose a one-to-one mapping of the set $N\backslash\{a\}$ to the set $M\backslash\{f(a)\}$. By the inductive assumption, there are $(m-1)(m-2)\ldots(m-n+1)$ possibilities for the latter choice. Altogether we have $m(m-1)(m-2)\ldots(m-n+1)$ one-to-one mappings $f: N \to M$. (Where is the picture? Well, these days, you can't expect to have everything in a book in this price category.) □

As we have noted for the postcards and 5-letter words, choosing a one-to-one mapping of an n-element set to an m-element set can also be viewed as selecting n objects from m distinct objects, where the order of the selected objects is important (i.e. we construct an ordered n-tuple). Such selections are sometimes called *variations of n elements from m elements without repetition.*

Exercises

1. Let $X = \{x_1, x_2, \ldots, x_n\}$ be an n-element set. Describe how each subset of X can be encoded by an n-letter word consisting of the letters a and b. Infer that the number of subsets of X is 2^n. (This is very similar to the second proof of Proposition 2.1.2.)

2. Determine the number of ordered pairs (A, B), where $A \subseteq B \subseteq \{1, 2, \ldots, n\}$.

3. Let N be an n-element set and M an m-element set. Define a bijection between the set of all mappings $f \colon N \to M$ and the n-fold Cartesian product M^n.

4. Among the numbers $1, 2, \ldots, 10^{10}$, are there more of those containing the digit 9 in their decimal notation, or of those with no 9?

5. (a) How many $n \times n$ matrices are there with entries chosen from the numbers $0, 1, \ldots, q - 1$?

   (b) *Let q be a prime. How many matrices as in (a) have a determinant that is not divisible by q? (In other words, how many nonsingular matrices over the q-element field are there?)

6. *Show that a natural number $n \geq 1$ has an odd number of divisors (including 1 and itself) if and only if $\sqrt{n}$ is an integer.

2.2   Permutations and factorials

A bijective mapping of a finite set X to itself is called a *permutation* of the set X.

If the elements of X are arranged in some order, we can also imagine a permutation as rearranging the elements in a different order. For instance, one possible permutation p of the set $X = \{a, b, c, d\}$ is given by $p(a) = b$, $p(b) = d$, $p(c) = c$, and $p(d) = a$. This can also be written in a two-row form:

$$\begin{pmatrix} a & b & c & d \\ b & d & c & a \end{pmatrix}.$$

In the first row, we have listed the elements of X, and under each element $x \in X$ in the first row, we have written the element $p(x)$ into the second row. Most often one works with permutations of the set $\{1, 2, \ldots, n\}$. If we use the convention that the first row always lists the numbers $1, 2, \ldots, n$ in the natural order, then it suffices to write the second row only. For example, (2 4 3 1) denotes the permutation p with values $p(1) = 2$, $p(2) = 4$, $p(3) = 3$, and $p(4) = 1$.

In the literature, permutations of a set X are sometimes understood as arrangements of the elements of X in some order, i.e. as linear orderings on the set X. This may be a quite useful point of view, but we will mostly regard permutations as mappings. This has some formal advantages. For instance, permutations can be composed (as mappings).

Yet another way of writing permutations is to use their so-called cycles. Cycles are perhaps easiest to define if we depict a permutation using arrows, in the way we depicted relations using arrows in Section 1.5. In the case of a permutation $p: X \to X$, we draw the elements of the set X as dots, and we draw an arrow from each dot x to the dot $p(x)$. For example, for the permutation $p = (4\ 8\ 3\ 5\ 2\ 9\ 6\ 1\ 7)$ (this is the one-line notation introduced above!), such a picture looks as follows:

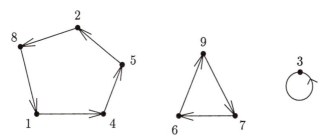

Each point has exactly one outgoing arrow and exactly one ingoing arrow. It is easy to see that the picture of a permutation consists of several disjoint groups of dots, where the dots in each group are connected by arrows into a cycle. One can walk around such a cycle in one direction following the arrows. The groups of elements connected together by these cycles are called the *cycles* of the considered permutation. (Any reader who is not satisfied with this pictorial definition can find a formal definition of a cycle in Exercise 2.) Using the cycles, the depicted permutation p can also be written $p = ((1,4,5,2,8)(3)(6,9,7))$. In each of the inner parentheses, the elements of one cycle are listed in the order along the arrows, starting with the smallest element in that cycle.

What can permutations be good for? They are studied, for instance, in the design and analysis of various sorting algorithms. Certain efficient algorithms for problems with graphs, or with geometric objects, start by rearranging the input objects into a random order, i.e. by performing a random permutation with them. In the so-called group theory, which is extremely important in almost all mathematics and also in many areas of modern physics, groups of permutations (with composition as the group operation) are one of the basic objects of study. An ultimate reason for the impossibility of a general algebraic solution of algebraic equations of degree 5 is in the properties of the group of all permutations on the 5-element set. Rubik's Cube, a toy which used to be extremely popular at the beginning of the 1980s, provides a pretty example of a complicated permutation group. Surprisingly involved properties of permutations are applied in a mathematical analysis of card shuffling. This is just a small sample of areas where permutations play a role.

According to Proposition 2.1.4, the number of permutations of an n-element set is $n(n-1)\cdot\ldots\cdot 2\cdot 1$. This number, regarded as a function of n, is denoted by $n!$ and is called n *factorial*. Hence we have

$$n! = n(n-1)\cdot\ldots\cdot 2\cdot 1 = \prod_{j=0}^{n-1}(n-j) = \prod_{i=1}^{n} i.$$

In particular, for $n = 0$ we have $0! = 1$ (because $0!$ is defined as the empty product).

Exercises

1. How many permutations of $\{1, 2, \ldots, n\}$ have a single cycle?

2. For a permutation $p\colon X \to X$, let p^k denote the permutation arising by a k-fold composition of p, i.e. $p^1 = p$ and $p^k = p \circ p^{k-1}$. Define a relation $\approx$ on the set X as follows: $i \approx j$ if and only if there exists a $k \geq 1$ such that $p^k(i) = j$. Prove that $\approx$ is an equivalence relation on X, and that its classes are the cycles of p.

3. Let p be a permutation, and let p^k be defined as in Exercise 2. By the *order* of the permutation p we mean the smallest natural number $k \geq 1$ such that $p^k = \mathrm{id}$, where id denotes the identity permutation (mapping each element onto itself).

   (a) Determine the order of the permutation (2 3 1 5 4 7 8 9 6).

   (b) Show that each permutation p of a finite set has a well-defined finite order, and *show how to compute the order using the lengths of the cycles of p.

4. ^{CS} Write a program that lists all permutations of the set $\{1, 2, \ldots, n\}$, each of them exactly once. Use a reasonable amount of memory even if $n!$, the number of permutations, is astronomically large. *Can you make the total number of operations of the algorithm proportional to $n!$, if the operations needed for the output (printing the permutations, say) are not counted?

5. (This is an exercise for those who are getting bored by the easy material covered in the first two sections of this chapter.) Let p be a permutation of the set $\{1, 2, \ldots, n\}$. Let us write it in the one-line notation, and let us mark the *increasing segments* in the resulting sequence of numbers, for example, (4 5 7 2 6 8 3 1). Let $f(n, k)$ denote the number of permutations of an n-element set with exactly k increasing segments.

(a) *Prove that $f(n, k) = f(n, n+1-k)$, and derive that the average number of increasing segments of a permutation is $(n+1)/2$ (the average is taken over all permutations of $\{1, 2, \ldots, n\}$).

(b) *Derive the following recurrent formula:
$f(n, k) = k \cdot f(n-1, k) + (n+1-k)f(n-1, k-1)$.

(c) Using (b), determine the number of permutations of $\{1, 2, \ldots, n\}$ with 2 increasing segments, with 3 increasing segments, and *with k increasing segments.

(d) *For a randomly chosen permutation π of the set $\{1, 2, \ldots, n\}$, calculate the probability that the first increasing segment has length k. Show that for n large, the average length of the first increasing segments approaches the number $e - 1$.

Remark. These and similar questions have been studied in the analysis of various algorithms for sorting.

6. Let π be a permutation of the set $\{1, 2, \ldots, n\}$. We say that an ordered pair $(i, j) \in \{1, 2, \ldots, n\} \times \{1, 2, \ldots, n\}$ is an *inversion* of π if $i < j$ and $\pi(i) > \pi(j)$.

(a) Prove that the set $I(\pi)$ of all inversions, regarded as a relation on $\{1, 2, \ldots, n\}$, is transitive.

(b) Prove that the complement of $I(\pi)$ is transitive too.

(c) ^{CS} Consider some sorting algorithm which rearranges n input numbers into a nondecreasing order, and in each step, it is only allowed to exchange two neighboring numbers (in the current order). Prove that there are input sequences whose sorting requires at least cn^2 steps of this algorithm, where $c > 0$ is some suitable constant.

(d) *,^{CS} Can you design an algorithm that calculates the number of inversions of a given permutation of $\{1, 2, \ldots, n\}$ using substantially less than n^2 steps? (See e.g. Knuth [38] for several solutions.)

7. (a) *Find out what is the largest power of 10 dividing the number 70! (i.e. the number of trailing zeros in the decimal notation for 70!).

(b) *Find a general formula for the highest power k such that $n!$ is divisible by p^k, where p is a given prime number.

8. Show that for every $k, n \geq 1$, $(k!)^n$ divides $(kn)!$.

2.3 Binomial coefficients

Let $n \geq k$ be nonnegative integers. The *binomial coefficient* $\binom{n}{k}$ (read "n choose k") is a function of the variables n, k defined by the formula

$$\binom{n}{k} = \frac{n(n-1)(n-2)\ldots(n-k+1)}{k(k-1)\cdot\ldots\cdot 2 \cdot 1} = \frac{\prod_{i=0}^{k-1}(n-i)}{k!}. \qquad (2.1)$$

The reader might know another formula, namely

$$\binom{n}{k} = \frac{n!}{k!(n-k)!}. \tag{2.2}$$

In our situation, this is equivalent to (2.1). Among these two possible definitions, the first one, (2.1), has some advantages. The numerical value of $\binom{n}{k}$ is more easily computed from it, and one also gets smaller intermediate results in the calculation. Moreover, (2.1) makes sense for an arbitrary real number n (more about this in Chapter 10), and, in particular, it defines the value of $\binom{n}{k}$ also for a natural number $n < k$; in such cases, $\binom{n}{k} = 0$.

The basic combinatorial meaning of the binomial coefficient $\binom{n}{k}$ is the *number of all k-element subsets of an n-element set*. We prove this in a moment, but first we introduce some notation.

2.3.1 Definition. *Let X be a set and let k be a nonnegative integer. By the symbol*

$$\binom{X}{k}$$

we denote the set of all k-element subsets of the set X.

For example, $\binom{\{a,b,c\}}{2} = \{\{a,b\}, \{a,c\}, \{b,c\}\}$. The symbol $\binom{x}{k}$ has two meanings now, depending on whether x is a number or a set. The following propositions put them into a close connection:

2.3.2 Proposition. *For any finite set X, the number of all k-element subsets equals $\binom{|X|}{k}$.*

In symbols, this statement can be written

$$\left| \binom{X}{k} \right| = \binom{|X|}{k}.$$

Proof. Put $n = |X|$. We will count all *ordered k-tuples* of elements of X (without repetitions of elements) in two ways. On the one hand, we know that the number of the ordered k-tuples is $n(n-1)\ldots(n-k+1)$ by Proposition 2.1.4 (see the remark following its proof). On the other hand, from one k-element subset $M \in \binom{X}{k}$, we can create $k!$ distinct ordered k-tuples, and each ordered k-tuple is obtained from exactly one k-element subset M in this way. Hence

$$n(n-1)\ldots(n-k+1) = k! \left| \binom{X}{k} \right|.$$

$\square$

Another basic problem leading to binomial coefficients. How many ways are there to write a nonnegative integer m as a sum of r nonnegative integer addends, where the order of addends is important? For example, for $n = 3$ and $r = 2$, we have the possibilities $3 = 0+3$, $3 = 1+2$, $3 = 2+1$, and $3 = 0+3$. In other words, we want to find out how many ordered r-tuples $(i_1, i_2, \ldots, i_r)$ of nonnegative integers there are satisfying the equation

$$i_1 + i_2 + \cdots + i_r = m. \tag{2.3}$$

The answer is the binomial coefficient $\binom{m+r-1}{r-1}$. This can be proved in various ways. Here we describe a proof almost in the style of a magician's trick.

Imagine that each of the variables $i_1, i_2, \ldots, i_r$ corresponds to one of r boxes. We have m indistinguishable balls, and we want to distribute them into these boxes in some way (we assume that each box can hold all the m balls if needed). Each possible distribution encodes one solution of Eq. (2.3). For example, for $m = 7$ and $r = 6$, the solution $0+1+0+3+1+2 = 7$ corresponds to the distribution

So we are interested in the number of distributions of the balls into boxes. We now let the bottoms and the leftmost and rightmost walls of the boxes disappear, so that only m balls and $r-1$ walls separating the boxes remain:

$$| \bullet | \: | \bullet \bullet \bullet | \bullet | \bullet \bullet |$$

(we have also moved the balls and walls for a better aesthetic impression). This situation still contains full information about the distribution of the balls into boxes. Hence, choosing a distribution of the balls means selecting the position of the internal walls among the balls. In other words, we have $m + r - 1$ objects, balls and internal walls, arranged in a row, and we determine which positions will be occupied by balls and which ones by walls. This corresponds to a selection of a subset of $r - 1$ positions from $m + r - 1$ positions, and this can be done in $\binom{m+r-1}{r-1}$ ways. $\qquad\square$

Simple properties of binomial coefficients. One well-known formula is

$$\binom{n}{k} = \binom{n}{n-k}.\tag{2.4}$$

Its correctness (for $n \geq k \geq 0$) can immediately be seen from the already-mentioned formula $\binom{n}{k} = \frac{n!}{k!(n-k)!}$. Combinatorially, Eq. (2.4) means that the number of k-element subsets of an n-element set is the same as the number of subsets with $n-k$ elements. This can be verified directly without referring to binomial coefficients—it suffices to assign to each k-element subset its complement.

Here is another important formula, attributed to Pascal:

$$\binom{n-1}{k-1} + \binom{n-1}{k} = \binom{n}{k}.\tag{2.5}$$

One elegant proof is based on a combinatorial interpretation of both sides of Eq. (2.5). The right-hand side is the number of k-element subsets of some n-element set X. Let us fix one element $a \in X$ and divide all k-element subsets of X into two groups depending on whether they contain a or not. The subsets not containing a are exactly all k-element subsets of $X \setminus \{a\}$, and so their number is $\binom{n-1}{k}$. If A is some k-element subset of X containing a, then we can assign the $(k-1)$-element set $A' = A \setminus \{a\}$ to A. It is easy to check that this assignment is a bijection between all k-element subsets of X containing the element a and all $(k-1)$-element subsets of $X \setminus \{a\}$. The number of the latter is $\binom{n-1}{k-1}$. Altogether, the number of all k-element subsets of X equals $\binom{n-1}{k} + \binom{n-1}{k-1}$. $\square$

The identity (2.5) is closely related to the so-called *Pascal triangle*:

$$
\begin{array}{ccccccccccc}
 & & & & & 1 & & & & & \\
 & & & & 1 & & 1 & & & & \\
 & & & 1 & & 2 & & 1 & & & \\
 & & 1 & & 3 & & 3 & & 1 & & \\
 & 1 & & 4 & & 6 & & 4 & & 1 & \\
1 & & 5 & & 10 & & 10 & & 5 & & 1 \\
 & & & \vdots & & & & \vdots & & &
\end{array}
$$

Every successive row in this scheme is produced as follows: under each pair of consecutive numbers in the preceding row, write their sum, and complement the new row by 1s on both sides. An induction using (2.5) shows that the $(n+1)$-st row contains the binomial coefficients $\binom{n}{0}$, $\binom{n}{1}, \ldots, \binom{n}{n}$.

Binomial theorem. Eq. (2.5) can be used for a proof of another well-known statement involving binomial coefficients: the binomial theorem.

2.3.3 Theorem (Binomial theorem). *For any nonnegative integer* n, *we have*

$$(1+x)^n = \sum_{k=0}^{n} \binom{n}{k} x^k \tag{2.6}$$

(this is an equality of two polynomials in the variable x, *so in particular it holds for any specific real number* x).

From the binomial theorem, we can infer various relations among binomial coefficients. Perhaps the simplest one arises by substituting $x = 1$, and it reads

$$\binom{n}{0} + \binom{n}{1} + \binom{n}{2} + \cdots + \binom{n}{n} = 2^n. \tag{2.7}$$

Combinatorially, this is nothing else than counting all subsets of an n-element set. On the left-hand side, they are divided into groups according to their size.

Second proof of Proposition 2.1.3 (about the number of odd-size subsets). By substituting $x = -1$ into the binomial theorem, we arrive at

$$\binom{n}{0} - \binom{n}{1} + \binom{n}{2} - \binom{n}{3} + \cdots = \sum_{k=0}^{n} (-1)^k \binom{n}{k} = 0. \tag{2.8}$$

Adding this equation to Eq. (2.7) leads to

$$2\left[\binom{n}{0} + \binom{n}{2} + \binom{n}{4} + \cdots \right] = 2^n.$$

The brackets on the left-hand side contain the total number of even-size subsets of an n-element set. Therefore, the number of even-size subsets equals 2^{n-1}. The odd-size subsets can be counted as a complement to 2^n. $\qquad\square$

Further identities with binomial coefficients. Literally thousands of formulas and identities with binomial coefficients are known and whole books are devoted to them. Here we present one more formula with a nice combinatorial proof. More formulas and methods on how to derive them will be given in the exercises and in Chapter 10.

2.3.4 Proposition.

$$\sum_{i=0}^{n} \binom{n}{i}^2 = \binom{2n}{n}.$$

Proof. The first trick is to rewrite the sum using the symmetry of binomial coefficients, (2.4), as

$$\sum_{i=0}^{n} \binom{n}{i}\binom{n}{n-i}.$$

Now we show that this sum expresses the number of n-element subsets of a $2n$-element set (and so it equals the right-hand side in the formula being proved). Consider a $2n$-element set X, and color n of its elements red and the remaining n elements blue. To choose an n-element subset of X now means choosing an i-element subset of the red elements plus an $(n-i)$-element subset of the blue elements, where $i \in \{0, 1, \ldots, n\}$:

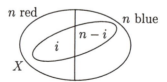

For a given i, there are $\binom{n}{i}$ possibilities to choose the red subset and, independently, $\binom{n}{n-i}$ possibilities for the blue subset. Altogether, an n-element subset of X can be selected in $\sum_{i=0}^{n} \binom{n}{i}\binom{n}{n-i}$ ways.   □

Multinomial coefficients and the multinomial theorem. Here is one of the favorite problems of American textbooks: how many distinct words, including nonsense ones, can be produced using all the letters of the word MISSISSIPPI? In other words, how many distinct ways are there to rearrange these letters? First, imagine that the letters in the name are distinguished somehow, so that we have 4 different Ss etc. In our text, we distinguish them by indices: $M_1 I_1 S_1 S_2 I_2 S_3 S_4 I_3 P_1 P_2 I_4$. So we have 11 distinct letters, and these can be permuted in 11! distinct ways. Now consider one (arbitrary) word produced from a "nonindexed" MISSISSIPPI, such as SIPISMSIPIS. From how many "indexed" words do we get this word by deleting the indices? The indices of the 4 letters S can be placed in 4! ways, the indices of the 4 letters I can be arranged (independently) in 4! ways, for the 2 letters P we have 2! possibilities, and finally for the single M we have 1 (or 1!) possibility. Thus, the word SIPISMSIPIS, and also any other word created from MISSISSIPPI, can be indexed in 4!4!2!1! ways. The number of nonindexed words, which is the answer to the problem, is $11!/(4!4!2!1!)$.

The same argument leads to the following general result: if we have objects of m kinds, k_i indistinguishable objects of the ith kind, where $k_1 + k_2 + \cdots + k_m = n$, then the number of distinct arrangements of these objects in a row is given by the expression

$$\frac{n!}{k_1! k_2! \ldots k_m!}.$$

This expression is usually written

$$\binom{n}{k_1, k_2, \ldots, k_m}$$

and is called a *multinomial coefficient*. In particular, for $m = 2$ we get a binomial coefficient, i.e. $\binom{n}{k, n-k}$ denotes the same thing as $\binom{n}{k}$. Why the name "multinomial coefficient"? It comes from the following theorem:

2.3.5 Theorem (Multinomial theorem). *For arbitrary real numbers $x_1, x_2, \ldots, x_m$ and any natural number $n \geq 1$, the following equality holds:*

$$(x_1 + x_2 + \cdots + x_m)^n = \sum_{\substack{k_1 + \cdots + k_m = n \\ k_1, \ldots, k_m \geq 0}} \binom{n}{k_1, k_2, \ldots, k_m} x_1^{k_1} x_2^{k_2} \ldots x_m^{k_m}.$$

The right-hand side of this formula usually has fairly many terms (we sum over all possible ways of writing n as a sum of m nonnegative integers). But the theorem is most often applied to determine the coefficient of some particular term. For example, it tells us that the coefficient of $x^2 y^3 z^5$ in $(x + y + z)^{10}$ is $\binom{10}{2,3,5} = 2520$.

The multinomial theorem can be proved by induction on n (see Exercise 26). A more natural proof can be given by the methods we discuss in Chapter 10.

Exercises

1. Formulate the problem of counting all k-element subsets of an n-element set as a problem with sending or buying postcards.

2. Prove the addition formula (2.5) by using the definition (2.1) of binomial coefficients and by manipulating expressions.

3. (a) Prove the formula

$$\binom{r}{r} + \binom{r+1}{r} + \binom{r+2}{r} + \cdots + \binom{n}{r} = \binom{n+1}{r+1} \qquad (2.9)$$

by induction on n (for r arbitrary but fixed). Note what the formula says for $r = 1$.

(b) *Prove the same formula combinatorially.

4. *For natural numbers $m \le n$ calculate (i.e. express by a simple formula not containing a sum) $\sum_{k=m}^{n} \binom{k}{m} \binom{n}{k}$.

5. Calculate (i.e. express by a simple formula not containing a sum)

(a) $\sum_{k=1}^{n} \binom{k}{m} \frac{1}{k}$,

(b) * $\sum_{k=0}^{n} \binom{k}{m} k$.

6. **Prove that

$$\sum_{k=0}^{m} \binom{m}{k}\binom{n+k}{m} = \sum_{k=0}^{m} \binom{n}{k}\binom{m}{k} 2^k.$$

7. *How many functions $f\colon \{1, 2, \ldots, n\} \to \{1, 2, \ldots, n\}$ are there that are *monotone*; that is, for $i < j$ we have $f(i) \le f(j)$?

8. How many terms are there in the sum on the right-hand side of the formula for $(x_1 + \cdots + x_m)^n$ in the multinomial theorem?

9. *How many k-element subsets of $\{1, 2, \ldots, n\}$ exist containing no two consecutive numbers?

10. (a) Using formula (2.9) for $r = 2$, calculate the sums $\sum_{i=2}^{n} i(i-1)$ and $\sum_{i=1}^{n} i^2$.

(b) Using (a) and (2.9) for $r = 3$, calculate $\sum_{i=1}^{n} i^3$.

(c) *Derive the result of (b) using Fig. 2.1 (the figure is drawn for the case $n = 4$).

11. Prove the binomial theorem by induction on n.

12. For a real number x and a natural number n, let the symbol $x^{\underline{n}}$ denote $x(x-1)(x-2)\ldots(x-n+1)$ (the so-called *nth factorial power* of x). Prove the following analog of the binomial theorem:

$$(x + y)^{\underline{n}} = \sum_{i=0}^{n} \binom{n}{i} x^{\underline{i}} y^{\underline{n-i}}.$$

Proceed by induction on n.

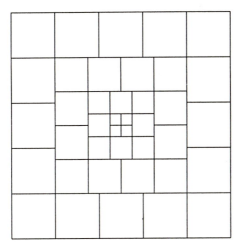

Fig. 2.1 A graphical derivation of the formula for $1^3 + 2^3 + \cdots + n^3$.

13. Prove the so-called *Leibniz formula* for the differentiation of a product. Let u, v be real functions of a single real variable, and let $f^{(k)}$ denote the kth derivative of a function f. Then

$$(uv)^{(n)} = \sum_{k=0}^{n} \binom{n}{k} u^{(k)} v^{(n-k)}$$

(supposing that all the derivatives in the formula exist). The case $n = 1$ is the formula for differentiating a product, $(uv)' = u'v + uv'$, which you may assume as being known.

14. ^{CS} Write a computer program that lists all k-element subsets of the set $\{1, 2, \ldots, n\}$, each of them exactly once. Use a reasonable amount of memory even if $\binom{n}{k}$, the number of such subsets, is very large. *Can you make the total number of operations of the algorithm proportional to $\binom{n}{k}$, if the operations needed for the output are not counted?

15. Let p be a prime and let n, k be natural numbers.

    (a) Prove that for $k < p$, $\binom{p}{k}$ is divisible by p.

    (b) Prove that $\binom{n}{p}$ is divisible by p if and only if $\lfloor n/p \rfloor$ is divisible by p.

16. (a) *Using the binomial theorem, derive a formula for the number of subsets of cardinality divisible by 4 of an n-element set.

    (b) *Count the subsets of size divisible by 3 of an n-element set.

17. We have n kinds of objects, and we want to determine the number of ways in which a k-tuple of objects can be selected. We consider variants: we may be interested in selecting ordered or unordered k-tuples,

and we may have either just 1 object of each kind or an unlimited supply of indistinguishable objects of each kind. Fill out the formulas in the following table:

	Only 1 object of each kind	Arbitrarily many objects of each kind
Ordered k-tuples		
Unordered k-tuples		

18. We have k balls, and we distribute them into n (numbered) bins. Fill out the formulas for the number of distributions for various variants of the problem in the following table:

	At most 1 ball into each bin	Any number of balls into each bin
Balls are distinguishable (have distinct colors)		
Balls are indistinguishable		

19. *How many ways are there to arrange 7 elves and 5 goblins in a row in such a way that no goblins stand next to each other?

20. A table is set with 13 large plates. We have 5 lobsters (indistinguishable ones) and 8 stuffed snails (also indistinguishable). We are interested in the number of ways to serve the snails and lobsters on the plates. The order of serving is important. Imagine we were writing a script for a movie: "Put a snail on plate no. 3, then serve a lobster on plate no. 11...". Only one item is served at a time. How many ways are there if

(a) if there are no restrictions, everything can come on the same plate, say, and

(b) if at least 1 item should come on each plate?

21. Draw a triangle ABC. Draw n points lying on the side AB (but different from A and B) and connect all of them by segments to the vertex C. Similarly, draw n points on the side AC and connect them to B.

(a) How many intersections of the drawn segments are there? Into how many regions is the triangle ABC partitioned by the drawn segments?

(b) *Draw n points also on the side BC and connect them to A. Assume that no 3 of the drawn segments intersect at a single point. How many intersections are there now?

(c) *How many regions are there in the situation of (b)?

22. Consider a convex n-gon such that no 3 diagonals intersect at a single point. Draw all the diagonals (i.e. connect every pair of vertices by a segment).

    (a) *How many intersections do the diagonals determine?

    (b) *Into how many parts is the polygon divided by the diagonals?

23. (Cayley's problem) *Consider a regular convex n-gon P with vertices $A_1, A_2, \ldots, A_n$. How many ways are there to select k of these n vertices, in such a way that no two of the selected vertices are consecutive (in other words, if we draw the polygon determined by the selected vertices, it has no side in common with P)? Hint: First, calculate the number of such selections including A_1.

24. *Consider a regular n-gon. We divide it by nonintersecting diagonals into triangles (i.e. we *triangulate* it), in such a way that each of the resulting triangles has at least one side in common with the original n-gon.

    (a) How many diagonals must we draw? How many triangles do we get?

    (b) *How many such triangulations are there?

25. (a) What is the coefficient of $x^2 y^3 z$ in the polynomial $(2x - y^2 + 3z)^6$? What about the coefficient of $x^2 y^2 z$?

    (b) Find the coefficient of $x^2 y^8 z$ in $(2x + y^2 - 5z)^7$.

    (c) What is the coefficient of $u^2 v^3 z^3$ in $(3uv - 2z + u + v)^7$?

26. (a) Prove the equality

$$
\binom{n}{k_1, k_2, \ldots, k_m} = \binom{n-1}{k_1 - 1, k_2, k_3, \ldots, k_m}
$$
$$
+ \binom{n-1}{k_1, k_2 - 1, k_3, \ldots, k_m}
$$
$$
+ \cdots + \binom{n-1}{k_1, k_2, \ldots, k_{m-1}, k_m - 1}
$$

$(n \geq 1, k_1 + \cdots + k_m = n, k_i \geq 1)$.

    (b) Prove the multinomial theorem by induction on n.

27. Count the number of linear extensions for the following partial orderings (a linear extension has been defined in Exercise 1.7.8):

    (a) X is a disjoint union of sets $X_1, X_2, \ldots, X_k$ of sizes $r_1, r_2, \ldots, r_k$, respectively. Each X_i is linearly ordered by $\preceq$, and no two elements from the different X_i are comparable.

    (b) *The Hasse diagram of $(X, \preceq)$ is a tree as in the following picture. The root has k sons, the ith son has r_i leaves.

2.4 Estimates: an introduction

If we are interested in some quantity and we ask the question "How much?", the most satisfactory answer seems to be one determining the quantity exactly. A millionairess may find some fascination in knowing that her account balance is 107,343,726.12 doublezons[2] at the moment. In mathematics, an answer to a counting problem is usually considered most satisfactory if it is given by an exact formula. But quite often we do not really *need* an exact result; for many applications it is enough to know a quantity approximately. For instance, many people may find it sufficient, although perhaps not comforting, to learn that their account balance is between 4000 and 4100 doublezons. Often even a one-sided inequality suffices: if we estimate that a program for finding an optimal project schedule by trying all possibilities would run for at least 10^{10} days, we probably need not put further effort into determining whether it would actually run for more than 10^{12} days or less than that.

Exact results may be difficult to find. Sometimes computing an exact result may be possible but laborious, and sometimes it is beyond our capabilities no matter how hard we try. Hence, heading for an estimate instead of the exact result may save us lots of work and considerably enlarge the range of problems we are able to cope with.

Another issue is that an exact answer may be difficult to grasp and relate to other quantities. Of course, if the answer is a single number, it is easy to compare it to other numbers, but the situation is more delicate if we have a formula depending on one or several variables. Such a formula defines a function, say a function of n, and we would like to understand "how big" this function is. The usual approach is to compare the considered function to some simple and well-known functions. Let us give a nontrivial example first.

2.4.1 Example (Estimating the harmonic numbers). The following sum appears quite often in mathematics and in computer science:

$$H_n = 1 + \frac{1}{2} + \frac{1}{3} + \cdots + \frac{1}{n} = \sum_{i=1}^{n} \frac{1}{i}.$$

---

[2]Doublezon is a currency unit taken from the book *L'Écume des jours* (English translation: *Froth on the Daydream*) by Boris Vian.

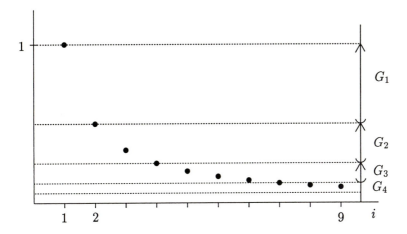

Fig. 2.2 Partitioning the sequence $(\frac{1}{1}, \frac{1}{2}, \frac{1}{3}, \ldots)$ into groups.

This H_n is called the nth *harmonic number*. It turns out that there is no way to simplify this sum (it has no "closed form"). We want to get some idea about the behavior of H_n for n growing to ∞. In particular, we want to decide whether $H_n \to \infty$ for $n \to \infty$.

A simple estimate. The idea is to divide the terms of the sequence $\frac{1}{1}, \frac{1}{2}, \frac{1}{3}, \ldots$ into groups, each group consisting of numbers that are roughly of the same magnitude. That is, we let the kth group G_k consist of the numbers $\frac{1}{i}$ with

$$\frac{1}{2^k} < \frac{1}{i} \le \frac{1}{2^{k-1}}$$

(see Fig. 2.2).

Hence G_k contains the 2^{k-1} numbers

$$\frac{1}{2^{k-1}}, \frac{1}{2^{k-1}+1}, \frac{1}{2^{k-1}+2}, \ldots, \frac{1}{2^k - 1}.$$

Therefore, the sum of the terms in each G_k satisfies

$$\sum_{x \in G_k} x \le |G_k| \max G_k = 2^{k-1} \frac{1}{2^{k-1}} = 1,$$

and similarly

$$\sum_{x \in G_k} x \ge |G_k| \min G_k > 2^{k-1} \frac{1}{2^k} = \frac{1}{2}.$$

A given term $\frac{1}{i}$ belongs to the group G_k with $2^{k-1} \leq i < 2^k$, i.e. with $k = \lfloor \log_2 i \rfloor + 1$. Therefore, H_n is no bigger than the sum of numbers in the first $\lfloor \log_2 n \rfloor + 1$ groups, and we get

$$H_n = \sum_{i=1}^{n} \frac{1}{i} \leq \sum_{k=1}^{\lfloor \log_2 n \rfloor + 1} 1 \leq \log_2 n + 1.$$

Similarly we can derive a lower bound

$$H_n > \sum_{k=1}^{\lfloor \log_2 n \rfloor} \frac{1}{2} \geq \frac{1}{2} \lfloor \log_2 n \rfloor.$$

We may conclude that H_n does grow to infinity but quite slowly, about as slowly as the logarithm function. Even for very large values of n, we can estimate the value of H_n by computing the logarithm. If n is large, the ratio of the upper and lower bounds is close to 2.

For somewhat more sophisticated and more precise estimates of H_n, see Exercise 2.5.13.

In this example, we seem to have been lucky. We could approximate the considered function H_n quite closely by suitable multiples of the logarithm function. But experience shows, and certain theoretical results confirm, that this is not exceptional luck, and that functions (of a single variable n) occurring in natural problems can usually be estimated fairly accurately by everyday functions like n, n^2, $n^{35/13}$, 2^n, 3.26^n, $3^{n^2/2}$, $\ln n$, $\frac{3}{11} n (\ln n)^2$, etc. But finding such estimates may often be quite tricky. In the subsequent sections, we will demonstrate several techniques which may be helpful in such an effort.

Asymptotic comparison of functions.   In the above example, we have shown that the function H_n is "smaller" than the function $\log_2 n + 1$, meaning that the inequality $H_n \leq \log_2 n + 1$ holds for all $n \in \mathbf{N}$. But if we consider the functions $f(n) = 5n$ and $g(n) = n^2$, then neither is smaller than the other, strictly speaking, since, for example, $f(1) = 5 > g(1) = 1$ but $f(6) = 30 < g(6) = 36$, so neither of the inequalities $f(n) \leq g(n)$ and $f(n) \geq g(n)$ is correct for all n. Yet we feel that g "grows much faster" than f: after some initial hesitation for small values of n, $g(n)$ exceeds $f(n)$ and remains above it for all the larger n.

In mathematics and in theoretical computer science, functions defined on the natural numbers are usually compared according to their behavior as n tends to infinity, while their behavior for small values n are ignored. This approach is usually called the *asymptotic analysis* of the considered functions. We also speak of the *asymptotic behavior* or *asymptotics* of some function, meaning its comparison to some simple functions for $n \to \infty$.

If f and g are real functions of a single variable n, we may introduce the symbol $f \preceq g$, meaning that there exists some number n_0 such that the inequality $f(n) \le g(n)$ holds for all $n \ge n_0$; that is, "g ultimately outgrows f". So, for the example in the preceding paragraph, we can write $5n \preceq n^2$.

It is useful to think a bit about the relation $\preceq$ just introduced. It can be viewed as a "soft" inequality between the considered functions. If $f \preceq g$ holds, we are sure that g outgrows f for large enough n but we generally do not know how large n must be. The notation $\preceq$ thus suppresses some information. This often makes it much easier to derive the $\preceq$ inequality between two functions than to prove the "hard" inequality $\le$ (which should hold for all n). But it may also make the $\preceq$ inequality treacherous for the "end-user". Suppose that someone sells us a black box that, for each input number n, computes and displays some value $f(n)$. We also get a guarantee that $f(n) \preceq n$. We can never really prove that the guarantee is invalid. No matter how many of the n we find with $f(n) > n$, the seller can always claim that the number n_0 implicit in the guarantee is still much bigger than our examples.

The notation $f \preceq g$ is not common in the literature (although we believe it has some didactical value for understanding the other notations to come). Instead, several other notations are used that suppress still somewhat more information, and thus may make the estimates yet more convenient to derive.

The "big-Oh" notation. The following notation is used quite often; for instance, it appears frequently in the analysis of algorithms.

2.4.2 Definition. *Let f, g be real functions of a single variable defined on the natural numbers (most often we assume that the values attained by both f and g are nonnegative). The notation*

$$f(n) = O(g(n))$$

means that there exists a constant C such that for all $n \in \mathbf{N}$, the inequality $|f(n)| \le C \cdot g(n)$ holds. If one has to read $f(n) = O(g(n))$ aloud one usually says "f is big-Oh of g".

Here the information suppressed by the notation $f(n) = O(g(n))$ is the value of the constant C. It may be 0.1, 10, or 10^{10}—we only learn that *some* constant C exists. The notation $f(n) = O(g(n))$ can intuitively be understood as saying that the function f doesn't grow much faster than g, i.e. that $f(n)/g(n)$ doesn't grow to infinity. Instead of $f(n)$ and $g(n)$, specific formulas may appear in this notation. For example, we may write $10n^2 + 5n = O(n^2)$.

We should warn that $f(n) = O(g(n))$ says that $f(n)$ is not too big, but it does *not* say anything about $f(n)$ not being very small. For example, $n+5 = O(n^2)$ is a true statement, although perhaps not as helpful as $n + 5 = O(n)$. Let us also emphasize that although the notation contains the equality sign "=", it is asymmetric (essentially, it is an inequality); one shouldn't write $O(f(n)) = g(n)$!

The $O()$ notation often allows us to simplify complicated expressions wonderfully. For example, we have

$$(7n^2 + 6n + 2)(n^3 - 3n + 2^8) = O(n^5). \qquad (2.10)$$

Why? We note the following two simple rules concerning the $O()$ notation: if we have $f_1(n) = O(g_1(n))$ and $f_2(n) = O(g_2(n))$ then $f_1(n) + f_2(n) = O(g_1(n) + g_2(n))$, and similarly for multiplication, $f_1(n)f_2(n) = O(g_1(n)g_2(n))$ (Exercise 6). Since obviously $n = O(n^2)$ and $1 = O(n^2)$, by a repeated application of the addition and multiplication rules we get $7n^2 + 6n + 2 = O(n^2 + n^2 + n^2) = O(n^2)$, and similarly $n^3 - 3n + 2^8 = O(n^3)$. A final application of the multiplication rule gives Eq. (2.10). A nice thing in this derivation is that we didn't need to multiply out the parentheses first!

After some practice, one can write estimates/simplifications like (2.10) right away without too much thinking, by quickly spotting the "main term" in an expression (the one that grows fastest) and letting all others disappear in the $O()$ notation. Such insight is usually based on a (maybe subconscious) use of the following simple rules:

2.4.3 Fact (Useful asymptotic inequalities). *In the following, let $C, a, \alpha, \beta > 0$ be some fixed real numbers independent of n. We have*

(i) $n^\alpha = O(n^\beta)$ *whenever* $\alpha \leq \beta$ *("a bigger power swallows a smaller one"),*

(ii) $n^C = O(a^n)$ *for any* $a > 1$ *("an exponential swallows a power"),*

(iii) $(\ln n)^C = O(n^\alpha)$ *for any* $\alpha > 0$ *("a power swallows a logarithm").*

(In fact, in all the inequalities above, we can write the $\preceq$ symbol instead of the $O()$ notation.)

Part (i) is trivial, proving part (ii) is a simple exercise in calculus, and part (iii) can be easily derived from (ii) by taking logarithms.

Using the symbol $O()$, we can also write a more exact comparison of functions. For example, the notation $f(n) = g(n) + O(\sqrt{n})$ means that the function f is the same as g up to an "error" of the order $\sqrt{n}$, i.e. that $f(n) - g(n) = O(\sqrt{n})$. A simple concrete example is $\binom{n}{2} = n(n-1)/2 = \frac{1}{2}n^2 + O(n)$.

The next example shows how to estimate a relatively complicated sum.

2.4.4 Example. Let us put $f(n) = 1^3 + 2^3 + 3^3 + \cdots + n^3$. We want to find good asymptotic estimates for $f(n)$.

In this case, it is possible to find an exact formula for $f(n)$ (see Exercise 2.3.10), but it is quite laborious.[3] But we can get reasonable asymptotic estimates for $f(n)$ in a less painful way. First, we may note that $f(n) \leq n \cdot n^3 = n^4$. On the other hand, at least $\frac{n}{2}$ addends in the sum defining $f(n)$ are bigger than $(n/2)^3$, and so $f(n) \geq (n/2)^4 = n^4/16$. As a first approximation, we thus see that $f(n)$ behaves like the function n^4, up to small multiplicative factors.

To get a more precise estimate, we can employ the summation formula (2.9) (Exercise 2.3.3) with $r = 3$:

$$\binom{3}{3} + \binom{4}{3} + \binom{5}{3} + \cdots + \binom{n}{3} = \binom{n+1}{4}.$$

Set $g(k) = \binom{k}{3}$. We find that $g(k) = \frac{k(k-1)(k-2)}{3!} = \frac{k^3}{6} + O(k^2)$. Hence we have

$$f(n) = \sum_{k=1}^{n} k^3 = \sum_{k=1}^{n} 6g(k) + \sum_{k=1}^{n} \left[k^3 - 6g(k) \right]$$

$$= 6\binom{n+1}{4} + O\left(\sum_{k=1}^{n} k^2 \right) = \frac{n^4}{4} + O(n^3).$$

In this derivation, we have used the following fact: if f, g are some functions such that $f(n) = O(g(n))$, then $\sum_{k=1}^{n} f(k) = O\left(\sum_{k=1}^{n} g(k) \right)$. It is a simple but instructive exercise to prove it.

A few more remarks. A similar "big-Oh" notation is also used for functions of several variables. For instance, $f(m, n) = O(g(m, n))$ means that for some constant C and for all values of m and n, we have $|f(n, m)| \leq C \cdot g(m, n)$.

A seemingly different definition of the notation $f(n) = O(g(n))$ says that there exist numbers C_0 and $n_0 \in \mathbf{N}$ such that for all $n \geq n_0$, we have $|f(n)| \leq C_0 g(n)$. In this definition, it doesn't matter if the function f or g is undefined for some small values of n. But if both f and g are

---

[3] At least by hand; many computer algebra systems can do it automatically.

defined for all natural numbers then this definition is equivalent to the former one (we leave this as an exercise).

In the literature, one frequently encounters several other symbols for expressing "inequality" between the order of magnitude of functions. They can be quite useful since once one gets used to them, they provide a convenient replacement for complicated phrases (such as "there exists a constant $c > 0$ such that for all $n \in \mathbf{N}$ we have ..." etc.). We will not discuss them in detail, but we will at least list the definitions of the most common symbols in the table below.

Notation	Definition	Meaning
$f(n) = o(g(n))$	$\lim_{n \to \infty} \frac{f(n)}{g(n)} = 0$	f grows much more slowly than g
$f(n) = \Omega(g(n))$	$g(n) = O(f(n))$	f grows at least as fast as g
$f(n) = \Theta(g(n))$	$f(n) = O(g(n))$ and $f(n) = \Omega(g(n))$	f and g have about the same order of magnitude
$f(n) \sim g(n)$	$\lim_{n \to \infty} \frac{f(n)}{g(n)} = 1$	$f(n)$ and $g(n)$ are almost the same

So, finally, it is natural to ask—what is a bound $f(n) = O(g(n))$ good for? Since it doesn't say anything about the hidden constant, we cannot deduce an estimate of $f(n)$ for any specific n from it! There are several answers to this question. In some mathematical considerations, we do not really care about any particular n, and it is enough to know that some function doesn't grow much faster than another one. This can be used, for example, to prove the existence of some object without actually constructing it (see Chapter 9 for such a proof method). A more practically oriented answer is that in most situations, the constant hidden in the $O()$ notation can actually be figured out if needed. One just has to go through a computation done with the $O()$ notation very carefully once more and track the constants used in all the estimates. This is usually tedious but possible. As a general rule of thumb (with many many exceptions), one can say that if a simple proof leads to an $O()$ estimate then the hidden constant is usually not too large, and so if we find that $f(n) = O(n)$ and $g(n) = \Omega(n^2)$ then typically $f(n)$ will be smaller than $g(n)$ even for moderate n. We add more remarks concerning the $O()$ notation in connection with algorithms in Section 4.3.

Exercises

1. Check that the relation $\preceq$ introduced in the text is a partial ordering on the set of all functions $f: \mathbf{N} \to \mathbf{R}$. Show that it is *not* a linear ordering.

2. Find positive and nondecreasing functions $f(n)$, $g(n)$ defined for all natural numbers such that neither $f(n) = O(g(n))$ nor $g(n) = O(f(n))$ holds.

3. Explain what the following notations mean, and decide which of them are true.

   (a) $n^2 = O(n^2 \ln n)$

   (b) $n^2 = o(n^2 \ln n)$

   (c) $n^2 + 5n \ln n = n^2 (1 + o(1)) \sim n^2$

   (d) $n^2 + 5n \ln n = n^2 + O(n)$

   (e) $\sum_{i=1}^{n} i^8 = \Theta(n^9)$

   (f) $\sum_{i=1}^{n} \sqrt{i} = \Theta(n^{3/2})$.

4. What is the meaning of the following notations: $f(n) = O(1)$, $g(n) = \Omega(1)$, $h(n) = n^{O(1)}$? How can they be expressed briefly in words?

5. *Order the following functions according to their growth rate, and express this ordering using the asymptotic notation introduced in this section: $n \ln n$, $(\ln \ln n)^{\ln n}$, $(\ln n)^{\ln \ln n}$, $n \cdot e^{\sqrt{\ln n}}$, $(\ln n)^{\ln n}$, $n \cdot 2^{\ln \ln n}$, $n^{1+1/(\ln \ln n)}$, $n^{1+1/\ln n}$, n^2.

6. Check that if we have $f_1(n) = O(g_1(n))$ and $f_2(n) = O(g_2(n))$ then $f_1(n) + f_2(n) = O(g_1(n) + g_2(n))$ and $f_1(n) f_2(n) = O(g_1(n) g_2(n))$.

2.5 Estimates: the factorial function

In this section, we are going to consider estimates of the function $n!$ (n factorial). At the first sight, it might seem that the definition of the factorial itself, i.e. the formula $n! = n(n-1) \cdot \ldots \cdot 2 \cdot 1$, tells us everything we may ever need to know. For small values of n, $n!$ can be very quickly evaluated by a computer, and for larger n, one might think that the values of the factorial are too large to have any significance in the "real world". For example, $70! > 10^{100}$, as many owners of pocket calculators with the $\boxed{n!}$ button may know. But in various mathematical considerations, we often need to compare the order of magnitude of the function $n!$ to other functions, even for very large values of n. For this purpose, the definition itself is not very suitable, and also an evaluation of $n!$ by a computer sometimes won't be of much help. What we need are good estimates that bound $n!$ by some "simpler" functions.

When approaching a problem, it is usually a good strategy to start looking for very simple solutions, and only try something complicated if simple things fail. For estimating $n!$, a very simple thing to try is the inequality

$$n! = \prod_{i=1}^{n} i \leq \prod_{i=1}^{n} n = n^n.$$

As for a very simple lower bound, we can write

$$n! = \prod_{i=2}^{n} i \geq \prod_{i=2}^{n} 2 = 2^{n-1}.$$

Hence, $n!$ is somewhere between the functions 2^{n-1} and n^n. In many problems, this may be all we need to know about $n!$. But in other problems, such as Example 2.5.1 below, we may start asking more sophisticated questions. Is $n!$ "closer" to n^n or to 2^{n-1}? Does the function $\frac{n^n}{n!}$ grow to infinity, and if so, how rapidly?

To some extent, this can be answered by still quite simple considerations (similar to the first part of the solution to Example 2.4.4). If n is even, then $\frac{n}{2}$ of the numbers in the set $\{1, 2, \ldots, n\}$ are at most $\frac{n}{2}$, and $\frac{n}{2}$ of them are larger than $\frac{n}{2}$. Hence, for n even, we have, on the one hand,

$$n! \geq \prod_{i=n/2+1}^{n} i > \prod_{i=n/2+1}^{n} \frac{n}{2} = \left(\frac{n}{2}\right)^{n/2} = \left(\sqrt{\frac{n}{2}}\right)^n, \qquad (2.11)$$

and on the other hand,

$$n! \leq \left(\prod_{i=1}^{n/2} \frac{n}{2}\right)\left(\prod_{i=n/2+1}^{n} n\right) = \frac{n^n}{2^{n/2}}. \qquad (2.12)$$

For n odd, one has to be slightly more careful, but it turns out that both the formulas $n! > \left(\sqrt{n/2}\right)^n$ and $n! \leq n^n/2^{n/2}$ can be derived for all odd $n \geq 3$ as well (Exercise 1). So, from Eq. (2.11) we see that $n!$ grows considerably faster than 2^n; in fact, sooner or later it outgrows any function C^n with a fixed number C. Eq. (2.12) tells us that n^n grows still faster than $n!$.

Here is a simple example where the question of comparing n^n and $n!$ arises naturally.

2.5.1 Example. Each of n people draws one card at random from a deck of n cards, remembers the card, and returns it back to the deck. What is the probability that no two of the people draw the same card? Is there some "reasonable chance", or is it a very rare event? Mathematically speaking, what is the probability that a mapping of the set $\{1, 2, \ldots, n\}$ to itself chosen at random is a permutation?

The number of all mappings is n^n, the number of permutations is $n!$, and so the required probability is $n!/n^n$. From the upper bound (2.12), we calculate

$$\frac{n!}{n^n} \leq \frac{n^n/2^{n/2}}{n^n} = 2^{-n/2}.$$

Therefore, the probability is no more than $2^{-n/2}$, and for n not too small, the considered event is extremely unlikely. From more precise estimates for $n!$ derived later on, we will see that the probability in question behaves roughly as the function e^{-n}.

A simple estimate according to Gauss. We show an elegant way of deriving estimates similar to (2.11) and (2.12) but a bit stronger. This proof is of some historical interest, since it was invented by the great mathematician Gauss (or, written in the German way, Gauß), and we also learn an important and generally useful inequality.

2.5.2 Theorem. *For every $n \geq 1$,*

$$n^{n/2} \leq n! \leq \left(\frac{n+1}{2}\right)^n.$$

We begin the proof with an inequality between the arithmetic and geometric mean of two numbers. For positive real numbers a, b, we define the *arithmetic mean* of a and b as $\frac{a+b}{2}$, and the *geometric mean*[4] of a and b as $\sqrt{ab}$.

2.5.3 Lemma (Arithmetic–geometric mean inequality). *For any pair of positive real numbers a, b, the geometric mean is no bigger than the arithmetic mean.*

Proof. The square of any real number is always positive, and so $(\sqrt{a} - \sqrt{b})^2 \geq 0$. By expanding the left-hand side we have $a - 2\sqrt{ab} + b \geq 0$, and by adding $2\sqrt{ab}$ to both sides of this inequality and dividing by 2 we get $\sqrt{ab} \leq \frac{a+b}{2}$. This is the desired inequality. $\square$

---

[4]If g denotes the geometric mean of a and b then the ratio $a : g$ is the same as $g : b$. From the point of view of the ancient Greeks, g is thus the appropriate segment "in the middle" between a segment of length a and a segment of length b, and that's probably why this mean is called geometric.

Proof of Theorem 2.5.2.   The idea is to pair up each number $i \in \{1, 2, \ldots, n\}$ with its "cousin" $n+1-i$ and estimate each of the products $i(n+1-i)$ from above and from below. If i runs through the values 1, 2, ..., n then $n+1-i$ runs through n, $n-1$, ..., 1. The product

$$\prod_{i=1}^{n} i(n+1-i)$$

thus contains each factor $j \in \{1, 2, \ldots, n\}$ exactly twice, and so it equals $(n!)^2$. Therefore we have

$$n! = \prod_{i=1}^{n} \sqrt{i(n+1-i)}. \qquad (2.13)$$

If we choose $a = i$ and $b = n+1-i$ in the arithmetic–geometric mean inequality, we get

$$\sqrt{i(n+1-i)} \leq \frac{i+n+1-i}{2} = \frac{n+1}{2},$$

and by (2.13)

$$n! = \prod_{i=1}^{n} \sqrt{i(n+1-i)} \leq \prod_{i=1}^{n} \frac{n+1}{2} = \left(\frac{n+1}{2}\right)^{n},$$

which proves the upper bound in Theorem 2.5.2.

In order to prove the lower bound for $n!$, it suffices to show that $i(n+1-i) \geq n$ for all $i = 1, 2, \ldots, n$. For $i = 1$ and $i = n$ we directly calculate that $i(n+1-i) = n$. For $2 \leq i \leq n-1$, we have a product of two numbers, the larger one being at least $\frac{n}{2}$ and the smaller one at least 2, and hence $i(n+1-i) \geq n$ holds for all i. Therefore, $n! \geq \sqrt{n^n} = n^{n/2}$ as was to be proved. $\qquad \square$

Of course, not everyone can invent such tricks as easily as Gauss did, but at least the arithmetic–geometric mean inequality is worth remembering.

Having learned some estimates of $n!$, we may keep asking more and more penetrating questions, such as whether $\left(\frac{n+1}{2}\right)^n / n!$ grows to infinity, and if so how fast, etc. We will now skip some stages of this natural evolution and prove bounds that estimate $n!$ up to a multiplicative factor of only n (note that in the preceding estimates, our uncertainty was still at least an exponential function of n). In these more sophisticated estimates, we encounter the so-called *Euler number* $e = 2.718281828\ldots$, the basis of the natural logarithms. The reader may learn much more about this remarkable constant in calculus. Here we need the following:

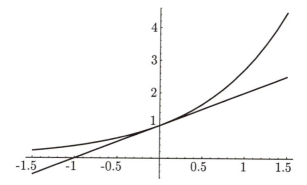

Fig. 2.3 The functions $y = 1 + x$ and $y = e^x$ in the vicinity of the origin.

2.5.4 Fact. *For every real number* x,

$$1 + x \leq e^x$$

holds (see Fig. 2.3).

This fact is something which should be marked in large bold-faced fluorescent letters in the notebook of every apprentice in asymptotic estimates. Here we use it to prove

2.5.5 Theorem. *For every* $n \geq 1$, *we have*

$$e \left(\frac{n}{e}\right)^n \leq n! \leq en \left(\frac{n}{e}\right)^n .$$

First proof (by induction). We only prove the upper bound $n! \leq en(n/e)^n$, leaving the lower bound as Exercise 9. For $n = 1$, the right-hand side becomes 1, and so the inequality holds. So we assume that the inequality has already been proved for $n - 1$, and we verify it for n. We have

$$n! = n \cdot (n-1)! \leq n \cdot e(n-1) \left(\frac{n-1}{e}\right)^{n-1}$$

by the inductive assumption. We further transform the right-hand side to

$$\left[en \left(\frac{n}{e}\right)^n \right] \cdot \left(\frac{n-1}{n}\right)^n e.$$

In the brackets, we have the upper bound for $n!$ we want to prove. So it suffices to show that the remaining part of the expression cannot

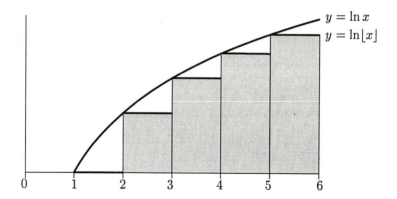

Fig. 2.4   Estimating the area below the step function by integration.

exceed 1. By an algebraic manipulation and by using Fact 2.5.4 with $x = -\frac{1}{n}$, we obtain

$$e\left(\frac{n-1}{n}\right)^n = e\left(1 - \frac{1}{n}\right)^n \le e\left(e^{-1/n}\right)^n = e \cdot e^{-1} = 1.$$

$\square$

Second proof (using an integral).   We again do the upper bound only. We begin with a formula for the factorial, $n! = 1 \cdot 2 \cdot \ldots \cdot n$, and we take natural logarithms on both sides. In this way, we get

$$\ln n! = \ln 1 + \ln 2 + \cdots + \ln n$$

(the function ln is the logarithm with base e). The expression on the right-hand side can be thought of as the area enclosed between the x-axis and the step function $x \mapsto \ln\lfloor x \rfloor$ on the interval $[1, n+1]$; see Fig. 2.4.

Since $\ln\lfloor x \rfloor \le \ln x$ on this interval, the area in question is no bigger than the area below the graph of the function $x \mapsto \ln x$ on the interval $[1, n+1]$. We express this latter area as an integral:

$$\ln n! \le \int_1^{n+1} \ln x \, dx = (n+1)\ln(n+1) - n,$$

as one can calculate as a simple exercise in integration. This estimate can be further manipulated into

$$n! \le e^{(n+1)\ln(n+1)-n} = \frac{(n+1)^{n+1}}{e^n}.$$

This is not yet the expression we want. But we can use this inequality for $n - 1$ instead of n, and this gives the formula in the theorem:

$$n! = n \cdot (n-1)! \le n \cdot \frac{n^n}{e^{n-1}} = en \left(\frac{n}{e}\right)^n.$$

□

For the reader's interest, let us mention a considerably more precise estimate for $n!$, known by the name *Stirling's formula*: If we define the function

$$f(n) = \sqrt{2\pi n} \left(\frac{n}{e}\right)^n,$$

where $\pi = 3.1415926535\ldots$ is the area of the unit disk, we have $f(n) \sim n!$. Recall that this means

$$\lim_{n\to\infty} \frac{f(n)}{n!} = 1.$$

So if we estimate $n!$ by $f(n)$ then the relative error of this estimate tends to 0 for n tending to infinity. For example, for $n = 8$, the error is about 1%. Let us note that Stirling's formula is approximately "in the middle" of the estimates from Theorem 2.5.5 (see also Exercise 10). Proving Stirling's formula requires somewhat more advanced tools from calculus and it doesn't quite fit into this book, so we omit it (see Knuth [38] for a proof).

Exercises

1. (a) Check that the formula $n! > \left(\sqrt{n/2}\right)^n$ is valid for all odd $n \ge 1$, by a consideration similar to Eq. (2.11).
   (b) Check that also $n! \le n^n/2^{n/2}$ holds for all odd $n \ge 3$.

2. Using Fact 2.5.4, prove that
   (a) $(1 + \frac{1}{n})^n \le e$ for all $n \ge 1$, and
   (b) * $(1 + \frac{1}{n})^{n+1} \ge e$ for all $n \ge 1$.
   (c) Using (a) and (b), conclude that $\lim_{n\to\infty}(1 + \frac{1}{n})^n = e$.
   (d) Prove $(1 - \frac{1}{n})^n \le \frac{1}{e} \le (1 - \frac{1}{n})^{n-1}$.

3. (Calculus required) *Prove Fact 2.5.4.

4. Show that $\sqrt[n]{n}$ tends to 1 for $n \to \infty$, and *use Fact 2.5.4 to prove that $\sqrt[n]{n} - 1 \ge \frac{\ln n}{n}$ for all $n \ge 1$.

5. Decide which of the following statements are true:
   (a) $n! \sim ((n+1)/2)^n$
   (b) $n! \sim ne(n/e)^n$
   (c) $n! = O((n/e)^n)$
   (d) $\ln(n!) = \Omega(n \ln n)$
   (e) $\ln(n!) \sim n \ln n$.

6. (a) For which pairs (a, b), $a, b > 0$, does the equality $\sqrt{ab} = (a + b)/2$ hold?

(b) The *harmonic mean* of positive real numbers a, b is defined by the expression $2ab/(a + b)$. Based on examples, suggest a hypothesis for the relation (inequality) of the harmonic mean to the arithmetic and geometric means, and *prove it.

7. Let $x_1, x_2, \ldots, x_n$ be positive reals. Their *arithmetic mean* equals $(x_1 + x_2 + \cdots + x_n)/n$, and their *geometric mean* is defined as $\sqrt[n]{x_1 x_2 \ldots x_n}$. Let $AG(n)$ denote the statement "for any n-tuple of positive reals $x_1, x_2, \ldots, x_n$, the geometric mean is less than or equal to the arithmetic mean". Prove the validity of $AG(n)$ for every n by the following strange induction:

(a) Prove that $AG(n)$ implies $AG(2n)$, for each n.

(b) *Prove that $AG(n)$ implies $AG(n - 1)$, for each $n > 1$.

(c) Explain why proving (a) and (b) is enough to prove the validity of $AG(n)$ for all n.

8. (Computation of the number π) *Define sequences $\{a_0, a_1, a_2, \ldots\}$ and $\{b_0, b_1, b_2, \ldots\}$ as follows: $a_0 = 2$, $b_0 = 4$, $a_{n+1} = \sqrt{a_n b_n}$, $b_{n+1} = 2a_{n+1}b_n/(a_{n+1} + b_n)$. Prove that both sequences converge to π. Hint: Find a relation of the sequences to regular polygons with 2^n sides inscribed in and circumscribed to the unit circle.

Remark. This method (of Archimedes) of calculation of π is not very efficient. Here is an example of a much faster algorithm: $x_1 = 2^{-3/4} + 2^{-5/4}$, $y_1 = 2^{1/4}$, $\pi_0 = 2 + \sqrt{2}$, $\pi_n = \pi_{n-1}(x_n + 1)/(y_n + 1)$, $y_{n+1} = (y_n\sqrt{x_n} + 1/\sqrt{x_n})/(y_n + 1)$, $x_{n+1} = (\sqrt{x_n} + 1/\sqrt{x_n})/2$. Then the π_n converge to π extremely fast. This and other such algorithms, as well as the remarkable underlying theory, can be found in Borwein and Borwein [15].

9. Prove the lower bound $n! \geq e(n/e)^n$ in Theorem 2.5.5

(a) by induction (use Fact 2.5.4 cleverly),

(b) via integration.

10. (Calculus required) *Prove the following upper bound for the factorial function (which is already quite close to Stirling's formula): $n! \leq e\sqrt{n}\,(n/e)^n$. Use the second proof of Theorem 2.5.5 as a starting point, but from the area below the curve $y = \ln x$, subtract the areas of suitable triangles.

11. Prove *Bernoulli's inequality*: for each natural number n and for every real $x \geq -1$, we have $(1 + x)^n \geq 1 + nx$.

12. Prove that for $n = 1, 2, \ldots$, we have

$$2\sqrt{n+1} - 2 < 1 + \frac{1}{\sqrt{2}} + \frac{1}{\sqrt{3}} + \cdots + \frac{1}{\sqrt{n}} \leq 2\sqrt{n} - 1.$$

13. Let H_n be as in Example 2.4.1: $H_n = \sum_{i=1}^{n} \frac{1}{i}$.
    (a) *Prove the inequalities $\ln n < H_n \leq \ln n + 1$ by induction on n (use Fact 2.5.4).
    (b) Solve (a) using integrals.

2.6 Estimates: binomial coefficients

Similar to the way we have been investigating the behavior of the function $n!$, we will now consider the function

$$\binom{n}{k} = \frac{n(n-1)\ldots(n-k+1)}{k(k-1)\cdot\ldots\cdot 2\cdot 1} = \prod_{i=0}^{k-1} \frac{n-i}{k-i}. \qquad (2.14)$$

From the definition of $\binom{n}{k}$, we immediately get

$$\binom{n}{k} \leq n^k,$$

and for many applications, this simple estimate is sufficient. For $k > \frac{n}{2}$, one should first use the equality $\binom{n}{k} = \binom{n}{n-k}$.

In order to derive some lower bound for $\binom{n}{k}$, we look at the definition of the binomial coefficient written as a product of fractions, as in (2.14). For $n \geq k > i \geq 0$ we have $\frac{n-i}{k-i} \geq \frac{n}{k}$, and hence

$$\binom{n}{k} \geq \left(\frac{n}{k}\right)^k.$$

Quite good upper and lower bounds for $\binom{n}{k}$ can be obtained from Stirling's formula, using the equality $\binom{n}{k} = \frac{n!}{k!(n-k)!}$. These bounds are somewhat cumbersome for calculation, however, and also we haven't proved Stirling's formula. We do prove good but less accurate estimates by different methods (the main goal is to demonstrate these methods, of course).

2.6.1 Theorem. *For every $n \geq 1$ and for every k, $1 \leq k \leq n$, we have*

$$\binom{n}{k} \leq \left(\frac{en}{k}\right)^k.$$

Proof. We in fact prove a stronger inequality:

$$\binom{n}{0} + \binom{n}{1} + \binom{n}{2} + \cdots + \binom{n}{k} \leq \left(\frac{en}{k}\right)^k.$$

We start from the binomial theorem, which claims that

$$\binom{n}{0} + \binom{n}{1}x + \binom{n}{2}x^2 + \cdots + \binom{n}{n}x^n = (1+x)^n$$

for an arbitrary real number x. Let us now assume $0 < x < 1$. Then by omitting some of the addends on the left-hand side, we get

$$\binom{n}{0} + \binom{n}{1}x + \cdots + \binom{n}{k}x^k \leq (1+x)^n,$$

and dividing this by x^k leads to

$$\frac{1}{x^k}\binom{n}{0} + \frac{1}{x^{k-1}}\binom{n}{1} + \cdots + \binom{n}{k} \leq \frac{(1+x)^n}{x^k}.$$

Each of the binomial coefficients on the left-hand side is multiplied by a coefficient that is at least 1 (since we assume $x < 1$), and so if we replace these coefficients by 1s the left-hand side cannot increase. We obtain

$$\binom{n}{0} + \binom{n}{1} + \cdots + \binom{n}{k} \leq \frac{(1+x)^n}{x^k}.$$

The number $x \in (0,1)$ can still be chosen at will, and we do it in such a way that the right-hand side becomes as small as possible. A suitable value, which can be discovered using some elementary calculus, is $x = \frac{k}{n}$. By substituting this value into the right-hand side, we find

$$\binom{n}{0} + \binom{n}{1} + \cdots + \binom{n}{k} \leq \left(1 + \frac{k}{n}\right)^n \left(\frac{n}{k}\right)^k.$$

Finally, by using Fact 2.5.4 we arrive at

$$\left(1 + \frac{k}{n}\right)^n \leq \left(e^{k/n}\right)^n = e^k,$$

and the inequality in Theorem 2.6.1 follows. $\qquad\square$

The trick used in this proof is a small glimpse into the realm of perhaps the most powerful known techniques for asymptotic estimates, using the so-called generating functions. We will learn something about generating functions in Chapter 10, but to see the full strength of this approach for asymptotic bounds, one needs to be familiar with the theory of functions of a complex variable.

The binomial coefficient $\binom{n}{\lfloor n/2 \rfloor}$**.** From the definition of the binomial coefficients, we can easily get the following formula:

$$\binom{n}{k} = \frac{n-k+1}{k} \binom{n}{k-1}.$$

Therefore, for $k \leq n/2$ we have $\binom{n}{k} > \binom{n}{k-1}$, and conversely, for $k \geq n/2$ we obtain $\binom{n}{k} > \binom{n}{k+1}$. Hence for a given n, the largest among the binomial coefficients $\binom{n}{k}$ are the middle ones: for n even, $\binom{n}{n/2}$ is bigger than all the others, and for n odd, the two largest binomial coefficients are $\binom{n}{\lfloor n/2 \rfloor}$ and $\binom{n}{\lceil n/2 \rceil}$.

The behavior of the binomial coefficient $\binom{n}{k}$ as a function of k, with n fixed as some large number and for k close to $n/2$, is illustrated in Fig. 2.5(a). The graph of the function $\binom{n}{k}$ isn't really a continuous curve (since $\binom{n}{k}$ is only defined for an integer k), but if n is very large, there are so many points that they visually blend into a curve. The "height" of this bell-shaped curve is exactly $\binom{n}{\lfloor n/2 \rfloor}$, and the "width" of the bell shape approximately in the middle of its height is about $1.5\sqrt{n}$. The scales on the vertical and horizontal axes are thus considerably different: the horizontal axis shows a range of k of length $3\sqrt{n}$, while the vertical range is $\binom{n}{\lfloor n/2 \rfloor}$ (which is nearly 2^n as we will soon see).

If you plot the function $x \mapsto e^{-x^2/2}$, you get a curve which looks exactly the same as the one we have plotted for binomial coefficients, up to a possibly different scaling of the axes. This is because the $e^{-x^2/2}$ curve, called the *Gauss curve*, is a limit of the curves for binomial coefficients for $n \to \infty$ (in a suitably defined precise sense). The Gauss curve is very important in probability theory, statistics, and other areas. For example, it describes a typical distribution of errors in physical measurements, the percentage of days with a given maximal temperature within a long time period, and so on. In statistics, the distribution given by the Gauss curve is called the *normal distribution*. The Gauss curve is one of the "ubiquitous" mathematical objects arising in many often unexpected contexts (another such omnipresent object is the Euler number e, and we will meet some others later in this book). You can learn more about the Gauss curve and related things in a probability theory textbook (Grimmett and Stirzaker [18] can be highly recommended).

How large is the largest binomial coefficient $\binom{n}{\lfloor n/2 \rfloor}$? A simple but often accurate enough estimate is

$$\frac{2^n}{n+1} \leq \binom{n}{\lfloor n/2 \rfloor} \leq 2^n.$$

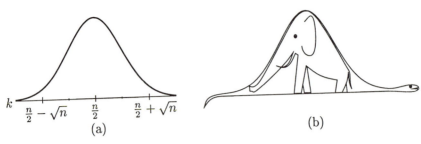

Fig. 2.5   A graph of $\binom{n}{k}$ as a function of k in the vicinity of $\frac{n}{2}$ (a), or perhaps a hat, or maybe a gigantic boa constrictor which has swallowed an elephant (b) (see [26]).

The upper bound is obvious from the equality $\sum_{k=0}^{n} \binom{n}{k} = 2^n$, and the lower bound follows from it as well, because $\binom{n}{\lfloor n/2 \rfloor}$ is largest among the $n+1$ binomial coefficients $\binom{n}{k}$ whose sum is 2^n.

We prove a considerably more precise estimate. For convenient notation, we will only work with even values of n, and so we write $n = 2m$.

2.6.2 Proposition. *For all $m \geq 1$ we have*

$$\frac{2^{2m}}{2\sqrt{m}} \leq \binom{2m}{m} \leq \frac{2^{2m}}{\sqrt{2m}}.$$

Proof. Both inequalities are proved similarly. Let us consider the number

$$P = \frac{1 \cdot 3 \cdot 5 \cdot \ldots \cdot (2m-1)}{2 \cdot 4 \cdot 6 \cdot \ldots \cdot 2m}$$

(the whole idea of the proof is hidden in this step). Since

$$P = \frac{1 \cdot 3 \cdot 5 \cdot \ldots \cdot (2m-1)}{2 \cdot 4 \cdot 6 \cdot \ldots \cdot 2m} \cdot \frac{2 \cdot 4 \cdot \ldots \cdot (2m)}{2 \cdot 4 \cdot \ldots \cdot (2m)} = \frac{(2m)!}{2^{2m}(m!)^2},$$

we get

$$P = \frac{1}{2^{2m}}\binom{2m}{m}.$$

Thus, we want to prove

$$\frac{1}{2\sqrt{m}} \leq P \leq \frac{1}{\sqrt{2m}}.$$

For the upper bound, consider the product

$$\left(1 - \frac{1}{2^2}\right)\left(1 - \frac{1}{4^2}\right)\cdots\left(1 - \frac{1}{(2m)^2}\right),$$

which can be rewritten as

$$\left(\frac{1\cdot 3}{2^2}\right)\left(\frac{3\cdot 5}{4^2}\right)\cdots\left(\frac{(2m-1)(2m+1)}{(2m)^2}\right) = (2m+1)P^2.$$

Since the value of the product is less than 1, we get $(2m+1)P^2 < 1$, and hence $P \le 1/\sqrt{2m}$.

For the lower bound we consider the product

$$\left(1-\frac{1}{3^2}\right)\left(1-\frac{1}{5^2}\right)\cdots\left(1-\frac{1}{(2m-1)^2}\right),$$

and we express it in the form

$$\left(\frac{2\cdot 4}{3^2}\right)\left(\frac{4\cdot 6}{5^2}\right)\cdots\left(\frac{(2m-2)(2m)}{(2m-1)^2}\right) = \frac{1}{2\cdot(2m)\cdot P^2},$$

which gives the lower bound in Proposition 2.6.2. □

Let us remark that by approximating both $(2m)!$ and $m!$ using Stirling's formula, we get a more precise result

$$\binom{2m}{m} \sim \frac{2^{2m}}{\sqrt{\pi m}}.$$

Such estimates have interesting relations with number theory, for example. One of the most famous mathematical theorems is the following statement about the density of primes:

2.6.3 Theorem (Prime number theorem). *Let $\pi(n)$ denote the number of primes not exceeding the number n. Then*

$$\pi(n) \sim \frac{n}{\ln n}$$

(i.e. $\lim_{n\to\infty} \pi(n)\ln n/n = 1$).

Several proofs of this theorem are known, all of them quite difficult (and a quest for interesting variations and simplifications still continues). Within the past century, Tschebyshev found a simple proof of the following weaker result:

$$\pi(n) = \Theta\left(\frac{n}{\ln n}\right),$$

i.e. $c_1 n/\ln n \le \pi(n) \le c_2 n/\ln n$ holds for all n and for certain constants $c_2 \ge c_1 > 0$. Part of the proof is based on the estimates $\frac{2^{2m}}{2m+1} \le \binom{2m}{m} \le 2^{2m}$ (see Exercise 2). Tschebyshev also proved the so-called Bertrand postulate: *For every $n \ge 1$, there exists a prime p with $n < p \le 2n$.* Perhaps the simplest known proof uses, among others, the estimates in Proposition 2.6.2. The reader can learn about these nice connections in Chandrasekhar [33], for example.

Exercises

1. (a) Prove the estimate $\binom{n}{k} \leq (en/k)^k$ by induction on k.

   (b) Prove the estimate in (a) directly from Theorem 2.5.5.

2. (Tschebyshev estimate of $\pi(n)$)

   (a) Show that the product of all primes p with $m < p \leq 2m$ is at most $\binom{2m}{m}$.

   (b) *Using (a), prove the estimate $\pi(n) = O(n/\ln n)$, where $\pi(n)$ is as in the prime number theorem 2.6.3.

   (c) *Let p be a prime, and let m, k be natural numbers. Prove that if p^k divides $\binom{2m}{m}$ then $p^k \leq 2m$.

   (d) Using (c), prove $\pi(n) = \Omega(n/\ln n)$.

2.7   Inclusion–exclusion principle

We begin with a simple motivating example. As many authors of examples with finite sets have already done, we resort to a formulation with clubs in a small town.

2.7.1 Example. The town of N. has 3 clubs. The lawn-tennis club has 20 members, the chandelier collectors club 15 members, and the membership of the Egyptology club numbers 8. There are 2 tennis players and 3 chandelier collectors among the Egyptologists, 6 people both play tennis and collect chandeliers, and there is even one especially eager person participating in all three clubs. How many people are engaged in the club life in N.?

As a warm-up, let us count the combined membership of tennis and Egyptology. We see that we have to add the number of tennis players and the Egyptology fans and subtract those persons who are in both these clubs, since they are accounted for twice in the sum. Written in symbols, we have $|T \cup E| = |T| + |E| - |T \cap E| = 20 + 8 - 2 = 26$. The reader who isn't discouraged by the apparent silliness of the whole problem[5] can probably find, with similar but more complicated considerations, that the answer for the 3 clubs is 33. To find the answer, it may be helpful to draw a picture:

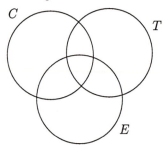

---

[5]Which may indicate mathematical inclinations.

The inclusion–exclusion principle mentioned in the section's title is
a formula which allows us to solve problems of a similar type for an
arbitrary number of clubs. It is used in situations where we want to
compute the size of the union of several sets, and we know the sizes
of all possible intersections of these sets. For 2 sets, T and E, such
a formula has been given above, and for 3 sets C, T, E it reads

$$|C \cup T \cup E| = |C| + |T| + |E| - |C \cap T| - |C \cap E| - |T \cap E| + |C \cap T \cap E|.$$

Expressed in words: in order to get the size of the union, we first add
up the sizes of all the sets, then we subtract the sizes of all pairwise
intersections, and finally we add the size of the intersection of all
the 3 sets. As will be shown in a moment, such a method also works
for an arbitrary number n of finite sets $A_1, A_2, \ldots, A_n$. The size of
their union, i.e. $|A_1 \cup A_2 \cup \cdots \cup A_n|$, is obtained as follows: we add
up the sizes of all the sets, then we subtract the sizes of all pairwise
intersections, add the sizes of all triple intersections, subtract the
sizes of all 4-wise intersections, etc.; as the last step, we either add
(for n odd) or subtract (for n even) the size of the intersection of all
the n sets.

How do we write this in a formula? One attempt might be

$$|A_1 \cup A_2 \cup \cdots \cup A_n| = |A_1| + |A_2| + \cdots + |A_n|$$
$$-|A_1 \cap A_2| - |A_1 \cap A_3| - \cdots - |A_1 \cap A_n| - |A_2 \cap A_3| - \cdots - |A_{n-1} \cap A_n|$$
$$+|A_1 \cap A_2 \cap A_3| + |A_1 \cap A_2 \cap A_4|$$
$$+ \cdots + (-1)^{n-1}|A_1 \cap A_2 \cap \cdots \cap A_n|.$$

This is a cumbersome and not very clear way of expressing such a
simple rule. Slightly better is a notation using sums:

$$|A_1 \cup A_2 \cup \cdots \cup A_n| = \sum_{i=1}^{n} |A_i| - \sum_{1 \le i_1 < i_2 \le n} |A_{i_1} \cap A_{i_2}|$$
$$+ \sum_{1 \le i_1 < i_2 < i_3 \le n} |A_{i_1} \cap A_{i_2} \cap A_{i_3}|$$
$$- \cdots + (-1)^{n-1}|A_1 \cap A_2 \cap \cdots \cap A_n|.$$

If we recall the notation $\binom{X}{k}$ for the set of all k-element subsets of
a set X, and if we use a notation similar to $\sum$ also for multiple
intersections and unions, we can write the same formula still more
elegantly:

2.7.2 Theorem (Inclusion–exclusion principle). *For any collection $A_1, A_2, \ldots, A_n$ of finite sets, we have*

$$\left| \bigcup_{i=1}^{n} A_i \right| = \sum_{k=1}^{n} (-1)^{k-1} \sum_{I \in \binom{\{1,2,\ldots,n\}}{k}} \left| \bigcap_{i \in I} A_i \right|. \tag{2.15}$$

In case you cannot see why this formula expresses the rule we have formulated in words, you may want to devote some time to it and work out the case $n = 3$ in detail. Many students have problems with this notation (or any mathematical notation) for the inclusion–exclusion principle, confusing numbers with sets and vice versa, and this makes a clean solution of more complicated problems very hard.

Finally, the shortest and almost devilish way of writing the inclusion–exclusion principle is

$$\left| \bigcup_{i=1}^{n} A_i \right| = \sum_{\emptyset \neq I \subseteq \{1,2,\ldots,n\}} (-1)^{|I|-1} \left| \bigcap_{i \in I} A_i \right|. \tag{2.16}$$

First proof of the inclusion–exclusion principle: by induction. The induction is on n, the number of sets. There is a small subtlety here: for the inductive step, we need the formula for the case $n = 2$, and so we use $n = 2$ as the basis for induction. For 2 sets, as we know, the formula holds. Assume its validity for arbitrary $n - 1$ sets. We have

$$\left| \bigcup_{i=1}^{n} A_i \right| = \left| \left(\bigcup_{i=1}^{n-1} A_i \right) \cup A_n \right| = \left| \bigcup_{i=1}^{n-1} A_i \right| + |A_n| - \left| \left(\bigcup_{i=1}^{n-1} A_i \right) \cap A_n \right|$$

(here we used inclusion-exclusion for 2 sets, i.e. the equality $|A \cup B| = |A| + |B| - |A \cap B|$ with $A = A_1 \cup \cdots \cup A_{n-1}$, $B = A_n$)

$$= \left| \bigcup_{i=1}^{n-1} A_i \right| + |A_n| - \left| \bigcup_{i=1}^{n-1} (A_i \cap A_n) \right|$$

(distributivity of the intersection: $X \cap (Y \cup Z) = (X \cap Y) \cup (X \cap Z)$; now we use the inductive hypothesis twice, once for $|A_1 \cup \cdots \cup A_{n-1}|$ and once for $|A_1' \cup \cdots \cup A_{n-1}'|$, where $A_i' = A_i \cap A_n$)

$$= \left(\sum_{k=1}^{n-1} (-1)^{k-1} \sum_{I \in \binom{\{1,2,\ldots,n-1\}}{k}} \left| \bigcap_{i \in I} A_i \right| \right) + |A_n| -$$

$$\left(\sum_{k=1}^{n-1} (-1)^{k-1} \sum_{I \in \binom{\{1,2,\dots,n-1\}}{k}} \left| \bigcap_{i \in I \cup \{n\}} A_i \right| \right).$$

We are nearly done. In the first sum, we add, with the proper signs, the sizes of all intersections not involving the set A_n. In the second sum, the sizes of all the intersections involving A_n appear, and the intersection of $k+1$ sets (i.e. some k sets among $A_1, \dots, A_{n-1}$ plus A_n) has the sign $-(-1)^{k-1} = (-1)^k$. The second sum doesn't include the term $|A_n|$, but this appears separately between both sums. Altogether, the size of the intersection of any k-tuple of sets among $A_1, \dots, A_n$ appears exactly once in the expression, with the sign $(-1)^{k-1}$. This agrees with Eq. (2.15), and the proof by induction is finished. Without a reasonable notation, we would easily get lost in this proof. □

Second proof of the inclusion–exclusion principle: by counting. Let us consider an arbitrary element $x \in A_1 \cup \dots \cup A_n$. It contributes exactly 1 to the size of the union on the left-hand side of (2.15). Let us look at how much x contributes to the various intersection sizes on the right-hand side. Let j be the number of sets among the A_i that contain x. We can rename the sets so that x is contained in $A_1, A_2, \dots, A_j$.

The element x now appears in the intersection of every k-tuple of sets among $A_1, A_2, \dots, A_j$ and in no other intersections. Since there are $\binom{j}{k}$ k-element subsets of a j-element set, x appears in $\binom{j}{k}$ intersections of k-tuples of sets. The sizes of k-wise intersections are counted with the sign $(-1)^{k-1}$, and so x contributes the quantity

$$j - \binom{j}{2} + \binom{j}{3} - \dots + (-1)^{j-1} \binom{j}{j}$$

to the right-hand side of the inclusion–exclusion formula (2.15). By the formula (2.8) for the sum of binomial coefficients with alternating signs, the above expression equals 1. The contribution of each element x to both sides of the inclusion–exclusion formula (2.15) is thus 1, and the formula is proved. □

And one more proof. If one looks at the inclusion–exclusion principle in a proper way, it is a consequence of the following formula for expanding a product:

$$(1+x_1)(1+x_2)\ldots(1+x_n) = \sum_{I\subseteq\{1,2,\ldots,n\}}\left(\prod_{i\in I}x_i\right). \qquad (2.17)$$

Contemplate what this formula says (write it out for $n = 1, 2, 3$, say) and why it holds.

In order to prove the inclusion–exclusion principle, let us denote $A = A_1 \cup A_2 \cup \cdots \cup A_n$, and let $f_i\colon A \to \{0,1\}$ be the characteristic function of the set A_i, which means that $f_i(a) = 1$ for $a \in A_i$ and $f_i(a) = 0$ otherwise. For every $a \in A$, we have $\prod_{i=1}^{n}(1 - f_i(a)) = 0$ (don't we?), and using (2.17) with $x_i = -f_i(a)$ we get

$$\sum_{I\subseteq\{1,2,\ldots,n\}}(-1)^{|I|}\prod_{i\in I}f_i(a) = 0.$$

By adding all these equalities together for all $a \in A$, and then by interchanging the summation order, we arrive at

$$0 = \sum_{a\in A}\left(\sum_{I\subseteq\{1,2,\ldots,n\}}(-1)^{|I|}\prod_{i\in I}f_i(a)\right)$$

$$= \sum_{I\subseteq\{1,2,\ldots,n\}}(-1)^{|I|}\left(\sum_{a\in A}\prod_{i\in I}f_i(a)\right). \qquad (2.18)$$

Now it suffices to note that the $\prod_{i\in I}f_i(a)$ is the characteristic function of the set $\bigcap_{i\in I}A_i$, and therefore $\sum_{a\in A}\prod_{i\in I}f_i(a) = \left|\bigcap_{i\in I}A_i\right|$. In particular, for $I = \emptyset$, $\prod_{i\in\emptyset}f_i(a)$ is the empty product, with value 1 by definition, and so $\sum_{a\in A}\prod_{i\in\emptyset}f_i(a) = \sum_{a\in A}1 = |A|$. Hence (2.18) means

$$|A| + \sum_{\emptyset\neq I\subseteq\{1,2,\ldots,n\}}(-1)^{|I|}\left|\bigcap_{i\in I}A_i\right| = 0,$$

and this is exactly the inclusion–exclusion principle. An expert in algebra can thus regard the inclusion–exclusion principle with mild contempt: a triviality, she might say. □

Bonferroni inequalities. Sometimes we can have the situation where we know the sizes of all the intersections up to m-fold ones, but we do not know the sizes of intersections of more sets than m. Then we cannot calculate the size of the union of all sets exactly. The so-called *Bonferroni inequalities* tell us that if we leave out all terms with $k > m$ on the right-hand side of the inclusion–exclusion principle (2.15) then the *error* that we make in this way in the calculation of the size of the union *has the same sign as the first omitted term*. Written as a formula, for every *even* q we have

$$\sum_{k=1}^{q}(-1)^{k-1}\sum_{I\in\binom{\{1,2,\ldots,n\}}{k}}\left|\bigcap_{i\in I}A_i\right| \le \left|\bigcup_{i=1}^{n}A_i\right| \qquad (2.19)$$

and for every *odd q* we have

$$\sum_{k=1}^{q}(-1)^{k-1}\sum_{I\in\binom{\{1,2,\ldots,n\}}{k}}\left|\bigcap_{i\in I}A_i\right| \geq \left|\bigcup_{i=1}^{n}A_i\right|. \qquad (2.20)$$

This means, for instance, that if we didn't know how many diligent persons are simultaneously in all the three clubs in Example 2.7.1, we could still estimate that the total number of members in all the clubs is at least 32. We do not prove the Bonferroni inequalities here.

Exercises

1. Explain why the formulas (2.15) and (2.16) express the same equality.

2. *Prove the Bonferroni inequalities. If you cannot handle the general case try at least the cases $q = 1$ and $q = 2$.

3. (Sieve of Eratosthenes) How many numbers are left in the set $\{1, 2, \ldots, 1000\}$ after all multiples of 2, 3, 5, and 7 are crossed out?

4. How many numbers $n < 100$ are not divisible by a square of any integer greater than 1?

5. *How many orderings of the letters A, B, C, D, E, F, G, H, I, J, K, L, M, N, O, P are there such that we cannot obtain any of the words BAD, DEAF, APE by crossing out some letters? What if we also forbid LEADING?

6. How many ways are there to arrange 4 Americans, 3 Russians, and 5 Chinese into a queue, in such a way that no nationality forms a single consecutive block?

2.8 The hatcheck lady & co.

2.8.1 Problem (Hatcheck lady problem). Honorable gentlemen, n in number, arrive at an assembly, all of them wearing hats, and they deposit their hats in a cloak-room. Upon their departure, the hatcheck lady, maybe quite absent-minded that day, maybe even almost blind after many years of service in the poorly lit cloak-room, issues one hat to each gentleman at random. What is the probability than none of the gentlemen receives his own hat?

As stated, this is a toy problem, but mathematically it is quite remarkable, and a few hundred years back, it occupied some of the best mathematical minds of their times. First we reformulate the problem using permutations. If we number the gentlemen (our apologies)

$1, 2, \ldots, n$, and their hats too, then the activity of the hatcheck lady results in a random permutation π of the set $\{1, 2, \ldots, n\}$, where $\pi(i)$ is the number of the hat returned to the ith gentleman. The question is, what is the probability of $\pi(i) \neq i$ holding for all $i \in \{1, 2, \ldots, n\}$? Call an index i with $\pi(i) = i$ a *fixed point* of the permutation π. So we ask: what is the probability that a randomly chosen permutation has no fixed point? Each of the $n!$ possible permutations is, according to the description of the hatcheck lady's method of working, equally probable, and so if we denote by $D(n)$ the number of permutations with no fixed point[6] on an n-element set, the required probability equals $D(n)/n!$.

Using the inclusion–exclusion principle, we derive a formula for $D(n)$. We will actually count the "bad" permutations, i.e. those with at least one fixed point. Let S_n denote the set of all permutations of $\{1, 2, \ldots, n\}$, and for $i = 1, 2, \ldots, n$, we define $A_i = \{\pi \in S_n : \pi(i) = i\}$. The bad permutations are exactly those in the union of all the A_i.

> Here we suggest that the reader contemplate the definition of the sets A_i carefully—it is a frequent source of misunderstandings (their elements are permutations, not numbers).

In order to apply the inclusion–exclusion principle, we have to express the size of the k-fold intersections of the sets A_i. It is easy to see that $|A_i| = (n-1)!$, because if $\pi(i) = i$ is fixed, we can choose an arbitrary permutation of the remaining $n - 1$ numbers. Which permutations lie in $A_1 \cap A_2$? Just those with both 1 and 2 as fixed points (and the remaining numbers can be permuted arbitrarily), and so $|A_1 \cap A_2| = (n-2)!$. More generally, for arbitrary $i_1 < i_2 < \cdots < i_k$ we have $|A_{i_1} \cap A_{i_2} \cap \cdots \cap A_{i_k}| = (n-k)!$, and substituting this into the inclusion–exclusion formula yields

$$|A_1 \cup \cdots \cup A_n| = \sum_{k=1}^{n} (-1)^{k-1} \binom{n}{k} (n-k)! = \sum_{k=1}^{n} (-1)^{k-1} \frac{n!}{k!}.$$

We recall that we have computed the number of bad permutations (with at least one fixed point), and so

$$D(n) = n! - |A_1 \cup \cdots \cup A_n| = n! - \frac{n!}{1!} + \frac{n!}{2!} - \cdots + (-1)^n \frac{n!}{n!},$$

which can still be rewritten as

---

[6]Such permutations are sometimes called *derangements*.

$$D(n) = n! \left(1 - \frac{1}{1!} + \frac{1}{2!} - \cdots + (-1)^n \frac{1}{n!} \right). \qquad (2.21)$$

As is taught in calculus, the series in parentheses converges to e^{-1} for $n \to \infty$ (where e is the Euler number), and it does so very fast. So we have the approximate relation $D(n) \approx n!/e$, and the probability in the hatcheck lady problem converges to the constant $e^{-1} = 0.36787\ldots$. This is what also makes the problem remarkable: the answer almost doesn't depend on the number of gentlemen!

The Euler function φ. A function denoted usually by φ and named after Leonhard Euler plays an important role in number theory. For a natural number n, the value of $\varphi(n)$ is defined as the number of natural numbers $m \leq n$ that are relatively prime to n; formally

$$\varphi(n) = |\{m \in \{1, 2, \ldots, n\}: \gcd(n, m) = 1\}|.$$

Here $\gcd(n, m)$ denotes the greatest common divisor of n and m; that is, the largest natural number that divides both n and m. As an example of application of the inclusion–exclusion principle, we find a formula which allows us to calculate $\varphi(n)$ quickly provided that we know the factorization of n into prime factors.

The simplest case is when $n = p$ is a prime. Then every $m < p$ is relatively prime to p, and so $\varphi(p) = p - 1$.

The next step towards the general solution is the case when $n = p^\alpha$ ($\alpha \in \mathbf{N}$) is a prime power. Then the numbers not relatively prime to p^α are multiples of p, i.e. $p, 2p, 3p, \ldots, p^{\alpha-1}p$, and there are $p^{\alpha-1}$ such multiples not exceeding p^α (in general, if d is an any divisor of some number n, then the number of multiples of d not exceeding n is n/d). Hence, there are $\varphi(p^\alpha) = p^\alpha - p^{\alpha-1} = p^\alpha(1 - 1/p)$ remaining numbers that are relatively prime to p^α.

An arbitrary n can be written in the form

$$n = p_1^{\alpha_1} p_2^{\alpha_2} \ldots p_r^{\alpha_r},$$

where $p_1, p_2, \ldots, p_r$ are distinct primes and $\alpha_i \in \mathbf{N}$. The "bad" $m \leq n$, i.e. those not contributing to $\varphi(n)$, are all multiples of some of the primes p_i. Let us denote by $A_i = \{m \in \{1, 2, \ldots, n\}: p_i | m\}$ the set of all multiples of p_i. We have $\varphi(n) = n - |A_1 \cup A_2 \cup \cdots \cup A_r|$. The inclusion–exclusion principle commands that we find the sizes of the intersections of the sets A_i. For example, the intersection $A_1 \cap A_2$ contains the numbers divisible by both p_1 and p_2, which are exactly the multiples of $p_1 p_2$, and hence $|A_1 \cap A_2| = n/(p_1 p_2)$. The same argument gives

$$|A_{i_1} \cap A_{i_2} \cap \cdots \cap A_{i_k}| = \frac{n}{p_{i_1} p_{i_2} \cdots p_{i_k}}.$$

Let us look at the particular cases $r = 2$ and $r = 3$ first. For $n = p_1^{\alpha_1} p_2^{\alpha_2}$ we have

$$\varphi(n) = n - |A_1 \cup A_2| = n - |A_1| - |A_2| + |A_1 \cap A_2|$$

$$= n - \frac{n}{p_1} - \frac{n}{p_2} + \frac{n}{p_1 p_2} = n \left(1 - \frac{1}{p_1} \right) \left(1 - \frac{1}{p_2} \right).$$

Similarly, for $n = p_1^{\alpha_1} p_2^{\alpha_2} p_3^{\alpha_3}$ we get

$$\varphi(n) = n - \frac{n}{p_1} - \frac{n}{p_2} - \frac{n}{p_3} + \frac{n}{p_1 p_2} + \frac{n}{p_1 p_3} + \frac{n}{p_2 p_3} - \frac{n}{p_1 p_2 p_3}$$

$$= n \left(1 - \frac{1}{p_1} \right) \left(1 - \frac{1}{p_2} \right) \left(1 - \frac{1}{p_3} \right).$$

This may raise a suspicion concerning the general formula.

2.8.2 Theorem. *For $n = p_1^{\alpha_1} p_2^{\alpha_2} \ldots p_r^{\alpha_r}$, we have*

$$\varphi(n) = n \left(1 - \frac{1}{p_1} \right) \left(1 - \frac{1}{p_2} \right) \ldots \left(1 - \frac{1}{p_r} \right). \qquad (2.22)$$

Proof. For an arbitrary r, the inclusion–exclusion principle (we use, to our advantage, the short formula (2.16)) gives

$$\varphi(n) = n - \sum_{\emptyset \neq I \subseteq \{1,2,\ldots,r\}} (-1)^{|I|-1} \frac{n}{\prod_{i \in I} p_i} = n \cdot \sum_{I \subseteq \{1,2,\ldots,r\}} \frac{(-1)^{|I|}}{\prod_{i \in I} p_i}.$$

We claim that this frightening formula equals the right-hand side of Eq. (2.22). This follows from the formula (2.17) for expanding the product $(1+x_1)(1+x_2)(1+x_3) \ldots$ by substituting $x_i = -1/p_i$, $i = 1, 2, \ldots, r$. □

Exercises

1. There are n married couples attending a dance. How many ways are there to form n pairs for dancing if no wife should dance with her husband?

2. (a) Determine the number of permutations with exactly one fixed point.

   (b) Count the permutations with exactly k fixed points.

3. What is wrong with the following inductive "proof" that $D(n) = (n-1)!$ for all $n \geq 2$? Can you find a false step in it? For $n = 2$, the formula holds, so assume $n \geq 3$. Let π be a permutation of $\{1, 2, \ldots, n-1\}$ with no fixed point. We want to extend it to a permutation π' of $\{1, 2, \ldots, n\}$ with no fixed point. We choose a number

$i \in \{1, 2, \ldots, n-1\}$, and we define $\pi'(n) = \pi(i)$, $\pi'(i) = n$, and $\pi'(j) = \pi(j)$ for $j \neq i, n$. This defines a permutation of $\{1, 2, \ldots, n\}$, and it is easy to check that it has no fixed point. For each of the $D(n-1) = (n-2)!$ possible choices of π, the index i can be chosen in $n-1$ ways. Therefore, $D(n) = (n-2)! \cdot (n-1) = (n-1)!$.

4. *Prove the equation

$$D(n) = n! - nD(n-1) - \binom{n}{2}D(n-2) - \cdots - \binom{n}{n-1}D(1) - 1.$$

5. (a) *Prove the recurrent formula $D(n) = (n-1)[D(n-1) + D(n-2)]$. Prove the formula (2.21) for $D(n)$ by induction.

(b) *Calculate the formula for $D(n)$ directly from the relation derived in (a). Use an auxiliary sequence given by $a_n = D(n) - n D(n-1)$.

6. How many permutations of the numbers $1, 2, \ldots, 10$ exist that map no even number to itself?

7. (Number of mappings onto) Now is the time to calculate the number of mappings of an n-element set *onto* an m-element set (note that we have avoided it so far). Calculate them

(a) for $m = 2$

(b) for $m = 3$.

(c) *Write a formula for a general m; check the result for $m = n = 10$ (what is the result for $n = m$?). Warning: The resulting formula is a sum, not a "nice" formula like a binomial coefficient.

(d) *Show, preferably without using part (c), that the number of mappings onto an m-element set is divisible by $m!$.

8. (a) *How many ways are there to divide n people into k groups (or: how many equivalences with k classes are there on an n-element set)? Try solving this problem for $k = 2, 3$ and $k = n - 1, n - 2$ first. For a general k, the answer is a sum.

(b) What is the total number of equivalences on an n-element set? (Here the result is a double sum.)

(c) *If we denote the result of (b) by B_n (the nth *Bell number*), prove the following (surprising) formula:

$$B_n = \frac{1}{e} \sum_{i=0}^{\infty} \frac{i^n}{i!}.$$

9. *Prove the formula (2.22) for the Euler function in a different way. Suppose it holds for $n = p^\alpha$ (a prime power). Prove the following *auxiliary claim*: if m, n are relatively prime, then $\varphi(mn) = \varphi(m)\varphi(n)$.

10. *For an arbitrary natural number n, prove that $\sum_{d\mid n} \varphi(d) = n$ (the sum is over all natural numbers d dividing n).

11. (a) How many divisors does the number $n = p_1^{\alpha_1} p_2^{\alpha_2} \ldots p_r^{\alpha_r}$ have ($p_1, p_2, \ldots, p_r$ are distinct primes)?

   (b) Show that the *sum* of all divisors of such a number n equals

   $$\prod_{i=1}^{r} \frac{p^{\alpha_i+1} - 1}{p_i - 1}.$$

   (c) **Call a number n *perfect* if it equals the sum of all its divisors (excluding itself). For example, $6 = 1 + 2 + 3$ is perfect. Prove that every even perfect number has the form $2^q(2^{q+1} - 1)$, where $q \geq 1$ is a natural number and $2^{q+1} - 1$ is a prime.

   Remark. No odd perfect numbers are known but no one can show that they don't exist.

12. (a) *For a given natural number N, determine the probability that two numbers $m, n \in \{1, 2, \ldots, N\}$ chosen independently at random are relatively prime.

   (b) *Prove that the limit of the probability in (a) for $N \to \infty$ equals the infinite product $\prod_p (1 - 1/p^2)$, where p runs over all primes. (Let us remark that the value of this product is $6/\pi^2$; this can be proved from Fact 10.7.1.)

13. (a) Determine the number of graphs with no vertices of degree 0 on a given n-element vertex set V (see Sections 3.1 and 3.3 for the relevant definitions).

   (b) Determine the number of all graphs with at least 2 vertices of degree 0 on V, and with exactly 2 vertices of degree 0.

14. *How many ways are there to seat n married couples at a round table with $2n$ chairs in such a way that the couples never sit next to each other?

3

Graphs: an introduction

3.1 The notion of a graph; isomorphism

Many situations in various practically motivated problems and also in mathematics and theoretical computer science can be captured by a scheme consisting of two things:

- a (finite) set of points, and
- lines joining some pairs of the points.

For example, the points may represent participants at a birthday party and the joins correspond to pairs of participants who know each other. Or the points can represent street crossings in a city and the joins the streets. Also a municipal transport network or a railway network is usually displayed as a scheme of this type (see Fig. 3.1), and electrotechnical schemes often have a similar character as well. In such cases, the points are commonly called *vertices* (or also *nodes*) and the joins are called *edges*.[1]

If we disregard the length, shape, and other properties of the joins and we only pay attention to which pairs of points are joined and which are not, we arrive at the mathematical notion of a graph. Although very simple, a graph is one of the key concepts in discrete mathematics. This is also illustrated in the subsequent sections.

3.1.1 Definition. *A graph[2] G is an ordered pair (V, E), where V is some set and E is a set of 2-point subsets of V. The elements of the set V are called* vertices *of the graph G and the elements of E* edges *of G.*

---

[1] The origins of this terminology are mentioned in Section 5.3.

[2] What we simply call a graph here is sometimes more verbosely called a *simple undirected graph*, in order to distinguish it from other, related notions. Some of them will be mentioned later.

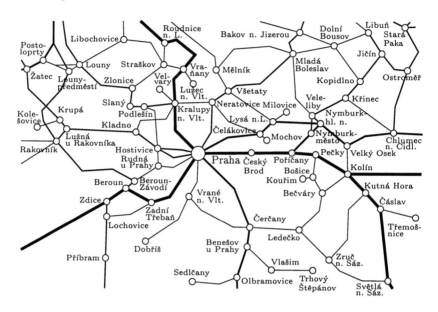

Fig. 3.1   A scheme of part of the Czech railway network in a region of about 150 × 100 km around Prague.

In this book, we almost always consider graphs with finite vertex sets. The few cases where we deal with infinite graphs too will be mentioned explicitly.

If we want to point out that some graph G has V as the vertex set and E as the edge set we write $G = (V, E)$. If we talk about some known graph G and we want to refer to its vertex set, we denote it by $V(G)$. Similarly we write $E(G)$ for the edge set of G. A useful notation is also $\binom{V}{2}$ for the set of all 2-element subsets of V (see Section 2.3.1 for a motivation of this symbol). We can briefly say that a graph is a pair (V, E), where $E \subseteq \binom{V}{2}$.

The following terminology is fairly self-explanatory: if $\{u, v\}$ is an edge of some graph G, we say that the vertices u and v are *adjacent* in G or that u is a *neighbor* of v (and v is a neighbor of u).

Graphs are usually depicted by drawings in the plane. The vertices of a graph are assigned points in the plane (drawn as dots, bullets, little circles, etc.) and the edges are expressed by connecting the corresponding pairs of points by straight or variously curved lines (these lines are called *arcs* in this context). In this way, we get pictures like this:

The word graph itself perhaps comes from the possibility of such a drawing. (We should emphasize that the word graph is used here with a different meaning than in "graph of a function"—we sincerely hope that the reader has noticed this by now.)

The role of the drawing of a graph is auxiliary, however. A graph can also be represented in many other ways, and, for instance, in a computer memory it is certainly not stored as a picture. One graph can be drawn in many different ways. For example, the first two of the above pictures show the same graph with vertex set $\{1, 2, 3, 4, 5\}$ and edges $\{1, 2\}$, $\{2, 3\}$, $\{3, 4\}$, $\{4, 5\}$, $\{5, 1\}$.

In a visually well-arranged drawing of a graph, the arcs should "cross" as little as possible. Crossings could possibly be mistaken for vertices, and also in some schemes of electronic circuits and in a number of other situations crossings are inadmissible. This leads to the study of an important class of the so-called planar graphs (see Chapter 5).

Drawing graphs is an important aid in the theory of graphs. Draw pictures for yourself whenever possible! Many notions are motivated "pictorially" and drawings can make such notions much more intuitive.

The railway scheme in Fig. 3.1 also illustrates that the notion of a graph is a nontrivial abstraction and simplification of real life situations. The arcane and historically developed railway network shown in the figure resembles a graph drawing, but, for example, there are few places where tracks branch outside of railway stations. Also, there are plenty of other types of information one might want to associate with a railway network scheme (track quality and number for each connection, which tracks go straight through a station and which form a side branch, station distances, train schedules, and so on). If one should make a mathematical model of a railway network and didn't know the notion of a graph, one would probably come up with something much more complex. In applications, graphs are indeed often augmented by further information. But the notion of a graph is very useful as a "skeleton" of such mathematical models, and once we use a graph as a significant part of the model, we immediately have a well-developed mathematical theory at our disposal, suggesting further notions, properties, and efficient algorithms to consider.

Important graphs. We introduce several types of specific graphs, which are quite often encountered in graph theory and for which standard notation and terminology have become customary.

The complete graph K_n: $V = \{1, 2, \ldots, n\}$, $E = \binom{V}{2}$.

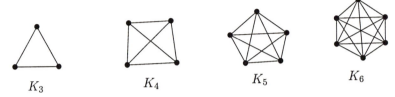

Mainly for aesthetic reasons, we also include a drawing of K_{23} with all of its 253 edges:

The cycle C_n: $V = \{1, 2, \ldots, n\}$, $E = \{\{i, i+1\}: i = 1, 2, \ldots, n-1\} \cup \{\{1, n\}\}$.

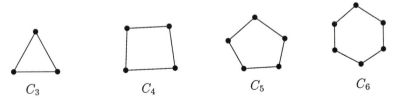

The path P_n: $V = \{0, 1, \ldots, n\}$, $E = \{\{i-1, i\}: i = 1, 2, \ldots, n\}$.

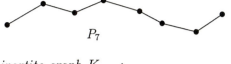

P_7

The complete bipartite graph $K_{n,m}$:
$$V = \{u_1, \ldots, u_n\} \cup \{v_1, \ldots, v_m\},$$
$$E = \{\{u_i, v_j\}: i = 1, 2, \ldots, n,\ j = 1, 2, \ldots, m\}.$$

$K_{1,1}$ $\qquad$ $K_{1,2}$ $\qquad$ $K_{1,3}$ $\qquad$ $K_{2,3}$ $\qquad$ $K_{3,3}$

A small explanation seems appropriate here. The word "bipartite" means "consisting of two parts". In general, a graph G is called *bipartite* if the set $V(G)$ can be divided into two disjoint sets V_1 and V_2 in such a way that each edge of G connects a vertex from V_1 to a vertex from V_2. Written in symbols, $E(G) \subseteq \{\{v, v'\}: v \in V_1, v' \in V_2\}$. Such sets V_1 and V_2 are sometimes called the *classes* of G, but *not* "partites", however tempting this neologism may be.

Graph isomorphism. Two graphs G and G' are considered identical (or equal) if they have the same set of vertices and the same set of edges, i.e. $G = G'$ means $V(G) = V(G')$ and $E(G) = E(G')$. But many graphs differ "only" by the names of their vertices and edges and have the same "structure". This is captured by the notion of isomorphism.

3.1.2 Definition. *Two graphs $G = (V, E)$ and $G' = (V', E')$ are called isomorphic if a bijection $f: V \to V'$ exists such that*

$$\{x, y\} \in E \quad \text{if and only if} \quad \{f(x), f(y)\} \in E'$$

holds for all $x, y \in V$, $x \neq y$. Such an f is called an isomorphism of the graphs G and G'. The fact that G and G' are isomorphic is written $G \cong G'$.

An isomorphism is thus required to map adjacent vertices to adjacent vertices and nonadjacent vertices to nonadjacent vertices, and it can be thought of as "renaming the vertices" of a graph. The relation $\cong$ ("to be isomorphic") is an equivalence, on any set of graphs (see Exercise 6).

Problem. The following three pictures show isomorphic graphs. Show this by finding suitable isomorphisms!

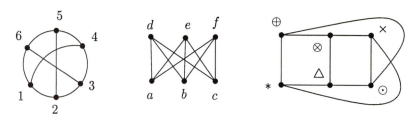

Solution. All the three graphs are isomorphic to $K_{3,3}$. An isomorphism of the first graph to the second graph: for instance, $1 \mapsto a$, $2 \mapsto d$, $3 \mapsto b$, $4 \mapsto e$, $5 \mapsto c$, $6 \mapsto f$ (several other possibilities exist!). The others are left to the reader.

Warning. The definition of isomorphism looks easy but many students tend to mess it up. They often think that an isomorphism of G to G' is any bijection between E and E' etc. As a precaution, it is best to try writing the definition down without looking in the textbook.

Testing isomorphism. For small pictures, it is usually not too difficult to find out whether they correspond to isomorphic graphs or not (although the preceding problem indicates that the pictures need not look the same at all). But the problem of deciding whether two given graphs are isomorphic or not is difficult in general, and no efficient algorithm is known for it (i.e. one working fast in all cases). It is even suspected that no such efficient algorithm exists. Roughly speaking, the difficulty lies in showing that two given graphs on n vertices are not isomorphic. To check this according to the definition, we must verify that none of the possible $n!$ bijections of the vertex sets is an isomorphism. Of course, often we can use a shortcut and exclude the possibility of an isomorphism right away. For instance, if the numbers of edges differ, the graphs cannot be isomorphic because isomorphism preserves the number of edges. More generally, if we can assign some number, vector, etc., to a graph in such a way that isomorphic graphs are always assigned the same value, we can sometimes use this to distinguish nonisomorphic graphs (examples will be discussed later). But so far no fast method has been found that would *always* succeed in distinguishing nonisomorphic graphs.

Number of nonisomorphic graphs. Let V be the set $\{1, 2, \ldots, n\}$. To choose a graph with vertex set V means choosing an arbitrary subset $E \subseteq \binom{V}{2}$. The set $\binom{V}{2}$ has $\binom{n}{2}$ elements, and thus the number of different graphs on V is exactly $2^{\binom{n}{2}}$. However, there are considerably fewer than $2^{\binom{n}{2}}$ pairwise nonisomorphic graphs with n vertices. For example, for $V = \{1, 2, 3\}$ we get the following $8 = 2^{\binom{3}{2}}$ distinct graphs:

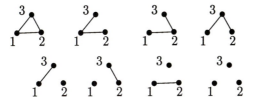

Among these 8 possibilities, only 4 nonisomorphic ones can be found:

How many pairwise nonisomorphic graphs on n vertices exist for a general n? In other words, how many classes of the equivalence relation $\cong$ on the set of all graphs with vertex set $V = \{1, 2, \ldots, n\}$ are there? Determining this number exactly is not easy (see e.g. Harary and Palmer [19]), but we can at least get a reasonable estimate by a simple (but clever!) trick. On the one hand, the number of nonisomorphic graphs on n vertices is certainly not larger than the number of all distinct graphs on the set V, i.e. $2^{\binom{n}{2}}$. On the other hand, consider a particular graph G with vertex set V. How many distinct graphs G' on V are isomorphic to it? For instance, if G is the graph

on the vertex set $\{1, 2, 3\}$, there are 3 such isomorphic graphs. By definition, if G' is such a graph isomorphic to G, there exists a bijection $f: V \to V$ that is an isomorphism of G and G'. The number of all possible bijections $f: V \to V$ is $n!$, and hence G is isomorphic to at most $n!$ distinct graphs on the set V. (We may be overcounting! For our specific example for $n = 3$, we had $3! = 6$ bijections but only 3 distinct graphs isomorphic to G—can you explain why?) In other words, each class of the equivalence $\cong$ on the set of all graphs with vertex set V consists of no more than $n!$ graphs, and therefore the number of equivalence classes is at least

$$\frac{2^{\binom{n}{2}}}{n!} .$$

Consequently, there is a collection of at least this many pairwise nonisomorphic graphs on n vertices.

We claim that this function of n doesn't grow much more slowly than $2^{\binom{n}{2}}$. To see this, we take the logarithms of both functions and manipulate the resulting expressions somewhat. We use the obvious estimate $n! \leq n^n$:

$$\log_2 \left[2^{\binom{n}{2}} \right] = \binom{n}{2} = \frac{n^2}{2} \left(1 - \frac{1}{n} \right),$$

$$\log_2 \frac{2^{\binom{n}{2}}}{n!} = \binom{n}{2} - \log_2 n! \geq \frac{1}{2}n^2 - \frac{1}{2}n - n\log_2 n$$
$$= \frac{n^2}{2}\left(1 - \frac{1}{n} - \frac{2\log_2 n}{n}\right).$$

We see that for large n, the logarithms of both functions behave "roughly as" the function $\frac{1}{2}n^2$: the relative error we would make by replacing their logarithms by $\frac{1}{2}n^2$ goes to 0 for $n \to \infty$. (Section 2.4 says more about estimating the growth of functions.) In particular, if n is sufficiently large, the number of nonisomorphic graphs on n vertices is much much larger than 2^n, say.

In the consideration just made, we have only shown that many non-isomorphic graphs exist, but, remarkably, we have constructed no specific collection of such graphs. Similar methods will be discussed more systematically in Chapter 9. Constructing many nonisomorphic graphs explicitly is not so easy—see Exercise 7.

Exercises

1. (a) Find an isomorphism of the following graphs:

(b) Show that both the graphs above are isomorphic to the following graph: the vertex set is $\binom{\{1,2,\dots,5\}}{2}$ (unordered pairs of numbers), and two vertices $\{i,j\}$ and $\{k,\ell\}$ ($i,j,k,\ell \in \{1,2,\dots,5\}$) form an edge if and only if $\{i,j\} \cap \{k,\ell\} = \emptyset$.

Remark. This graph is called the *Petersen graph* and it is one of the most remarkable small graphs (being a counterexample to numerous conjectures, an exceptional case in many theorems, etc.). It is the smallest nontrivial member of a family of the so-called *Kneser graphs*, which supplies many more examples of graphs with interesting properties. The vertex set of a Kneser graph is $\binom{\{1,2,\dots,n\}}{k}$ for natural numbers $n > k \geq 1$, and edges correspond to empty intersections.

2. An *automorphism* of a graph $G = (V, E)$ is any isomorphism of G and G, i.e. any bijection $f: V \to V$ such that $\{u, v\} \in E$ if and only if $\{f(u), f(v)\} \in E$. A graph is called *asymmetric* if its only automorphism is the identity mapping (each vertex is mapped to itself).

(a) Find an example of an asymmetric graph with at least 2 vertices.

(b) Show that no asymmetric graph G exists with $1 < |V(G)| \leq 5$.

3. Show that a graph G with n vertices is asymmetric (see Exercise 2) if and only if $n!$ distinct graphs on the set $V(G)$ are isomorphic to G.

4. Call a graph $G = (V, E)$ *vertex-transitive* if for any two vertices $v, v' \in V$ an automorphism $f : V \to V$ of G exists (see Exercise 2) with $f(v) = v'$. Similarly, G is *edge-transitive* if for any two edges $e, e' \in E$ an automorphism $f : V \to V$ exists with $f(e) = e'$ (if $e = \{u, v\}$ then the notation $f(e)$ stands for the set $\{f(u), f(v)\}$).

   (a) Prove that the graph in Exercise 1 is vertex-transitive.

   (b) Decide whether each vertex-transitive graph is edge-transitive as well.

   (c) Find a graph that is edge-transitive but not vertex-transitive.

   (d) *Show that any graph as in (c) is necessarily bipartite.

5. How many graphs on the vertex set $\{1, 2, \ldots, 2n\}$ are isomorphic to the graph consisting of n vertex-disjoint edges (i.e. with edge set $\{\{1, 2\}, \{3, 4\}, \ldots, \{2n - 1, 2n\}\}$?

6. Let V be a finite set. Let $\mathcal{G}$ denote the set of all possible graphs with vertex set V. Verify that $\cong$ ("to be isomorphic") is an equivalence relation on $\mathcal{G}$.

7. Construct as many pairwise nonisomorphic graphs with vertex set $\{1, 2, \ldots, n\}$ as possible (suppose that n is a very large number). Can you find more than n^2 of them? At least $2^{n/10}$, or even substantially more?

8. (a) ^{CS} Plot the logarithms of the functions $2^{\binom{n}{2}}$ and $2^{\binom{n}{2}}/n!$ in a suitable range.

   (b) ^{*,CS} Write a computer program for calculating the number of non-isomorphic graphs on n vertices for a given n. (Warning: Unless you devise a clever method, you will only be able to deal with very small values of $n!$). For the values of n you can handle, draw the actual numbers on the plot made in (a).

   (c) *If you were able to solve (b) cleverly, the numbers should indicate that the lower bound $2^{\binom{n}{2}}/n!$ for the number of nonisomorphic graphs is much closer to the truth than the upper bound $2^{\binom{n}{2}}$. The upper bound was gained by a quite trivial method anyway; can you improve it?

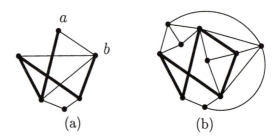

Fig. 3.2   An example of a subgraph (a), and of an induced subgraph (b).

3.2   Subgraphs, components, adjacency matrix

The next definition captures the intuitive notion of "one graph being contained in another graph". It turns out there are at least two ways of making this precise.

3.2.1 Definition. *Let G and G' be graphs. We say that G is a subgraph of G' if $V(G) \subseteq V(G')$ and $E(G) \subseteq E(G')$.*

   We say that G is an induced subgraph of G' if $V(G) \subseteq V(G')$ and $E(G) = E(G') \cap \binom{V(G)}{2}$.

   This definition can also be rephrased as follows: an induced subgraph of a graph G' arises by deleting some vertices of G' and all edges containing a deleted vertex. To get a subgraph, we can also delete some more edges although none of their end-vertices has been deleted. Fig. 3.2(a) shows a graph and its subgraph isomorphic to the path P_4 drawn by a thick line. This subgraph is not induced (because of the edge $\{a, b\}$). Fig. 3.2(b) shows an induced subgraph, isomorphic to the cycle C_5; this is a subgraph as well, of course.

Paths and cycles. A subgraph of a graph G isomorphic to some path P_t is called a *path in the graph* G; see Fig. 3.2(a). A path in a graph G can also be understood as a sequence

$$(v_0, e_1, v_1, \ldots, e_t, v_t),$$

where $v_0, v_1, \ldots, v_t$ are mutually distinct vertices of the graph G, and for each $i = 1, 2, \ldots, t$ we have $e_i = \{v_{i-1}, v_i\} \in E(G)$. This is like the log of a traveler who followed the path from one end to the other end and recorded the visited vertices and edges.[3] We also say

---

[3]Since we only consider simple graphs, recording the edges is not really necessary, as they can be reconstructed from the sequence of vertices. The definition we use is advantageous if multiple edges connecting two vertices are allowed.

that the path $(v_0, e_1, v_1, \ldots, e_t, v_t)$ is a *path from v_0 to v_n of length t*. Let us remark that we also allow $t = 0$, i.e. a path of zero length consisting of a single vertex.

Similarly, a subgraph of G isomorphic to some cycle C_t ($t \geq 3$) is called a *cycle in the graph G*; see Fig. 3.2(b). (An alternative name used in the literature is a *circuit*.) A cycle in a graph G can also be understood as a sequence

$$(v_0, e_1, v_1, e_2, \ldots, e_{t-1}, v_{t-1}, e_t, v_0)$$

(the initial and final points coincide), where $v_0, v_1, \ldots, v_{t-1}$ are pairwise distinct vertices of the graph G, and $e_i = \{v_{i-1}, v_i\} \in E(G)$ for $i = 1, 2, \ldots, t - 1$, and also $e_t = \{v_{t-1}, v_0\} \in E(G)$. The number $t \geq 3$ is the length of the cycle.

Connectedness, components. We say that a graph G is *connected* if for any two vertices $x, y \in V(G)$, G contains a path from x to y. Diagram (a) shows an example of a connected graph,

(a)　　　　　　　　(b)

while (b) is a drawing of a disconnected graph.

The notion of connectedness can also be defined slightly differently. First we define a notion similar to a path in a graph. Let $G = (V, E)$ be a graph. A sequence $(v_0, e_1, v_1, e_2, \ldots, e_t, v_t)$ is called a *walk* in G (more verbosely, a *walk of length t from v_0 to v_t*) if we have $e_i = \{v_{i-1}, v_i\} \in E$ for all $i = 1, \ldots, t$. In a walk some edges and vertices may be repeated, while for a path this was forbidden. A walk is the log of a leisurely traveler who doesn't mind visiting edges or vertices several times.

Next, we define a relation $\sim$ on the set $V(G)$ by letting $x \sim y$ if and only if there exists a walk from x to y in G. It is a fairly easy exercise to check that $\sim$ is an equivalence relation. Let $V = V_1 \dot\cup V_2 \dot\cup \cdots \dot\cup V_k$ be a partition of the vertex set V into classes of the equivalence $\sim$. The subgraphs of G induced by the sets V_i are called the *components* of the graph G. The following observation relates the definition of components to the previous definition of a connected graph.

3.2.2 Observation. *Each component of any graph is connected, and a graph is connected if and only if it has a single component.*

Proof. Clearly, a connected graph has a single component. On the other hand, any two vertices x, y in the same component of a graph G can be connected by a walk. Any walk from x to y of the shortest possible length must be a path. □

Why did we choose the somewhat roundabout definition of components using walks, rather than using paths? We could define the components using the relation $\sim'$, where $x \sim' y$ if a path from x to y exists. The above considerations show that $\sim'$ is in fact the same relation as $\sim$. But showing directly that $\sim'$ is an equivalence is a bit messy, and the approach via walks seems cleaner.

It is relatively easy to decide whether a given graph is connected, or to find the components. We aren't going to describe such algorithms here; they can be found in almost any textbook on algorithms. They are usually presented as algorithms for searching a graph, or a maze. One such algorithm is the so-called *depth-first search*.

Distance in graphs. Let $G = (V, E)$ be a connected graph. We define the *distance* of two vertices $v, v' \in V(G)$, denoted by $d_G(v, v')$, as the length of a shortest path from v to v' in G.

Hence d_G is a function, $d_G : V \times V \to \mathbf{R}$, and it is called the *distance function* or the *metric* of the graph G. The metric of G has the following properties:

1. $d_G(v, v') \geq 0$, and $d_G(v, v') = 0$ if and only if $v = v'$;
2. (symmetry) $d_G(v, v') = d_G(v', v)$ for any pair of vertices v, v';
3. (triangle inequality) $d_G(v, v'') \leq d_G(v, v') + d_G(v', v'')$ for any three vertices $v, v', v'' \in V(G)$.

The validity of these statements can be readily checked from the definition of the distance function $d_G(v, v')$. Each mapping $d : V \times V \to \mathbf{R}$ with properties 1–3 is called a *metric* on the set V, and the set V together with such a mapping d is called a *metric space*. The distance function d_G of a graph has, moreover, the following special properties:

4. $d_G(v, v')$ is a nonnegative integer for any two vertices v, v';
5. if $d_G(v, v'') > 1$ then there exists a vertex v', $v \neq v' \neq v''$, such that $d_G(v, v') + d_G(v', v'') = d_G(v, v'')$.

Conditions 1–5 already characterize functions arising as distance functions of graphs with vertex set V (see Exercise 7).

Graph representations. We have seen representations of graphs by drawings, and also by writing out a list of vertices and edges.

Graphs can also be represented in many other ways. Some of them become particularly important if we want to store and manipulate graphs in a computer. A very basic and very common representation is by an adjacency matrix:

3.2.3 Definition. *Let $G = (V, E)$ be a graph with n vertices. Denote the vertices by $v_1, v_2, \ldots, v_n$ (in some arbitrary order). The adjacency matrix of G, with respect to the chosen vertex numbering, is an $n \times n$ matrix $A_G = (a_{ij})_{i,j=1}^n$ defined by the following rule:*

$$a_{ij} = \begin{cases} 1 & \text{if } \{v_i, v_j\} \in E \\ 0 & \text{otherwise.} \end{cases}$$

This is very similar to the adjacency matrix of a relation defined in Section 1.5. The adjacency matrix of a graph is always a symmetric square matrix with entries 0 and 1, with 0s on the main diagonal. Conversely, each matrix with these properties is the adjacency matrix of some graph.

Example. The graph $G =$

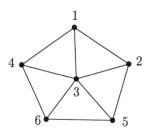

has the adjacency matrix

$$A_G = \begin{pmatrix} 0 & 1 & 1 & 1 & 0 & 0 \\ 1 & 0 & 1 & 0 & 1 & 0 \\ 1 & 1 & 0 & 1 & 1 & 1 \\ 1 & 0 & 1 & 0 & 0 & 1 \\ 0 & 1 & 1 & 0 & 0 & 1 \\ 0 & 0 & 1 & 1 & 1 & 0 \end{pmatrix}.$$

Let us emphasize that the adjacency matrix also depends on the chosen numbering of the vertices of a graph!

It might seem that we gain nothing new by viewing a graph as a matrix since the graph and the adjacency matrix both encode the same information. To illustrate that we can profit from this alternative representation, let us show a simple connection between matrix multiplication and the graph metric.

3.2.4 Proposition. *Let $G = (V, E)$ be a graph with vertex set $V = \{v_1, v_2, \ldots, v_n\}$ and let $A = A_G$ be its adjacency matrix. Let A^k denote the kth power of the adjacency matrix (the matrices are multiplied as is usual in linear algebra, i.e. if we put $B = A^2$, we have $b_{ij} = \sum_{k=1}^{n} a_{ik}a_{kj}$). Let $a_{ij}^{(k)}$ denote the element of the matrix A^k at position (i, j). Then $a_{ij}^{(k)}$ is the number of walks of length exactly k from the vertex v_i to the vertex v_j in the graph G.*

Proof. This is easy but very instructive. We proceed by induction on k. A walk of length 1 between two vertices means exactly that these vertices are connected by an edge, and hence for $k = 1$ the proposition just reformulates the definition of the adjacency matrix.

Next, let $k > 1$, and let v_i, v_j be two arbitrary vertices (possibly identical). Any walk of length k from v_i to v_j consists of an edge from v_i to some neighbor v_ℓ of v_i and a walk of length $k - 1$ from v_ℓ to v_j:

By the inductive hypothesis, the number of walks of length $k - 1$ from v_ℓ to v_j is $a_{\ell j}^{(k-1)}$. Hence the number of walks of length k from v_i to v_j is

$$\sum_{\{v_i, v_\ell\} \in E(G)} a_{\ell j}^{(k-1)} = \sum_{\ell=1}^{n} a_{i\ell} a_{\ell j}^{(k-1)}.$$

But this is exactly the element at position (i, j) in the product of the matrices A and A^{k-1}, i.e. $a_{ij}^{(k)}$. $\square$

3.2.5 Corollary. *The distance of any two vertices v_i, v_j satisfies*

$$d_G(v_i, v_j) = \min\{k \geq 1 \colon a_{ij}^{(k)} \neq 0\}.$$

This result has surprising applications. For instance, if one wants to find the distance for all pairs of vertices in a given graph, one can apply sophisticated algorithms for matrix multiplication (some of them are described in Aho, Hopcroft, and Ullman [9]) and some other ideas and get unexpectedly fast methods for computing the function d_G. Exercises 10 and 11 indicate other algorithmic applications.

Let us remark that the adjacency matrix is not always the best computer representation of a graph. Especially if a graph has few edges (much fewer than $\binom{n}{2}$), it is usually better to store the list of neighbors for every vertex. For a fast implementation of certain algorithms, other, more complicated representations are used as well.

Exercises

1. Prove that the complement of a disconnected graph G is connected. (The *complement* of a graph $G = (V, E)$ is the graph $(V, \binom{V}{2} \setminus E)$.)

2. What is the maximum possible number of edges of a graph with n vertices and k components?

3. ^{CS} Design an algorithm that finds the decomposition of a given graph G into its components. (Try to get an algorithm which needs at most $O(n + m)$ steps for a graph with n vertices and m edges.)

4. *Prove that a graph is bipartite if and only if it contains no cycle of an odd length.

5. *Describe all graphs containing no path (not necessarily induced!) of length 3.

6. *Having solved the preceding exercise, describe all graphs containing no path of length 4.

7. Show that if a function $d: V \times V \to \mathbf{N}$ satisfies conditions 1–5 above then a graph $G = (V, E)$ exists such that $d_G(v, v') = d(v, v')$ for any pair of elements of V.

8. Define the "diameter" and "radius" of a graph (in analogy with the intuitive meaning of these notions).

9. (a) Find a connected graph of n vertices for which each of the powers $A_G^1, A_G^2, \dots$ of the adjacency matrix contains some zero elements.

   (b) Let G be a graph on n vertices, $A = A_G$ its adjacency matrix, and I_n the $n \times n$ identity matrix (with 1s on the diagonal and 0s elsewhere). Prove that G is connected if and only if the matrix $(I_n + A)^{n-1}$ has no 0s.

   (c) Where are the 0s in the matrix $(I_n + A)^{n-1}$ if the graph G is not connected?

10. Show that a graph G contains a triangle (i.e. a K_3) if and only if there exist indices i and j such that both the matrices A_G and A_G^2 have the entry (i, j) nonzero, where A_G is the incidence matrix of G.

   Remark. In connection with algorithms for fast matrix multiplication, this observation gives the fastest known method for deciding whether a given graph contains a triangle, substantially faster than the obvious $O(n^3)$ algorithm.

11. *Let G be a graph. Prove the following formula for the number of cycles of length 4 in G (not necessarily induced):

$$\frac{1}{8}\left(\text{trace}\left(A_G^4\right) - 2|E(G)| - 4\sum_{v \in V(G)} \binom{\deg_G(v)}{2} \right).$$

Here A_G^4 is the 4th power of the incidence matrix, and $\text{trace}\left(A_G^4\right)$ denotes the sum of the elements on the main diagonal of A_G^4. For the definition of $\deg_G(v)$, see the next section. Note that this gives an $O(n^3)$ algorithm for counting the number of cycles of length 4, or even a faster algorithm using algorithms for fast matrix multiplication.

12. Prove that G and G' are isomorphic if and only if a permutation matrix P exists such that
$$A_{G'} = PA_GP^T.$$

Here A_G is the adjacency matrix of G and P^T denotes the transposed matrix P. A matrix P is called a *permutation matrix* if its entries are 0 and 1 and each row and each column contain precisely one 1.

3.3 Graph score

Let G be a graph, and let v be a vertex of G. The number of edges of G containing the vertex v is denoted by the symbol $\deg_G(v)$. The number $\deg_G(v)$ is called the *degree* of v in the graph G.

Let us denote the vertices of G by $v_1, v_2, \ldots, v_n$ (in some arbitrarily chosen order). The sequence

$$(\deg_G(v_1), \deg_G(v_2), \ldots, \deg_G(v_n))$$

is called a *degree sequence* of the graph G, or a *score* of G. By choosing different numberings of the vertices of the same graph, we usually obtain several different sequences of numbers differing by the order of their terms. Thus, we will not distinguish two scores if one of them can be obtained from the other by rearranging the order of the numbers. We will usually write scores in nondecreasing order, with the smallest degree coming first.

It is easy to see that two isomorphic graphs have the same scores, and thus two graphs with different scores are necessarily nonisomorphic. On the other hand, graphs with the same score need not be isomorphic! For example, the graphs

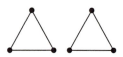

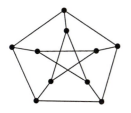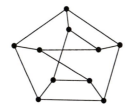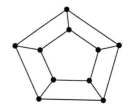

Fig. 3.3 Three connected nonisomorphic graphs with the same score.

both have score $(2, 2, 2, 2, 2, 2)$, but they cannot be isomorphic since one of them is connected while the other one is not. All the three graphs in Fig. 3.3 have score $(3, 3, 3, 3, 3, 3, 3, 3, 3, 3)$ but no two of them are isomorphic (to prove this is a bit harder; see Exercise 1). In spite of these negative examples, the score is an important and easily computable characteristic of a graph, and it can often help to distinguish nonisomorphic graphs in practice.

A problem studied in graph theory, although not really one of the most important ones, is to characterize sequences of numbers that can appear as scores of graphs. One such basic and often quite significant property is a consequence of the following observation:

3.3.1 Proposition. *For each graph $G = (V, E)$ we have*

$$\sum_{v \in V} \deg_G(v) = 2|E|.$$

Proof. The degree of a vertex v is the number of edges containing v. Each edge contains 2 vertices, and hence by summing up all degrees we get twice the number of edges. □

3.3.2 Corollary (Handshake lemma). *The number of odd-degree vertices is even, in any graph. (Or: the number of participants at a birthday party who shook hands with an odd number of other participants is always even—for any finite party.)*

Let us remark that the handshake lemma is not true for infinite parties. A one-sided infinite path has a single odd-degree vertex:

The handshake lemma (Corollary 3.3.2), and some other simple necessary conditions, are not sufficient to characterize sequences that can show up as graph scores (see Exercise 2). A full characterization of scores is not quite simple, and it is related to so-called network flows which are not treated in this book. Here we explain a simple algorithm for deciding whether a given sequence of integers is a graph score or not. The algorithm is an easy consequence of the following result.

3.3.3 Theorem (Score theorem). *Let $D = (d_1, d_2, \ldots, d_n)$ be a sequence of natural numbers, $n > 1$. Suppose that $d_1 \leq d_2 \leq \cdots \leq d_n$, and let the symbol D' denote the sequence $(d'_1, \ldots, d'_{n-1})$, where*

$$
d'_i = \begin{cases} d_i & \text{for } i < n - d_n \\ d_i - 1 & \text{for } i \geq n - d_n. \end{cases}
$$

For example, for $D = (1, 1, 2, 2, 2, 3, 3)$, we have $D' = (1, 1, 2, 1, 1, 2)$. Then D is a graph score if and only if D' is a graph score.

Proof.   One implication is easy. Suppose that D' is a score of a graph $G' = (V', E')$, where $V' = \{v_1, v_2, \ldots, v_{n-1}\}$ and $\deg_G(v_i) = d'_i$ for all $i = 1, 2, \ldots, n - 1$. Fix a new vertex v_n distinct from all of $v_1, \ldots, v_{n-1}$, and define a new graph $G = (V, E)$ where

$$
V = V' \cup \{v_n\}
$$
$$
E = E' \cup \{\{v_i, v_n\}: i = n - d_n, n - d_n + 1, \ldots, n - 1\}.
$$

Expressed less formally, the new vertex v_n is connected to the d_n last vertices of the graph G'. Clearly G has score D. (Remembering this construction is the best way to remember the statement of the theorem.)

It is more difficult to prove the reverse implication, i.e. if D is a score then D' is a score. So assume that D is a score of some graph. The trouble is that in general, we cannot reverse the construction by which we passed from D' to D, i.e. to tear off a largest-degree vertex, since such a vertex can be connected to other vertices than we would need. An example is shown in the following picture:

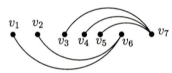

We thus consider the set $\mathcal{G}$ of all graphs on the vertex set $\{v_1, \ldots, v_n\}$ in which the degree of each vertex v_i equals d_i, $i = 1, 2, \ldots, n$. We prove the following:

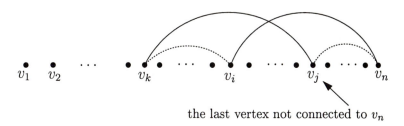

the last vertex not connected to v_n

Fig. 3.4 Illustration for the proof of the score theorem.

Claim. The set $\mathcal{G}$ contains a graph G_0 in which the vertex v_n is adjacent exactly to the vertices $v_{n-d_n}, v_{n-d_n+1}, \ldots, v_{n-1}$, i.e. to the last d_n vertices.

Having a graph G_0 as in the claim, it is already clear that the graph $G' = (\{v_1, \ldots, v_{n-1}\}, E')$, where $E' = \{e \in E(G_0) \colon v_n \notin e\}$, has score D' (i.e. we can remove the vertex v_n in G_0), and this proves the score theorem. It remains to establish the claim.

If $d_n = n - 1$, i.e. v_n is connected to all other vertices, then any graph from $\mathcal{G}$ satisfies the claim. So suppose $d_n < n - 1$ and define, for a graph $G \in \mathcal{G}$, a number $j(G)$ as the largest index $j \in \{1, 2, \ldots, n-1\}$ such that $\{v_j, v_n\} \notin E(G)$. Let $G_0 \in \mathcal{G}$ be a graph with the smallest possible value of $j(G)$. We prove that $j(G_0) = n - d_n - 1$, and from this one can already see that G_0 satisfies the claim.

For contradiction, let us suppose that $j = j(G_0) > n - d_n - 1$. The vertex v_n has to be adjacent to d_n vertices, and at most $d_n - 1$ of them can have a larger index than v_j. Hence there exists some index $i < j$ such that v_i is adjacent to v_n, and so we have $\{v_j, v_n\} \notin E(G_0)$, $\{v_i, v_n\} \in E(G_0)$ (refer to Fig. 3.4). Since $\deg_{G_0}(v_i) \leq \deg_{G_0}(v_j)$, there exists a vertex v_k adjacent to v_j but not to v_i. In this situation, we consider a new graph $G' = (V, E')$, where

$$E' = (E(G_0) \setminus \{\{v_i, v_n\}, \{v_j, v_k\}\}) \cup \{\{v_j, v_n\}, \{v_i, v_k\}\}.$$

It is easy to check that the graph G' has score D too, and at the same time $j(G') \leq j(G_0) - 1$, which contradicts the choice of G_0. This proves the claim, and hence also Theorem 3.3.3 is proved. □

As we have said, the theorem just proved gives an easy method for deciding whether a given sequence is a graph score. We illustrate the procedure with a concrete example.

Problem.   Decide whether there exists a graph with score $(1, 1, 1,$
$2, 2, 3, 4, 5, 5)$.

Solution.   We reduce the given sequence by a repeated use of the
score theorem 3.3.3:

   $(1, 1, 1, 2, 2, 3, 4, 5, 5)$
   $(1, 1, 1, 1, 1, 2, 3, 4)$
   $(1, 1, 1, 0, 0, 1, 2)$; rearranged nondecreasingly $(0, 0, 1, 1, 1, 1, 2)$
   $(0, 0, 1, 1, 0, 0)$; rearranged $(0, 0, 0, 0, 1, 1)$
   $(0, 0, 0, 0)$.

Since the last sequence is a graph score (of the graph with 4
vertices and no edges), we get that the original sequence is a score
of some graph as well. Construct an example of such a graph!

Exercises

1. Prove that the three graphs in Fig. 3.3 are pairwise nonisomorphic.

2. Construct an example of a sequence of length n in which each term is
    some of the numbers $1, 2, \ldots, n - 1$ and which has an even number of
    odd terms, and yet the sequence is not a graph score.

3. Where was the assumption $d_1 \leq d_2 \leq \cdots \leq d_n$ used in the proof of
    the score theorem? Show that the statement is not true if we omit this
    assumption.

4. Find a smallest possible example (with the smallest number of vertices)
    of two connected nonisomorphic graphs with the same score.

5. Draw all nonisomorphic graphs with score $(6, 3, 3, 3, 3, 3, 3)$. Prove that
    none was left out!

6. Find an example, as small as possible, of a graph with 6 vertices of
    degree 3, other vertices of degree ≤ 2, and with 12 edges.

7. Let G be a graph with 9 vertices, each of degree 5 or 6. Prove that it
    has at least 5 vertices of degree 6 or at least 6 vertices of degree 5.

8. (a) Decide for which $n \geq 2$ there exists a graph whose score consists
    of n distinct numbers.

    (b) *For which n does there exist a graph on n vertices whose score has
    $n - 1$ distinct numbers (i.e. exactly 2 vertices have the same degree)?

9. Let G be a graph in which all vertices have degree at least d. Prove
    that G contains a path of length d (not necessarily an induced one).

10. *Let G be a graph with maximum degree 3. Prove that its vertices can
    be colored by 2 colors (each vertex gets one color) in such a way that
    there is no path of length 2 whose 3 vertices all have the same color.

11. **Let G be a graph with all vertices of degree at least 3. Show that G contains a cycle which is not induced (i.e. it has a "diagonal").

12. A graph G is called *k-regular* if all its vertices have degree exactly k. Determine all pairs (k, n) such that there exists a k-regular graph on n vertices.

13. Draw all nonisomorphic 3-regular graphs on 6 vertices.

14. Find a 3-regular asymmetric graph (see Exercise 3.1.2).

15. (a) Show that the following graph and the graph in Fig. 8.3 are isomorphic:

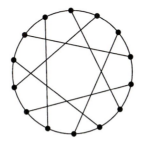

(b) *Prove that any 3-regular graph with 14 vertices containing no cycle of length 5 or smaller is isomorphic to the graph in (a) (this graph is called the *Heawood graph*). Reading Chapter 8 first might help.

16. *Let G be a connected graph in which any two distinct vertices u, v have either 0 or 5 common neighbors. Prove that G is k-regular for some k.

17. **Prove that each graph with an even number of vertices has two vertices with an even number of common neighbors.

3.4 Eulerian graphs

Here is one of the oldest problems concerning graph drawing.

Problem. Draw a given graph $G = (V, E)$ with a single closed line, without lifting the pencil from the paper (and drawing each edge only once).

Mathematically, this can be formalized as follows: find a closed walk $(v_0, e_1, v_1, \ldots, e_{m-1}, v_{m-1}, e_m, v_0)$ containing all the vertices and all the edges, each edge exactly once (while vertices can be repeated). (Note that the first vertex and the last vertex coincide.) Such a walk is called a *closed Eulerian tour* in G, and a graph possessing a closed Eulerian tour is called *Eulerian*.

Here is an example of drawing a graph with a single line:

and here is another one:

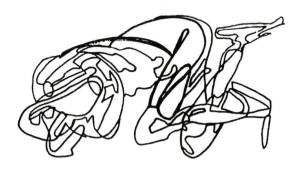

It turns out that Eulerian graphs can be characterized nicely by a "local" condition, using vertex degrees.

3.4.1 Theorem (Characterization of Eulerian graphs). *A graph $G = (V, E)$ is Eulerian if and only if it is connected and each vertex has an even degree.*

Proof. It is rather easy to show that the condition is necessary for G to be Eulerian. Clearly, an Eulerian graph must be connected. The reason for each degree being even is that whenever a closed Eulerian tour enters a vertex it must also leave it. In more detail: if we fix some direction of traversing the closed Eulerian tour and consider some vertex $v \in V(G)$, the edges incident to v can be classified as either *ingoing* or *outgoing*, and the tour defines a bijection between the set of the ingoing edges and the set of the outgoing edges.

Proving that a connected graph with all degrees even has a closed Eulerian tour is a bit more demanding. For brevity, define a *tour* in G as a walk in which no edge is repeated (vertices can be repeated, though). Consider a tour $T = (v_0, e_1, v_1, \ldots, e_m, v_m)$ in G of the largest possible length, m. We prove that

(i) $v_0 = v_m$, and

(ii) $\{e_i \colon i = 1, 2, \ldots, m\} = E$.

Ad (i). If $v_0 \neq v_m$ then the vertex v_0 is incident to an odd number of edges of the tour T. But since the degree $\deg_G(v_0)$ is even, there exists an edge $e \in E(G)$ not contained in T. Hence we could extend T by this edge—a contradiction.

Ad (ii). So, assume that $v_0 = v_m$. Write $V(T)$ for the set of vertices occurring in T and $E(T)$ for the set of edges in T. First assume $V(T) \neq V$. Thanks to the connectedness of G, an edge exists of the form $e = \{v_k, v'\} \in E(G)$, where $v_k \in V(T)$ and $v' \notin V(T)$. In this case, the tour

$$(v', e, v_k, e_{k+1}, v_{k+1}, \ldots, v_{m-1}, v_0, e_1, v_1, \ldots, e_k, v_k)$$

has length $m + 1$ and thus leads to a contradiction. Pictorially:

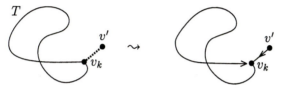

If $V(T) = V$ and $E(T) \neq E$, consider an edge $e \in E \setminus E(T)$, and write $e = \{v_k, v_\ell\}$. Analogously to the previous case, a new tour

$$(v_k, e_{k+1}, v_{k+1}, \ldots, v_{m-1}, e_m, v_0, e_1, v_1, \ldots, e_k, v_k, e, v_\ell)$$

leads to a contradiction. Pictorially:

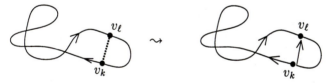

This proves Theorem 3.4.1. □

Note that the main trick in the proof is to look at the *longest possible* tour; the rest is more or less routine. This trick is worth remembering, since similar twists occur in numerous graph theory proofs (in this book see e.g. Exercise 3.3.11 or Lemma 4.1.3).

Remark about multiple edges. So far we have defined an edge of a graph as a 2-point set of vertices, and we will stick to this definition in most of the rest of this book. This means, among other things, that two vertices can be connected by at most one edge. In some applications, it is natural to admit two vertices to be connected by several distinct edges. We thus get graphs with *multiple edges*, also called *multigraphs*. Mathematically, this notion can be formalized in several different ways, some of them being more handy than others.

For instance, we could assign a nonnegative integer $m(u, v)$ to each pair $\{u, v\}$ of vertices. This $m(u, v)$ would be the *multiplicity* of the edge $\{u, v\}$. Hence $m(u, v) = 0$ would mean that the edge is not present in the graph, $m(u, v) = 1$ would denote an "ordinary" (simple) edge, and $m(u, v) > 1$ would say that the graph contains $m(u, v)$ "copies" of the edge $\{u, v\}$. A multigraph would then be an ordered pair (V, m), where $m: \binom{V}{2} \to \{0, 1, 2, \ldots\}$.

Another common way of introducing multiple edges, a more elegant one from a certain point of view, is to consider edges as "abstract" objects, i.e. to take E as some finite set disjoint from the vertex set V. For each edge $e \in E$, we then determine the pair of end-vertices of e. The same pair of vertices can occur for several edges. Formally, a graph with multiple edges is an ordered triple (V, E, ε), where V and E are disjoint sets and $\varepsilon: E \to \binom{V}{2}$ is a mapping determining the end-vertices of edges. Imagine that we have a graph-building kit with a supply of vertices and edges; then the mapping ε tells us how to assemble a particular graph from the vertices and edges:

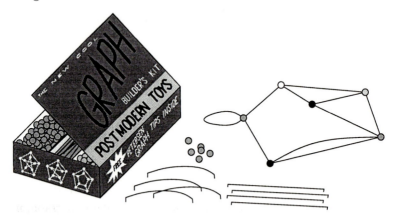

Sometimes it is useful also to admit *loops* in a graph, i.e. edges beginning and ending in the same vertex. Formally, loops can again be introduced in a number of ways. The simplest way is to represent a loop attached to a vertex v as the 1-element set $\{v\}$ in the edge set E (while the ordinary edges are 2-element sets). If multiple edges are introduced using the mapping ε, loops can be added by letting ε be a mapping into the set $\binom{V}{2} \cup V$, and a loop e is mapped by ε to its single end-vertex.

Yet another modification of the notion of a graph, the so-called directed graphs, will be discussed in Section 3.6.

For simple applications, the formal way of introducing multiple edges and loops doesn't really matter a great deal, provided that we choose a single way and keep to it consistently, and if this chosen way is not too clumsy.

Exercises

1. The following sketch of a city plan depicts 7 bridges:

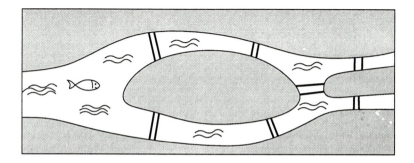

(a) Show that one cannot start walking from some place, cross each of the bridges exactly once, and come back to the starting place (no swimming please). Can one cross each bridge exactly once if it is not required to return to the starting position?

This is a historical motivation for the notion of the Eulerian graphs. The scheme (loosely) corresponds to a part of the city of Königsberg, Královec, Królewiec, or Kaliningrad—that's what it was variously called during its colorful history—and the problem was solved by Euler in 1736. Can you find the city on a modern map?

(b) How many bridges need to be added (and where) so that a closed tour exists?

Remark. Many people, not armed with the notion of a graph, might try to solve the Königsberg bridges practically, by actually walking through the city. If you have ever tried to find your way in a foreign city, you will probably agree that the chance of finding the negative solution in this way is negligible. From this point of view, one can appreciate Euler's genius and the simplicity of the graph model of the situation.

2. Characterize graphs that have a tour, not necessarily a closed one, covering all edges.

3. Draw the following graphs with a single line:

   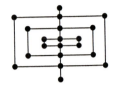

4. (a) Formulate an algorithm for finding a closed Eulerian tour in a given graph, based on the above proof of Theorem 3.4.1.

(b) ^{CS} How fast can you implement it (i.e. how many steps are needed for a graph with n vertices and m edges in the worst case)?

5. Give an alternative proof of Theorem 3.4.1, based on the following outline:

(a) Prove that if a graph G has all degrees even, its edge set can be expressed as a disjoint union of edge sets of cycles in G.

(b) Show that two edge-disjoint closed tours with at least one vertex in common can be combined into a single closed tour. Prove Theorem 3.4.1 using (a) and (b).

6. Check that Theorem 3.4.1 also holds for graphs with loops and multiple edges (what is the correct way to define the degree of a vertex for such graphs?).

7. Characterize the sequences of nonnegative integers that can appear as scores of graphs possibly with loops and multiple edges. (A loop increases the vertex degree by 2.)

8. A *Hamiltonian cycle* in a graph G is a cycle containing all vertices of G. This notion may seem quite similar to an Eulerian tour but it turns out that it is much more difficult to handle. For instance, no efficient algorithm is known for deciding whether a graph has a Hamiltonian cycle or not. This and the next two exercises are a microscopic introduction to this notion (another nice result is mentioned in Exercise 4.3.3).

(a) Decide which of the graphs drawn in Fig. 5.1 has a Hamiltonian cycle. Try to prove your claims!

(b) Construct two connected graphs with the same score, one with and one without a Hamiltonian cycle.

9. For a graph G, let $L(G)$ denote the so-called *line graph* of G, given by $L(G) = (E, \{\{e, e'\}: e, e' \in E(G), e \cap e' \neq \emptyset\})$. Decide whether the following is true for every graph G:

(a) G is connected if and only if $L(G)$ is connected.

(b) G is Eulerian if and only if $L(G)$ has a Hamiltonian cycle (see Exercise 8 for a definition).

10. (a) *Prove that every graph G with n vertices and with all vertices having degree at least $\frac{n}{2}$ has a Hamiltonian cycle (see Exercise 8 for a definition).

    (b) *Is it sufficient to assume degrees at least $\lfloor n/2 \rfloor$ in (a)?

3.5 An algorithm for an Eulerian tour

Theorem 3.4.1 provides a very simple method for showing the nonexistence of an Eulerian tour in a connected graph: find an odd-degree vertex. If all degrees are even, the proof given in the preceding section actually allows us to construct an Eulerian tour, i.e. to draw a graph with a single line: start with an arbitrary tour and keep extending it. In this section we give another proof of the theorem, formulated directly as an algorithm for finding a suitable way of drawing. This algorithm is more suitable for "manual" use; having already drawn some edges, the algorithm tells us how to continue. A more detailed analysis, which we do not include here, also shows that the new algorithm is faster, provided that all operations are implemented cleverly.

We need some preparation. Let $G = (V, E)$ be a graph. Call an edge $e \in E$ a *bridge*[4] of G if the graph $(V, E \setminus \{e\})$ has more components than G. This definition has a quite intuitive meaning:

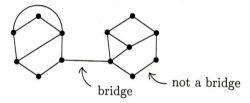

bridge   not a bridge

3.5.1 Lemma. *If all vertex degrees in a graph G are even then G has no bridge.*

Proof. We may assume that G is connected, for otherwise the lemma could be applied to each component separately. For contradiction, suppose that an edge $\{v_1, v_2\} = e$ is a bridge of G. The graph $(V, E \setminus \{e\})$ has exactly 2 components. Let G_1 be the component containing v_1. All vertices of G_1 have an even degree except for v_1 whose degree in G_1 is odd. But this is impossible by the handshake lemma (Corollary 3.3.2). $\qquad\square$

---

[4]Another common term, more appropriate for a sailor's mentality, is *isthmus*.

3.5.2 Algorithm (How to draw a graph with a single line).

The algorithm can be summarized as follows: keep drawing and never use an edge whose deletion creates a new component if this can be avoided. A detailed description follows.

Let $G = (V, E)$ be a connected graph with all vertex degrees even, and put $|E| = m$.

Step 1. Choose $v_0 \in V$ arbitrarily. Set $T_0 = (v_0)$.

Step 2. Repeat step 3 below for $i = 0, 1, 2, \ldots$, until it is no longer possible. If step 3 cannot be executed anymore, then $i = m$ and T_m is the required Eulerian tour.

Step 3. (Extending the tour) Let $T_i = (v_0, e_1, v_1, \ldots, e_i, v_i)$ be the tour constructed so far. Choose an edge $e_{i+1} \in E \backslash \{e_1, e_2, \ldots, e_i\}$ containing the vertex v_i. Moreover, whenever possible, choose e_{i+1} in such a way that the graphs $(V, E \setminus \{e_1, \ldots, e_i\})$ and $(V, E \setminus \{e_1, \ldots, e_i, e_{i+1}\})$ have the same number of components.

Proof of correctness of the algorithm. Let $T_k = (v_0, e_1, v_1, \ldots, e_k, v_k)$ be the result of the computation of the algorithm. For $i = 1, 2, \ldots, k$, let the symbol G_i denote the graph G after edges e_1 through e_i have been removed. Clearly $\deg_{G_k}(v_k) = 0$, for otherwise we could extend the tour in step 3, and hence $v_0 = v_k$.

For contradiction, suppose that the tour T_k does not pass through all the edges of G, i.e. that $E(G_k) \neq \emptyset$, and let W be the set of vertices whose degree in G_k is nonzero.

Since G is a connected graph, the tour T_k visits the (nonempty) set W at least once. Let v_t be the vertex in which the tour T_k visits W for the last time. This means that t is the maximum index with $v_t \in W$. Since $v_k \notin W$, we have $v_{t+1} \notin W$. Because G_k's edges all live on the set W and the edge e_{t+1} is the last edge of the tour T_k incident to W, the edge e_{t+1} is the only edge of the graph G_t joining a vertex of W to a vertex outside W. In other words, e_{t+1} is a bridge in the graph G_t and G_{t+1} has more components than G_t. The following figure illustrates the situation in the graph G_t:

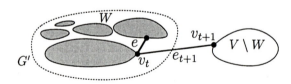

Let e be an edge of G_t containing the vertex v_t and different from e_{t+1} (such an edge exists since $\deg_{G_{t+1}}(v_t) \geq \deg_{G_k}(v_t) > 0$). By the rule in step 3, this e must also be a bridge in G_t, otherwise we would have selected it instead of e_{t+1}. The other endpoint of e also lies in W (no edge in G_t leaves W except for e_{t+1}), and so e is even a bridge in $G' = (W, E(G_t) \cap \binom{W}{2})$, the subgraph induced by the set W in the graph G_t.

At the same time, the subgraph induced by W doesn't change by removing the subsequent edges of the tour T_k. Written in symbols, $E(G') = E(G_k) \cap \binom{W}{2} = E(G_k)$. The degrees of all vertices in the graph G_k, and thus also in G', are even, and this contradicts Lemma 3.5.1 applied to G'. □

Example. This graph G

can be drawn with a single line like this

or like this

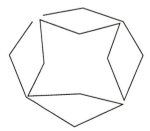

but not like this

wrong step

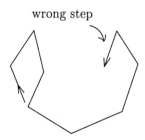

Exercises

1. ^{CS} Design the details of the algorithm described above for finding an Eulerian tour (the main problem is how to check the condition in step 3 efficiently). Try to estimate the number of steps of the algorithm for a graph with n vertices and m edges in the worst case.

2. We say that a graph $G = (V, E)$ is *randomly Eulerian* from a vertex v_0 if every maximal tour starting at v_0 is already a closed Eulerian tour in G. That is, if we start at v_0 and draw edges one by one, choosing a continuation arbitrarily among the yet unused edges, we can never get stuck. (It would be nice if art galleries or zoos were randomly Eulerian, but unfortunately they seldom are. The result in part (c) below indicates why.)

   (a) Prove that the following graphs are randomly Eulerian:

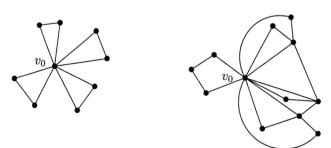

   (b) Show that the graphs below are not randomly Eulerian:

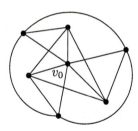

(c) *Prove the following characterization of randomly Eulerian graphs. A connected graph $G = (V, E)$ all of whose vertices have even degree is randomly Eulerian from a vertex v_0 if and only if the graph $(V \setminus \{v_0\}, \{e \in E: v_0 \notin e\})$ contains no cycle.

3. Let G be a bipartite k-regular graph for some $k \geq 2$. Prove that G has no bridge.

3.6 Eulerian directed graphs

All graphs considered so far were "undirected"—their edges were unordered pairs of vertices. Many situations involve one-way streets and schemes similar to the following:

 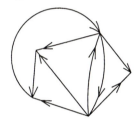

To reflect such situations, one introduces directed graphs, where every edge has a direction.

3.6.1 Definition. *A directed graph G is a pair (V, E), where E is a subset of the Cartesian product $V \times V$. The ordered pairs $(x, y) \in E$ are called directed edges. We say that a directed edge $e = (x, y)$ has head y and tail x, or[5] that e is an edge from x to y.*

Further we could introduce directed graphs with multiple edges, and also for each notion or problem for undirected graphs, we could look at its directed analogy. Sometimes the results for directed graphs are simple modifications of the results for undirected graphs. In some other problems, the directed and undirected cases differ substantially, and as a rule, the directed version is then more difficult to handle. In this book, we deal almost exclusively with undirected graphs. Let us make an exception here and introduce Eulerian directed graphs and describe one of their cute applications.

An attentive reader might have noticed that a directed graph $G = (V, E)$ is the same object as a relation on the set V. Nevertheless, both

---

[5] Here is some alternative terminology. The artificial word *digraph* is often used for a directed graph. A directed edge is sometimes called an *arrow*, a (directed) *arc*, etc. An *oriented graph* is a special type of directed graph, where we do not admit directed edges (x, y) and (y, x) simultaneously.

these notions are introduced, since directed graphs are investigated in different contexts than relations.

It is quite natural to define a *directed tour* in a directed graph $G = (V, E)$ as a sequence

$$(v_0, e_1, v_1, e_2, \ldots, e_m, v_m)$$

such that $e_i = (v_{i-1}, v_i) \in E$ for each $i = 1, 2, \ldots, m$ and, moreover, $e_i \neq e_j$ whenever $i \neq j$. Similarly we can define a directed walk, directed path, and directed cycle[6].

We say that a directed graph (V, E) is *Eulerian* if it has a closed directed tour containing all vertices and passing each directed edge exactly once. Eulerian directed graphs can again be characterized nicely. Before stating the theorem, we should add a few more notions.

For a given vertex v in a directed graph $G = (V, E)$, let us denote the number of directed edges ending in v (i.e. having v as the head) by $\deg_G^+(v)$. Similarly, $\deg_G^-(v)$ stands for the number of directed edges originating in v. The number $\deg_G^+(v)$ is called the *indegree* of v, and $\deg_G^-(v)$ is the *outdegree* of v.

Each directed graph $G = (V, E)$ can be assigned an undirected graph $\mathrm{sym}(G) = (V, \bar{E})$, where

$$\bar{E} = \{\{x, y\} : (x, y) \in E \text{ or } (y, x) \in E\}.$$

The graph $\mathrm{sym}\,(G)$ is called the *symmetrization* of the graph G.

Now we can formulate the promised characterization of Eulerian directed graphs.

3.6.2 Proposition. *A directed graph is Eulerian if and only if its symmetrization is connected[7] and $\deg_G^+(v) = \deg_G^-(v)$ holds for each vertex $v \in V(G)$.*

A proof of this proposition is very similar to the proof of Theorem 3.4.1, and we leave it as an exercise.

---

[6] A directed cycle is sometimes simply called a cycle in the literature, or the neologism *dicycle* is also occasionally used, which may sound more like a name for some obscure vehicle.

[7] A directed graph whose symmetrization is connected is called *weakly connected* (a policeman who can ignore one-way street signs can get from any vertex to any other one). On the other hand, in a *strongly connected* directed graph, any two vertices can be connected by a directed path, in both directions.

An application. A wheel has a sequence of n digits 0 and 1 written along its circumference. We can read k consecutive digits through a slot:

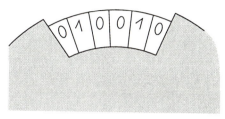

The sequence of n digits should be such that the position of the wheel can always be detected unambiguously from the k digits in the slot, no matter how the wheel is rotated. (Imagine a device for controlling the angular position of a radar or something else your fantasy can envisage.) For a given k, we want to manufacture a wheel with n as large as possible (so that the angular position can be controlled fairly precisely). A mathematical formulation of the problem is the following:

Problem. Find a cyclic sequence of digits 0 and 1, as long as possible, such that no two k-tuples of consecutive digits coincide (here a cyclic sequence means positioning the digits on the circumference of a circle).

Let $\ell(k)$ denote the maximum possible number of digits in such a sequence for a given k. We prove the following surprising result:

Proposition. *For each $k \geq 1$ we have $\ell(k) = 2^k$.*

Proof. Since the number of distinct k-digit sequences made of digits 0 and 1 is 2^k, the length of the cyclic sequence cannot be longer than 2^k. It remains to construct a cyclic sequence of length 2^k with the required property. The case $k = 1$ is easy, so let us assume $k \geq 2$.

Define a graph $G = (V, E)$ in the following manner:

- V is the set of all sequences of 0s and 1s of length $k - 1$ (so $|V| = 2^{k-1}$).
- The directed edges are all pairs of $(k-1)$-digit sequences of the form

$$((a_1, \ldots, a_{k-1}), (a_2, \ldots, a_k)).$$

Directed edges are in a bijective correspondence with k-digit sequences

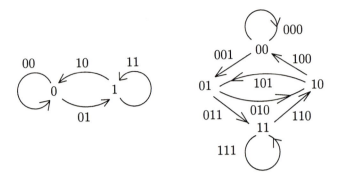

Fig. 3.5   The directed graph in the wheel problem.

$$(a_1, a_2, \ldots, a_k),$$

and hence $|E| = 2^k$. For brevity, let us denote the directed edge $((a_1, \ldots, a_{k-1}), (a_2, \ldots, a_k))$ by $(a_1, a_2, \ldots, a_k)$. No confusion should arise.

The reader will probably agree that $\deg_G^-(v) = \deg_G^+(v) = 2$ for each vertex $v \in V$. The symmetrization of G is connected, because by repeatedly omitting the last term of a $(k-1)$-digit sequence and adding 0s to its beginning we can convert any sequence to the sequence of 0s. Hence G is an Eulerian directed graph. Examples for $k = 2$ and $k = 3$ are shown in Fig. 3.5.

Set $|E| = 2^k = K$, and let $(e^1, \ldots, e^K)$ be the sequence of edges in some directed Eulerian tour in G. Each edge e^i has the form $e^i = (a_1^i, \ldots, a_k^i)$. The desired cyclic sequence of digits 0 and 1 of length K for our wheel can be defined as $(a_1^1, a_1^2, \ldots, a_1^K)$. That is, we take the first element from each e^i. Each subsequence of k consecutive digits in this sequence corresponds to traversing one directed edge of the Eulerian tour, and since no directed edge is repeated in the tour, no two k-digit segments coincide. This proves $\ell(k) = 2^k$.

For example, for $k = 2$, from the graph in Fig 3.5 we can find a tour 00, 01, 11, 10 and the corresponding cyclic sequence 0011, and for $k = 3$ we get a tour 000, 001, 011, 111, 110, 101, 010, 100 and the corresponding cyclic sequence 00011101. $\square$

Let us remark that the noteworthy graphs constructed in the preceding proof are called the *De Bruijn graphs*. Although they are exponentially large in k, the neighbors of a given vertex can be found quickly. They are sometimes used as interconnecting networks in parallel computing. Other graphs with similar properties are the *k-dimensional cubes*: the vertex set is again all sequences of 0s and 1s of length k, and two sequences are adjacent if and only if they differ in exactly one coordinate.

Exercises

1. Prove Proposition 3.6.2.

2. Design an algorithm for finding an Eulerian directed tour in a directed graph.

3. When can a directed graph be drawn with a single line (not necessarily a closed one)? Each directed edge must be drawn exactly once and in the direction from its tail to its head.

4. Let $G = (V, E)$ be a graph. An *orientation of* G is any oriented graph $G' = (V, E')$ arising by replacing each edge $\{u, v\} \in E$ either by the directed edge (u, v) or by the directed edge (v, u).

    (a) Prove that if all degrees of G are even then an orientation H of G exists with $\deg_H^+(v) = \deg_H^-(v)$ for all vertices $v \in V(G)$.

    (b) Prove that a directed graph G satisfying $\deg_G^+(v) = \deg_G^-(v)$ for all vertices v is strongly connected if and only if it is weakly connected.

5. *Let $G = (V, E)$ be a directed graph, and let $w: E \rightarrow \mathbf{R}$ be a function assigning a real number to each edge. A function $p: V \rightarrow \mathbf{R}$ defined on vertices is called a *potential* for w if $w(e) = p(v) - p(u)$ holds for every directed edge $e = (u, v)$. Prove that a potential for w exists if and only if the sum of the values of w over the edges of any directed cycle in G is 0.

6. *Prove that the following two conditions for a strongly connected directed graph G are equivalent:

    (i) G contains a directed cycle of an even length.

    (ii) The vertices of G can be colored by 2 colors (each vertex receives one color) in such a way that for each vertex u there exists a directed edge (u, v) with v having the color different from the color of u.

7. **Knights from two enemy castles are sitting at a round table and negotiating for peace. The number of knights with an enemy sitting on their right-hand side is the same as the number of knights with an ally on their right-hand side. Prove that the total number of knights is divisible by 4.

8. *A *tournament* is a directed graph such that for any two distinct vertices u, v, exactly one of the directed edges (u, v) and (v, u) is present in the graph. Prove that each tournament has a directed path passing through all vertices (such a path is called *Hamiltonian*).

9. *Prove that in any tournament (see Exercise 8 for a definition), there exists a vertex v that can be reached from any other vertex by a directed path of length at most 2.

3.7   2-connectivity

A graph G is called *k-vertex-connected* if it has at least $k + 1$ vertices and it remains connected after removing any $k-1$ vertices. A graph G is called *k-edge-connected* if we obtain a connected graph by deleting any $k - 1$ edges of G. The maximum k such that G is k-vertex-connected is called the *vertex connectivity* of G, and similarly for *edge connectivity*.

If a graph is a scheme of a city public transport network, a railway network, telephone cables, etc., its higher connectivity gives hope for a reasonable functioning of the network even in critical conditions, when one or several nodes or connections of the network fail. The notions of vertex connectivity and edge connectivity are theoretically and practically quite important in graph theory. They are related to so-called *network flows*, which are not treated in this book. Here we restrict our attention to 2-vertex-connectivity, which will be needed in a chapter on planar graphs, and which will also serve as an illustration for some proof methods and constructions.

Instead of vertex 2-vertex-connectivity we will briefly say 2-connectivity. To be on the safe side, let us give the definition once more:

3.7.1 Definition (2-connectivity). *A graph G is called* 2-connected *if it has at least 3 vertices, and by deleting any single vertex we obtain a connected graph.*

It is easy to check that a 2-connected graph is also connected (here we need the assumption that a 2-connected graph has at least 3 vertices—we recommend the reader to think this over). In this section we give alternative descriptions of 2-connected graphs. Before we begin with this, we introduce the notation for several graph-theoretic operations. It simplifies formulas considerably and will also be useful later on.

3.7.2 Definition (Some graph operations). *Let $G = (V, E)$ be a graph. We define various new graphs created from G:*

- *(Edge deletion)*

$$G - e = (V, E \setminus \{e\}),$$

where $e \in E$ is an edge of G;

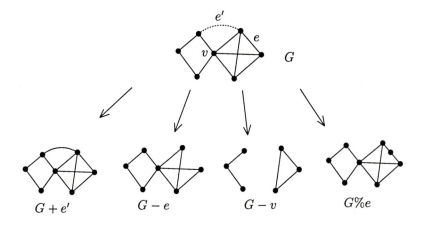

Fig. 3.6 Examples of graph operations.

- *(Adding a new edge)*

$$G + \bar{e} = (V, E \cup \{\bar{e}\}),$$

where $\bar{e} \in \binom{V}{2} \setminus E$ is a pair of vertices that is not an edge of G;
- *(Vertex deletion)*

$$G - v = (V \setminus \{v\}, \{e \in E \colon v \notin e\}),$$

where $v \in V$ *(we delete the vertex v and all edges having v as an endpoint);*
- *(Edge subdivision)*

$$G\%e = \Big(V \cup \{z\}, (E \setminus \{\{x,y\}\}) \cup \{\{x,z\}, \{z,y\}\}\Big),$$

where $e = \{x, y\} \in E$ is an edge, and $z \notin V$ is a new vertex (we "draw a new vertex z" on the edge $\{x, y\}$).

We say that a graph G' is a *subdivision* of the graph G if G' is isomorphic to a graph created from G by successive operations of edge subdivision.

Examples of the operations just defined are shown in Fig. 3.6.

Let us go back to 2-connectivity. Here is the first remarkable equivalent characterization:

3.7.3 Theorem. *A graph G is 2-connected if and only if there exists, for any two vertices of G, a cycle in G containing these two vertices.*

Let us remark that this theorem is a particular case of a very important result called *Menger's theorem*, which says the following. If x and y are two vertices in a k-vertex-connected graph, then there exist k paths from x to y that are mutually disjoint except for sharing the vertices x and y.

Proof. The given condition is, no doubt, sufficient, since if two vertices v, v' lie on a common cycle then there exist two paths connecting them having no common vertices except for the end-vertices, and so v and v' can never fall into distinct components by removing a single vertex.

We now prove the reverse implication. The existence of a common cycle for v, v' will be established by induction on $d_G(v, v')$, the distance of the vertices v and v'.

First let $d_G(v, v') = 1$. This means that $\{v, v'\} = e \in E(G)$. By 2-connectivity of G, the graph $G - e$ is connected (if it were disconnected, at least one of the graphs $G - v$, $G - v'$ would also be disconnected). Therefore there exists a path from v to v' in the graph $G - e$, and this path together with the edge e forms the required cycle containing both v and v'.

Next, suppose that any pair of vertices at distance less than k lies on a common cycle, for some $k \geq 2$. Consider two vertices $v, v' \in V$ at distance k. Let $P = (v = v_0, e_1, v_1, \ldots, e_k, v_k = v')$ be a shortest path from v to v'. Since $d_G(v, v_{k-1}) = k - 1$, a cycle exists containing both v and v_{k-1}. This cycle consists of two paths, P_1 and P_2, from v to v_{k-1}. Now consider the graph $G - v_{k-1}$. It is connected, and hence it has a path $\check{P}$ from v to v'. This path thus doesn't contain v_{k-1}. Let us look at the last vertex on the path $\check{P}$ (when going from v to v') belonging to one of the paths P_1, P_2, and denote this vertex by w, as in the illustration:

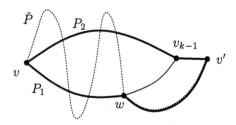

Without loss of generality, suppose that w is a vertex of P_1. Then the desired cycle containing v and v' is formed by the path P_2, by the portion of the path P_1 between v and w, and by the portion of the path $\check{P}$ between w and v' (drawn by a thick line). □

3.7.4 Observation. *A graph G is 2-connected if and only if any subdivision of G is 2-connected.*

Proof. It is enough to show that, for any edge $e \in E(G)$, G is 2-connected if and only if $G\%e$ is 2-connected. If $v \in V(G)$ is a vertex of G, it is easy to see that $G - v$ is connected if and only if $(G\%e) - v$ is connected. Therefore, if $G\%e$ is 2-connected then so is G. For the reverse implication, it remains to show that for a 2-connected G, the graph $(G\%e) - z$ is connected, where z is the newly added vertex. This follows from the fact (observed in the previous proof) that $G - e$ is connected for a 2-connected G. □

The next characterization of 2-connected graphs is particularly suitable for proofs. We show how 2-connected graphs are built from simpler graphs.

3.7.5 Theorem (2-connected graph synthesis). *A graph G is 2-connected if and only if it can be created from a triangle (i.e. from K_3) by a sequence of edge subdivisions and edge additions.*

Such a synthesis is illustrated below:

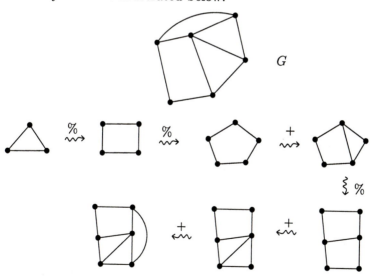

Proof. Every graph that can be produced from K_3 by the above-mentioned operations is obviously 2-connected. So, we need to prove that we can construct each 2-connected graph.

Actually, we show the possibility of creating any 2-connected graph by a somewhat different construction. We start with a cycle G_0, and if a graph G_{i-1} has already been built, a graph G_i arises by adding a path P_i connecting two vertices of the graph G_{i-1}. The path P_i only shares its end-vertices with G_{i-1}, while all edges and all inner vertices are new. As illustrated in the following drawing,

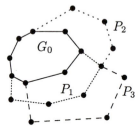

we successively glue "ears" to the graph G (and, indeed, the decomposition is commonly called an *ear decomposition*).

Since adding a path can be simulated by an edge addition and edge subdivisions,[8] it suffices to show that every 2-connected graph G can be produced by a repeated ear addition.

Let us pick a cycle G_0 in the graph G arbitrarily. Suppose that graphs $G_j = (E_j, V_j)$ for $j \leq i$ have already been defined, with properties as described above. If $G_i = G$ the proof is over, so let us assume that $E_i \neq E(G)$. Since G is connected, there exists an edge $e \in E(G) \setminus E_i$ such that $e \cap V_i \neq \emptyset$.

If both vertices of e lie in V_i then we put $G_{i+1} = G_i + e$. In the other case, let $e = \{v, v'\}$, where $v \in V_i$, $v' \notin V_i$:

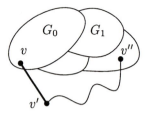

Consider the graph $G - v$. This is connected (since G is 2-connected), and therefore a path P exists connecting the vertex v' to some vertex $v'' \in V_i$, where v'' is the only vertex of the path P belonging to V_i (to

---

[8]One has to be careful here: if $v, v' \in V(G_{i-1})$ are already connected by an edge and if we want to connect them by a new path, we cannot start by adding the edge $\{v, v'\}$ (at least if we do not allow multiple edges). We have to subdivide the edge $\{v, v'\}$ first, then again add the edge $\{v, v'\}$, and then continue extending the path by subdivisions if needed.

this end, take the shortest path joining v' to V_i in the graph $G - v$). Then we can define the graph G_{i+1} by adding the edge e and the path P to G_i, i.e. $V_{i+1} = V_i \cup V(P)$, $E_{i+1} = E_i \cup \{e\} \cup E(P)$. $\quad\square$

Exercises

1. Prove that for any two edges of a 2-connected graph, a cycle exists containing both of them.

2. Let G be a *critical 2-connected graph*; this means that G is 2-connected but no graph $G - e$ for $e \in E(G)$ is 2-connected.

   (a) Prove that at least one vertex of G has degree 2.

   (b) For each n, find an example of a critical 2-connected graph with a vertex of degree at least n.

   (c) *For each n, give an example of a critical 2-connected graph with a vertex of degree $\geq n$, which is at distance at least n from each vertex of degree 2.

3. (a) Is it true that any critical 2-connected graph (see Exercise 2) can be obtained from a cycle by successive gluing of "ears" (paths) of length at least 2?

   (b) Is it true that any critical 2-connected graph can be obtained from a cycle by a successive gluing of "ears" in such a way that each of the intermediate graphs created along the way is also critical 2-connected?

4. Prove that any 2-connected graph has a strongly connected orientation (see Section 3.6 for these notions).

5. *Determine the vertex connectivity of the k-dimensional cube. The k-dimensional cube was defined at the end of Section 3.6.

6. *Let $d \geq 3$ be an integer, and let G be a d-regular graph (every vertex has degree d) which is d-edge-connected. Prove that such a G is *tough*, meaning that removing any k vertices disconnects G into at most k connected components (for all $k \geq 1$).

7. (Mader's theorem) **Let G be a graph on n vertices such that $|E(G)| \geq (2k - 3)(n - k + 1) + 1$, where k is natural number with $2k - 1 \leq n$. By induction on n, prove that G has a k-vertex-connected subgraph.

4

Trees

4.1 Definition and characterizations of trees

Even very abstract and elusive concepts in mathematics are often given names from common language. Similarly as mathematical definitions are not at all arbitrary, the name for a notion may be quite important. Sometimes the name helps to communicate, on an intuitive level, some key property of the object which is not easy to notice behind the formal definition. Other names may sound quite illogical in today's context without knowing their often convoluted history. For example, fields in mathematics are neither green fields nor strawberry fields nor any other kind of fields most people may be used to, and it is hard to imagine why something in the plane should be called a vertex (see Chapter 5 for the origin of this name). And, an unfitting name may lead one completely astray. In this chapter, the reader can judge how well mathematicians succeeded in choosing a name for a simple and fundamental graph-theoretic concept—a tree. While trees in mathematics perhaps cannot match those in nature for beauty or diversity, they still form quite a rich area in graph theory, and closely related concepts of trees appear in other branches of mathematics and of computer science as well.

In graph theory, a tree is a graph similar to the ones in the following drawings:

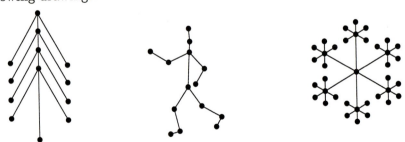

From these nature-inspired pictures, one may get quite a good idea of what is meant by a tree in graph theory. But how do we define this notion rigorously? Try it by yourself before reading further!

Perhaps the most usual definition is the following:

4.1.1 Definition. *A tree is a connected graph containing no cycle.*

Don't worry if your definition doesn't match ours. A tree can be defined in several rather different ways. Soon we will give even four more equivalent definitions.

Why should one define a tree in several different ways? First of all, the definition just given is somewhat unsuitable from several points of view. For instance, it is not clear from it how we can check whether a given graph is a tree or not. The connectedness can be verified by a simple algorithm, but deciding the existence of a cycle seems to be a problem. The alternative descriptions given below give very straightforward algorithmic ways of recognizing trees, and they tell us several interesting properties of trees which may be useful in applications.

Second, while a proof of the equivalence of the various definitions is probably not very exciting reading, it provides good exercise material for students' own proofs (according to our experience). The various implications in the proof are not formidably difficult, but they are not completely easy either, and a student trying to prove one such implication has plenty of opportunities to make and discover errors or gaps in the proof. The proof given below also illustrates how one can proceed in proving the equivalence of several statements.

Third, the equivalent characterizations of trees below are simple but they show samples of "good" characterizations of a mathematical object one should look for; difficult major theorems in several areas have a formally similar pattern.

Here are the promised equivalent definitions of a tree. The most remarkable of them is perhaps the one saying that among connected graphs, a tree can be recognized simply by counting its edges and vertices.

4.1.2 Theorem (Tree characterizations). *The following conditions are all equivalent for a graph $G = (V, E)$:*

(i) G is a tree.

(ii) (Path uniqueness)
For every two vertices x, $y \in V$, there exists exactly one path from x to y.

(iii) (Minimal connected graph)
The graph G is connected, and deleting any of its edges gives rise to a disconnected graph.

(iv) (Maximal graph without cycles)
  The graph G contains no cycle, and any graph arising from G by adding an edge (i.e. a graph of the form $G+e$, where $e \in \binom{V}{2} \setminus E$) already contains a cycle.

(v) (Euler's formula)
  G is connected and $|V| = |E| + 1$.

Note that this theorem not only describes various properties of trees, such as "any tree on n vertices has $n-1$ edges", but also lists properties equivalent to Definition 4.1.1, so for instance it says "A graph on n vertices is a tree if and only if it is connected and has $n-1$ edges".

Proving equivalences of various pairs of statements in Theorem 4.1.2 seems far from trivial for beginners, and all sorts of shortcomings appear in attempts at such proofs. Here is a (hypothetical) example. Consider the following implication: "If G is a tree, then any two vertices of G can be connected by exactly one path." Most people quickly notice that since G is connected, any two vertices can be connected by *at least one* path. Then, if the proof is being created from the interaction of a student with the teacher, dialog of the following sort often develops:

S.: "We proceed by contradiction. If u and v are two vertices in G that can be connected by two different paths, then G contains a cycle. Hence the implication holds."

T.: "But why must there be a cycle in G if u and v are connected by two distinct paths?"

S.: "Well, the two paths together contain a cycle."

For the teacher, this is not at all easy to argue with, in particular because it's true and, moreover, "obvious from a picture".

T: "But why? Can you prove it rigorously? In your picture, there is indeed a cycle, and I can't show you a graph with no cycle and with two paths between u and v either, but isn't it possible that some extraterrestrians, much more clever than me and you and all other people together, can construct such a graph?"

(Suggestions of didactically more convincing ways of arguing are welcome.) In fact, a rigorous proof is not entirely trivial, but it seems that the message most difficult to get through is that there really *is* something to prove. As you will see, in our proof of Theorem 4.1.2 below we preferred to avoid this direct argument.

We now begin with the proof of Theorem 4.1.2. Since there are many implications to prove, it is important to organize the proof suitably. The basic idea in all steps is to proceed by induction on the number of vertices of the considered graph, and to "tear off" a vertex of degree 1 in the inductive step. In few simple lemmas below, we prepare the ground for this method.

Let us call a vertex of degree 1 in a graph G an *end-vertex* of G or a *leaf* of G. We begin with the following almost obvious observation:

4.1.3 Lemma (End-vertex lemma). *Each tree with at least 2 vertices contains at least 2 end-vertices.*

Proof. Let $P = (v_0, e_1, v_1, \ldots, e_t, v_t)$ be a path of the maximum possible length in a tree $T = (V, E)$. Clearly the length of the path P is at least 1, and so in particular $v_0 \neq v_t$. We claim that both v_0 and v_t are end-vertices. This can be shown by contradiction: if, for example, v_0 is not an end-vertex, then there exists an edge $e = \{v_0, v\}$ containing the vertex v_0 and different from the first edge $e_1 = \{v_0, v_1\}$ of the path P. Then either v is one of the vertices of the path P, i.e. $v = v_i$, $i \geq 2$ (in this case the edge e together with the portion of the path P from v_0 to v_i form a cycle), or $v \notin \{v_0, \ldots, v_t\}$—in that case we could extend P by adding the edge e. In both cases we thus get a contradiction. $\qquad\square$

Let us remark that the end-vertex lemma does not hold for infinite trees (the proof just given fails because a path of the maximum length need not exist). For instance, the "one-sided infinite path" has only one end-vertex

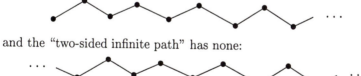

and the "two-sided infinite path" has none:

We only consider finite graphs here, however.

Next, we recall a notation from Section 3.7: if $G = (V, E)$ is a graph and v is a vertex of G, then $G - v$ stands for the graph arising from G by deleting the vertex v and all edges containing it. In case v is an end-vertex of a tree T, the graph $T - v$ arises by deleting the vertex v and the single edge containing v.

4.1.4 Lemma (Tree-growing lemma). *The following two statements are equivalent for a graph G and its end-vertex v:*

(i) *G is a tree*
(ii) *$G - v$ is a tree.*

Proof. This is also quite easy. First we prove the implication (i) ⇒ (ii). The graph G is a tree, and we want to prove that $G - v$ is a tree as well. Consider two vertices x, y of $G - v$. Since G is connected, x and y are connected by a path in G. This path cannot contain a vertex of degree 1 different from both x and y, and so it doesn't contain v. Therefore it is completely contained in $G - v$, and we conclude that $G - v$ is connected. Since G has no cycle, obviously $G - v$ cannot contain a cycle, and thus it is a tree.

It remains to prove the implication (ii) ⇒ (i). Let $G - v$ be a tree. By adding the end-vertex v back to it, no cycle can be created. We must also check the connectedness of G, but this is obvious too: any two vertices distinct from v were connected already in $G - v$, and a path to v from any other vertex x is obtained by considering the (single) neighbor v' of v in G, connecting it to x by a path in $G - v$, and extending this path by the edge $\{v', v\}$. □

This lemma allows us to reduce a given tree to smaller and smaller trees by removing end-vertices successively. Now we are going to apply this device.

Proof of Theorem 4.1.2. We prove that each of the statements (ii)–(v) is equivalent to (i). This, of course, proves the mutual equivalence of all the statements. The proofs go by induction on the number of vertices of G using the tree-growing lemma 4.1.4. As for the induction basis, we note that all the statements are valid for the graph with a single vertex.

First let us see that (i) implies all of (ii)–(v). To this end, let G be a tree with at least 2 vertices, let v be one of its end-vertices, and let v' be the single neighbor of v in G. Suppose that the graph $G - v$ already satisfies (ii)–(v); this is our inductive hypothesis.

In this situation, the validity of (ii), (iii), and (v) for G can be considered obvious (we leave a detailed argument to the reader).

As for (iv), we do not even need the inductive hypothesis for $G - v$. Since G is connected, any two vertices $x, y \in V(G)$ can be connected by a path, and if $\{x, y\} \notin E(G)$ then the edge $\{x, y\}$ together with the just-mentioned path creates a cycle. This proves the implication (i) ⇒ (iv).

Let us now prove that each of the conditions (ii)–(v) implies (i). In (ii) and (iii) we already assume connectedness. Moreover, a graph satisfying (ii) or (iii) cannot contain a cycle. For (ii), this is because two vertices in a cycle can be connected by two distinct paths, and for (iii), the reason is that by omitting an edge in a cycle we obtain a connected graph. Thus we have already proved the equivalence of (i)–(iii).

In order to verify the implication (iv)⇒(i), it suffices to check that G is connected. For this, we can use the argument by which we have proved (i)⇒(iv) turned upside down. If $x, y \in V(G)$ are two vertices, either they are connected by an edge, or the graph $G+\{x,y\}$ contains a cycle, and removing the edge $\{x, y\}$ from this cycle gives a path from x to y in G.

Finally the implication (v)⇒(i) is again proved by induction on the number of vertices. Let us consider a connected graph G satisfying $|V| = |E| + 1 \geq 2$. The sum of the degrees of all vertices is thus $2|V| - 2$ (why?). This means that not all vertices can have degree 2 or larger, and since all the degrees are at least 1 (by connectedness!), there exists a vertex v of degree exactly 1, i.e. an end-vertex of the graph G. The graph $G' = G - v$ is again connected and it satisfies $|V(G')| = |E(G')| + 1$. Hence it is a tree by the inductive hypothesis, and thus G is a tree as well. □

Exercises

1. Draw all trees with vertex set $\{1, 2, 3, 4\}$, and all pairwise nonisomorphic trees on 6 vertices.

2. Prove that any graph $G = (V, E)$ having no cycles and satisfying $|V| = |E| + 1$ is a tree.

3. *Let $n \geq 3$. Prove that a graph G on n vertices is a tree if and only if it is not isomorphic to K_n and adding any edge (on the same vertex set) not present in G creates exactly one cycle.

4. Prove that a graph on n vertices with c components has at least $n - c$ edges.

5. Suppose that a tree contains a vertex of degree k. Show that it has at least k end-vertices.

6. Let T be a tree with n vertices, $n \geq 2$. For a positive integer i, let p_i be the number of vertices of T of degree i. Prove

$$p_1 - p_3 - 2p_4 - \cdots - (n - 3)p_{n-1} = 2.$$

(This provides an alternative proof of the end-vertex lemma.)

7. King Uxamhwiashurh had 4 sons, 10 of his male descendants had 3 sons each, 15 had 2 sons, and all others died childless. How many male descendants did King Uxamhwiashurh have?

8. Consider the following two conditions for a sequence $(d_1, d_2, \ldots, d_n)$ of (strictly) positive integers (where $n \geq 1$):

   (i) There exists a tree with score $(d_1, d_2, \ldots, d_n)$.
   (ii) $\sum_{i=1}^{n} d_i = 2n - 2$.

   The goal is to show that these conditions are equivalent (and so there is a very simple way to tell sequences that are scores of trees from those that aren't).

   (a) Why does (i) imply (ii)?

   (b) Why is the following "proof" of the implication (ii)$\Rightarrow$(i) insufficient (or, rather, makes no sense)? We proceed by induction on n. The base case $n = 1$ is easy to check, so let us assume that the implication holds for some $n \geq 1$. We want to prove it for $n + 1$. If $D = (d_1, d_2, \ldots, d_n)$ is a sequence of positive integers with $\sum_{i=1}^{n} d_i = 2n - 2$, then we already know that there exists a tree T on n vertices with D as a score. Add another vertex v to T and connect it to any vertex of T by an edge, obtaining a tree T' on $n + 1$ vertices. Let D' be the score of T'. We know that the number of vertices increased by 1, and the sum of degrees of vertices increased by 2 (the new vertex has degree 1 and the degree of one old vertex increased by 1). Hence the sequence D' satisfies condition (ii) and it is a score of a tree, namely of T'. This finishes the inductive step.

   (c) *Prove (ii)$\Rightarrow$(i).

9. Suppose we want to prove that any connected graph $G = (V, E)$ with $|V| = |E| + 1$ is a tree, i.e. the implication (v)$\Rightarrow$(i) in Theorem 4.1.2. What is wrong with the following proof?

   We already assume that the considered graph is connected, so all we need to prove is that it has no cycle. We proceed by induction on the number of vertices. For $|V| = 1$, we have a single vertex and no edge, and the statement holds. So assume the implication holds for any graph $G = (V, E)$ on n vertices. We want to prove it also for a graph $G' = (V', E')$ arising from G by adding a new vertex. In order that the assumption $|V'| = |E'| + 1$ holds for G', we must also add one new edge, and because we assume G' is connected, this new edge must connect the new vertex to some vertex in V. Hence the new vertex has degree 1 and so it cannot be contained in a cycle. And because G has no cycle (by the inductive hypothesis), we get that neither does G' have a cycle, which finishes the induction step.

4.2   Isomorphism of trees

As we have mentioned in Section 3.1, no fast algorithm is known for deciding whether two given graphs are isomorphic or not. For some special classes of graphs efficient algorithms exist, however. One of these classes is trees. In fact, many (perhaps most) algorithmic problems which are intractable for general graphs can be solved relatively easily for trees.

In this section we demonstrate a fast and simple algorithm for testing the isomorphism of two given trees T and T'. For a given tree T, the algorithm computes a sequence of 0s and 1s of length $2n$, called the *code* of the tree T. Isomorphic trees yield identical sequences, while nonisomorphic ones receive distinct sequences. In this way, the testing for isomorphism is reduced to a simple sequence comparison.

Next, we introduce a number of specialized notions, each with a particular name. This is quite usual in algorithmic graph theory, where many notions arose from various applications and where the terminology is quite diverse.

A *rooted tree* is a pair (T, r), where T is a tree and $r \in V(T)$ is a distinguished vertex of T called the *root*. If $\{x, y\} \in E(T)$ is an edge and the vertex x lies on the unique path from y to the root, we say that x is the *father* of y (in the considered rooted tree) and y is a *son* of x.

A *planted tree* is a rooted tree (T, r) plus some its fixed drawing in the plane. In this drawing, the root is marked by an arrow pointing downwards, and the sons of each vertex all lie above that vertex.

For those who don't like this definition, note that a planar drawing of a tree is fully described, up to a suitable continuous deformation of the plane, by the left-to-right order of the sons of each vertex. A planted tree is thus a rooted tree in which every vertex v is assigned a linear ordering of its sons. Thus, we can formally write a planted tree as a triple (T, r, ν), where ν is a collection of linear orderings, one linear ordering for the set of sons of each vertex.

For each of the above-defined types of trees, an isomorphism is defined in a slightly different way. Let us recall that a mapping $f: V(T) \to V(T')$ is an isomorphism of trees T and T' if f is a bijection (i.e. it is one-to-one and onto) satisfying $\{x, y\} \in E(T)$ if and only if $\{f(x), f(y)\} \in E(T')$. The existence of such an isomorphism is written $T \cong T'$. An *isomorphism of rooted trees* (T, r) and (T', r') is an isomorphism f of the trees T and T' for which we have, moreover, $f(r) = r'$. This is denoted by $(T, r) \cong' (T', r')$. An *isomorphism of planted trees* is an isomorphism of rooted trees that preserves the left-to-right ordering of the sons of each vertex. The fact that two planted trees are isomorphic in this sense is denoted by $(T, r, \nu) \cong'' (T', r', \nu')$.

The definitions of $\cong$, $\cong'$, and $\cong''$ are successively stronger and stronger. This can be best understood in the following small examples:

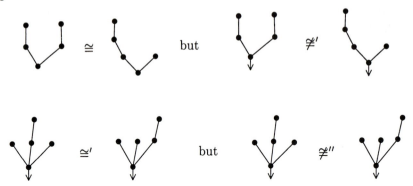

The definition of the isomorphism of planted trees is most restrictive, and thus the coding of these trees is easiest. The following method assigns a certain code to each vertex of a planted tree. The code of the whole tree is then defined as the code of the root.

K1. Each end-vertex distinct from the root is assigned 01.
K2. Let v be a vertex with sons $v_1, v_2, \ldots, v_t$ (written in the left-to-right order). If A_i is the code of the son v_i, then the vertex v receives the code $0A_1 A_2 \ldots A_t 1$.

The process of a successive building of a code is illustrated below:

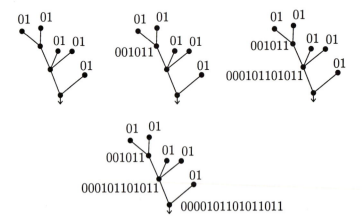

Clearly, isomorphic planted trees have been assigned the same codes, because in the code construction, we only used properties of a planted tree preserved by the isomorphism of planted trees.

Now we show how the original planted tree can be reconstructed from the resulting code. In this way, we prove that nonisomorphic planted trees are assigned distinct codes. We proceed by induction on the length of the code.

The shortest possible code, 01, corresponds to the planted tree with a single vertex. In an induction step, suppose that we are given a code k of length $2(n + 1)$. This code has the form $0\,A\,1$, where $A = A_1 A_2 \ldots A_t$ is a concatenation of codes of several planted trees. The part A_1 can be identified as the shortest initial segment of the sequence A containing the same number of 0s and 1s. Similarly, A_2 is the next shortest segment with the number of 0s and 1s balanced, and so on. By the inductive hypothesis, each A_i corresponds to a unique planted tree. The planted tree coded by the code k has a single root r, and this root has as sons the roots $r_1, r_2, \ldots, r_t$ of the planted trees coded by $A_1, A_2, \ldots, A_t$, respectively (in the left-to-right order). Hence the code uniquely determines a planted tree.

Decoding by the arrow method. We present an intuitive (pictorial) procedure for reconstructing a planted tree from its code.

In a given code sequence, we replace every 0 by the arrow "↑" and every 1 by the arrow "↓". Next, we take this sequence of arrows as instructions for drawing a tree. When encountering an "↑", we draw an edge from the current point upwards (and to the right of the parts already drawn from that point). For a "↓", we follow an already drawn edge downwards. The whole procedure is illustrated in the following picture:

0000101101011011

↓

↑↑↑↑↓↑↓↓↑↓↑↓↓↑↓↓ ⟶

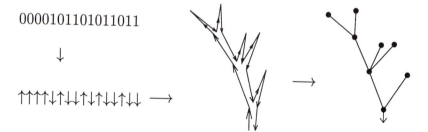

(The procedure draws the root's arrow as well, since a one-point tree has the code 01.)

We have studied the isomorphism of planted trees in some detail, since the algorithm for planted trees easily leads to an algorithm for rooted trees. For a rooted tree (T, r), we build its code in a similar way to the method for planted trees, but the rule K2 is replaced by the following modification:

K2′. Suppose that each son w of a vertex v has already been assigned a code $A(w)$. Let the sons of v be denoted by $w_1, w_2, \ldots, w_t$, in such a way that $A(w_1) \leq A(w_2) \leq \cdots \leq A(w_t)$. The vertex v receives the code $0\, A_1\, A_2 \ldots A_t\, 1$, where $A_i = A(w_i)$.

Here $A \leq B$ means that the sequence A is less than or equal to the sequence B in some fixed linear ordering of all finite sequences of 0s and 1s. For definiteness, we can use the so-called *lexicographic ordering*. Two (distinct) sequences $A = (a_1, a_2, \ldots, a_n)$ and $B = (b_1, b_2, \ldots, b_m)$ are compared as follows:

- If A is an initial segment of B then $A < B$. If B is an initial segment of A then $B < A$. (For example, we have $0010 < 00100$ and $0 < 0111$.)
- Otherwise, let j be the smallest index with $a_j \neq b_j$. Then if $a_j < b_j$ we let $A < B$, and if $a_j > b_j$ we let $A > B$. (For example, we have $011 < 1$ and $10011 < 10110$.)

We have to check that two rooted trees are isomorphic if and only if they have the same codes. This can be done quite similarly as for planted trees, and we leave it to the reader.

We now turn our attention to coding trees without a root.[1] Our task would be greatly simplified if we could identify a vertex which would play the role of the root, and which would be preserved by any isomorphism. For trees, such a distinguished vertex can indeed be found (well, not always, but the exceptional cases can be characterized and they can be handled slightly differently). The relevant definitions can be useful also in other contexts, so we formulate them for general graphs rather than just for trees.

For a vertex v of a graph G, let the symbol $\mathrm{ex}_G(v)$ denote the maximum of the distances of v from other vertices of G. The number $\mathrm{ex}_G(v)$ is called the *excentricity* of the vertex v in the graph G. We can imagine the vertices with a large excentricity as lying on the periphery of G.

Now let $C(G)$ denote the set of all vertices of G with the minimum excentricity. The set $C(G)$ is called the *center*[2] of G. The example

---

[1]In the literature, the term *free tree* is sometimes used if one speaks about a tree and wants to emphasize that it is not considered as a rooted tree.

[2]The excentricity of the vertices of the center is called the *radius* of the graph G.

of a cycle (and many other graphs) shows that sometimes the center may coincide with the whole vertex set. However, for trees, we have

Proposition. *For any tree T, $C(T)$ has at most 2 vertices. If $C(T)$ consists of 2 vertices x and y then $\{x, y\}$ is an edge.*

Proof. We describe a procedure for finding the center of a tree. Let $T = (V, E)$ be a given tree. If T has at most 2 vertices, then its center coincides with the vertex set and the proposition holds. Otherwise let $T' = (V', E')$ be the tree arising from T by removing all leaves. Explicitly,

$$V' = \{x \in V \colon \deg_T(x) > 1\},$$
$$E' = \{\{x, y\} \in E \colon \deg_T(x) > 1 \text{ and } \deg_T(y) > 1\}.$$

We clearly have $V(T') \neq \emptyset$, since not all vertices of T can be leaves. Further, for any vertex v, the vertices most distant from v are necessarily leaves, and hence for each $v \in V'$ we get

$$\mathrm{ex}_T(v) = \mathrm{ex}_{T'}(v) + 1.$$

In particular, we have $C(T') = C(T)$. If T' has at least 3 vertices we repeat the construction just described, otherwise we have found the center of T.
$\square$

We can now specify the coding of a tree T.

- If the center of T is a single vertex, v, then we define the code of T to be the code of the rooted tree (T, v).
- If the center of T consists of an edge $e = \{x_1, x_2\}$, we consider the graph $T - e$. This graph has exactly 2 components T_1 and T_2; the notation is chosen in such a way that $x_i \in V(T_i)$. Let the letter A denote the code of the rooted tree (T_1, x_1) and the letter B the code of the rooted tree (T_2, x_2). If $A \leq B$ in the lexicographic ordering, the tree T is coded by the code of the rooted tree (T, x_1), and for $A \geq B$ it is coded by the code of (T, x_2).

This finishes the coding procedure for trees.

The decoding is done in exactly the same way as for planted trees (we then obtain a "canonical" drawing of the considered tree). Since an isomorphism maps a center to a center and since we have already seen that the coding works properly for rooted trees, it is easy to see that two trees have the same code if and only if they are isomorphic.

The algorithms for the isomorphism of various types of trees explained in this section can be implemented in such a way that the number of elementary steps (computer instructions, say) is bounded by a linear function of the total number of vertices of the input trees.

Several more classes of graphs are known for which the isomorphism problem can be solved efficiently. Perhaps the most important examples are the class of all planar graphs and the class of all graphs with maximum degree bounded by a small constant. Here the known algorithms for isomorphism testing are fairly complicated.

Exercises

1. (a) Find an *asymmetric* tree, i.e. a tree with a single (identical) automorphism (see Exercise 3.1.2), with at least 2 vertices.

   (b) Find the smallest possible number of vertices a tree as in (a) can have (i.e. prove that no smaller tree can be asymmetric).

2. Find two nonisomorphic trees with the same score.

3. A rooted tree is called *binary* if each nonleaf vertex has exactly 2 sons.

   (a) Draw all nonisomorphic binary trees with 9 vertices.

   (b) Characterize the codes of binary trees.

4. Prove in detail that isomorphic trees (not rooted) receive the same code by the explained procedure, and nonisomorphic trees get distinct codes.

5. *,CS Let $A_1, \ldots, A_t$ be sequences of 0s and 1s (possibly of different lengths). Let n denote the sum of their lengths. Devise an algorithm that sorts these sequences lexicographically in $O(n)$ steps. One step may only access one term of some A_i; it is not possible to manipulate a whole sequence of 0s and 1s at once.

6. Prove that there exist at most 4^n pairwise nonisomorphic trees on n vertices.

7. Let $T = (V, E)$ be a tree and v some vertex of T. Put

$$\tau(v) = \max(|V(T_1)|, |V(T_2)|, \ldots, |V(T_k)|),$$

where $T_1, \ldots, T_k$ are all the components of the graph $T - v$. The *centroid* of the tree T is the set of all vertices $v \in V$ with the minimum value of $\tau(v)$.

   (a) *Prove that the centroid of any tree is either a single vertex or 2 vertices connected by an edge.

   (b) Is the centroid always identical to the center?

   (c) Prove that if v is a vertex in the centroid then $\tau(v) \le \frac{2}{3}|V(T)|$.

4.3 Spanning trees of a graph

A spanning tree is one of the basic graph constructions:

4.3.1 Definition. *Let $G = (V, E)$ be a graph. An arbitrary tree of the form (V, E'), where $E' \subseteq E$, is called a* spanning tree *of the graph G. So a spanning tree is a subgraph of G that is a tree and contains all vertices of G.*

Obviously, a spanning tree may only exist for a connected graph G. It is not difficult to show that every connected graph has a spanning tree. We prove it by giving two (fast) algorithms for finding a spanning tree of a given connected graph. In the subsequent sections we will need variants of these algorithms, so let us study them carefully.

4.3.2 Algorithm (Spanning tree). Let $G = (V, E)$ be a graph with n vertices and m edges. We order the edges of G arbitrarily into a sequence $(e_1, e_2, \ldots e_m)$. The algorithm successively constructs sets of edges $E_0, E_1, \ldots \subseteq E$.

We let $E_0 = \emptyset$. If the set E_{i-1} has already been found, the set E_i is computed as follows:

$$E_i = \begin{cases} E_{i-1} \cup \{e_i\} & \text{if the graph } (V, E_{i-1} \cup \{e_i\}) \text{ has no cycle} \\ E_{i-1} & \text{otherwise.} \end{cases}$$

The algorithm stops either if E_i already has $n - 1$ edges or if $i = m$, i.e. all edges of the graph G have been considered. Let E_t denote the set for which the algorithm has stopped, and let T be the graph (V, E_t).

4.3.3 Proposition (Correctness of Algorithm 4.3.2). *If Algorithm 4.3.2 produces a graph T with $n-1$ edges then T is a spanning tree of G. If T has $k < n - 1$ edges then G is a disconnected graph with $n - k$ components.*

Proof. According to the way the sets E_i are created, the graph G contains no cycle. If $k = |E(T)| = n - 1$ then T is a tree by Theorem 4.1.2(v), and hence it is a spanning tree of the graph G. If $k < n - 1$, then T is a disconnected graph whose every component is a tree (such a graph is called a *forest*). It is easy to see that it has $n - k$ components.

We prove that the vertex sets of the components of the graph T coincide with the vertex sets of the components of the graph G. For contradiction, suppose this is not true, and let x and y be vertices lying in the same component of G but in distinct components of T. Let C denote the component of T containing the vertex x, and consider some path $(x = x_0, e_1, x_1, e_2, \ldots, e_\ell, x_\ell = y)$ from x to y in the graph G, as in the following picture:

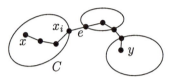

Let i be the last index for which x_i is contained in the component C. Obviously $i < \ell$, and hence $x_{i+1} \notin C$. The edge $e = \{x_i, x_{i+1}\}$ thus does not belong to the graph T, and so it had to form a cycle with some edges already selected into T at some stage of the algorithm. Therefore the graph $T + e$ also contains a cycle, but this is impossible as e connects two distinct components of T. This provides the desired contradiction. $\qquad\square$

Complexity of the algorithm. We have just shown that Algorithm 4.3.2 always computes what it is supposed to compute, i.e. a spanning tree of the input graph. But if we really needed to find spanning trees for some large graphs, should we choose this algorithm and spend our time programming it, or our money by buying some existing code?

To answer such a question is no simple matter, and algorithms are compared according to different, and often contradictory, criteria. For instance, it is important to consider the clarity and simplicity of the algorithm (a complicated or obscure algorithm easily leads to programming errors), the robustness (how do rounding errors or small changes in the input data influence the correctness of the output?), memory requirements, and so on. Perhaps the most common measure of complexity of an algorithm is its *time complexity*, which means the number of elementary operations (such as additions, multiplications, comparisons of two numbers, etc.) the algorithm needs for solving the input problem. Most often the *worst-case* complexity is considered, i.e. the number of operations needed to solve the worst possible problem, one expressly chosen to make the algorithm slow, for a given size of input. For computing a spanning tree, the input size can be measured as the number of vertices plus the number of edges of the input graph. Instead of "worst-case time complexity" we will speak briefly of "complexity", since we do not discuss other types of complexity.

The complexity of an algorithm can seldom be determined precisely. In order that we could even think of doing it, we would have to determine exactly what the allowed primitive operations are (so, in principle, we would restrict ourselves to a specific computer), and also we would have to describe the algorithm in the smallest details including various routine steps; that is, essentially look at a concrete program. Even if we did both these things, determining the precise complexity is quite laborious even for very simple algorithms. For these reasons, the complexity of algorithms is only analyzed asymptotically in most cases. We could thus say that some algorithm has complexity $O(n^{3/2})$, another one $O(n \log n)$, and so on (here n is a parameter measuring the size of the input).

For a real assessment of algorithms, it is usually necessary to complement such a theoretical analysis by testing the algorithm for various input data on a particular computer. For example, if the asymptotic analysis yields complexity $O(n^2)$ for one algorithm and $O(n \log^4 n)$ for another then the second algorithm looks clearly better at first sight because the function $n \log^4 n$ grows much more slowly than n^2. But if the exact complexity of the first algorithm were, say, $n^2 - 5n$ and of the second one $20n(\log_2 n)^4$, the superiority of the second algorithm will only show for $n > 5 \cdot 10^6$, and such a superiority is quite illusory from a practical point of view.

Let us try to estimate the asymptotic complexity of Algorithm 4.3.2. We have described the algorithm on a "high level", however. This doesn't refer to a prestigious social position but to the fact that we have used, for instance, a test of whether a given set of edges contains a cycle, which cannot be considered an elementary operation even with a very liberal approach. The complexity of the algorithm will thus depend on our ability to realize such a complex operation by elementary operations.

For our Algorithm 4.3.2, we may note that it is not necessary to store all the edge sets E_i, and that all of them can be represented by a single variable (say, a list of edges) which successively takes values $E_0, E_1, \ldots$. The only significant question is how to test efficiently whether adding a new edge e_i creates a cycle or not. Here is a crucial observation: a cycle arises if and only if the vertices of the edge e_i belong to the same connected component of the graph (V, E_{i-1}). Hence we need to solve the following problem:

4.3.4 Problem (UNION–FIND problem). Let $V = \{1, 2, \ldots, n\}$ be a set of vertices. Initially, the set V is partitioned into 1-element equivalence classes; that is, no distinct vertices are considered equivalent. Design an algorithm which maintains an equivalence relation on V (in other words, a partition of V into classes) in a suitable data structure, in such a way that the following two types of operations can be executed efficiently:

(i) (UNION) Make two given nonequivalent vertices $i, j \in V$ equivalent, i.e. replace the two classes containing them by their union.

(ii) (Equivalence testing—FIND) Given two vertices $i, j \in V$, decide whether they are currently equivalent.

A new request for an operation is input to the algorithm only after it has executed the previous operation.

Our Algorithm 4.3.2 for finding a spanning tree needs at most $n - 1$ operations UNION and at most m operations FIND.

We describe a quite simple solution of Problem 4.3.4. In the beginning, we assign distinct marks to the vertices of V, say the marks $1, 2, \ldots, n$. During the computation, the marks will always be assigned so that two vertices are equivalent if and only if they have the same mark. Thus, equivalence testing (FIND) is a trivial comparison of marks. For replacing two classes by their union, we have to change the marks for the elements of one of the classes. So, if the elements of each class are also stored in a list, the time needed for the mark-changing operation is proportional to the size of the class whose marks are changed.

For a very rough estimate of the running time, we can say that no class has more than n elements, so a single UNION operations never needs more than $O(n)$ time. For $n - 1$ UNION operations and m FIND operations we thus get the bound $O(n^2 + m)$. One inconspicuous improvement is to maintain also the size of each class and to change marks always for the smaller class. For such an algorithm, one can show a much better total bound: $O(n \log n + m)$ (Exercise 1). The best known solution of Problem 4.3.4, due to Tarjan, needs time at most $O(n\alpha(n) + m)$ for m FIND and $n - 1$ UNION operations (see e.g. Aho, Hopcroft, and Ullman [9]), where $\alpha(n)$ is a certain function of n. We do not give the definition of $\alpha(n)$ here; we only remark that $\alpha(n)$ does grow to infinity with $n \to \infty$ but extremely slowly, much more slowly than functions like $\log \log n$, $\log \log \log n$, etc. For practical purposes, the solution described above (with re-marking the smaller class) may be fully satisfactory.

Let us present one more algorithm for spanning trees, perhaps even a simpler one.

4.3.5 Algorithm (Growing a spanning tree). Let a given graph $G = (V, E)$ have n vertices and m edges. We will successively construct sets $V_0, V_1, V_2, \ldots \subseteq V$ of vertices and sets $E_0, E_1, E_2, \ldots \subseteq E$ of edges. We let $E_0 = \emptyset$ and $V_0 = \{v\}$, where v is an arbitrary vertex.

Having already constructed V_{i-1} and E_{i-1}, we find an edge $e_i = \{x_i, y_i\} \in E(G)$ such that $x_i \in V_{i-1}$ and $y_i \in V \setminus V_{i-1}$, and we set $V_i = V_{i-1} \cup \{y_i\}$, $E_i = E_{i-1} \cup \{e_i\}$. If no such edge exists, the algorithm finishes and outputs the graph constructed so far, $T = (V_t, E_t)$.

4.3.6 Proposition (Correctness of Algorithm 4.3.5). *If the algorithm finishes with a graph T with n vertices, then T is a spanning* tree of G. Otherwise G is a disconnected graph and T is a spanning tree of the component of G containing the initial vertex v.

Proof. The graph T is a tree because it is connected and has the right number of edges and vertices. If T has n vertices, it is a spanning tree, so let us assume that T has $\bar{n} < n$ vertices. It remains to show that $V(T)$ is the vertex set of a component of G.

Let us suppose the contrary: let there be an $x \in V(T)$ and a $y \notin V(T)$ connected by a path in the graph G. As in the proof of Proposition 4.3.3, we find an edge $e = \{x_j, y_j\} \in E(G)$ on this path such that $x_j \in V(T)$ and $y_j \in V \setminus V(T)$. The algorithm could thus have added the edge e and the vertex y_j to the tree, and should not have finished with the tree T. This contradiction concludes the proof. $\square$

Remark. The details of the algorithm just considered can be designed in such a way that the running time is $O(n + m)$ (see Exercise 2).

Exercises

1. Prove that if Problem 4.3.4 is solved by the described method (always changing the marks for the smaller class), then the total complexity of $n - 1$ UNION operations is at most $O(n \log n)$.

2. $*,CS$ Design the details of Algorithm 4.3.5 is such a way that the running time is $O(n + m)$ in the worst case. (This may require some knowledge of simple list-like data structures.)

3. From Exercise 3.4.8, we recall that a Hamiltonian cycle in a graph G is a cycle containing all vertices of G. For a graph G and a natural number $k \geq 1$, define the graph $G^{(k)}$ as the graph with vertex set $V(G)$ and two (distinct) vertices connected by an edge if and only if their distance in G is at most k.

   (a) *Prove that for each tree T, the graph $T^{(3)}$ has a Hamiltonian cycle.

   (b) Using (a), conclude that $G^{(3)}$ has a Hamiltonian cycle for any connected graph G.

   (c) Find a connected graph G such that $G^{(2)}$ has no Hamiltonian cycle.

4.4 The minimum spanning tree problem

Imagine a map of your favorite region of countryside with some 30–40 villages. Some pairs of villages are connected by gravel roads, in such a way that each village can be reached from any other along these roads. The county's council decides to modernize some of these roads to highways suitable for fast car driving, but it wants to invest as little money as possible under the condition that it will be possible to travel between any two villages along a highway. In this way, we arrive at a fundamental problem called the *minimum spanning tree problem*. This section is devoted to its solution.

Of course, you may object that your favorite part of the countryside was full of first-class expressways a long time ago. So you may consider some less advanced country, or find another natural formulation of the underlying mathematical problem.

We also assume that the existing roads have no branchings outside the villages, and that the new roads can only be built along the old ones (because of proprietary rights, say). Otherwise, it may be cheaper for instance to connect four places like this

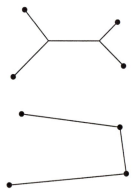

instead of like this

If we allowed the former kind of connection, we would arrive at a different algorithmic problem (called the *Steiner tree problem*), which turns out to be much less tractable than the minimum spanning tree problem.

A mathematical formulation of the minimum spanning tree problem requires that the notion of graph is enriched a bit: we will consider graphs with *weighted edges*. This means that for every edge $e \in E$ we are given a number $w(e)$, called the *weight* of the edge e. The weight of an edge is usually a nonnegative integer or real number. A graph $G = (V, E)$ together with a weight function w on its edges, $w: E \to \mathbf{R}$, is sometimes called a *network*.

Let us formulate the above road-building problem in graph-theoretic terms:

Problem. Given a connected graph $G = (V, E)$ with a nonnegative weight function w on the edges, find a spanning connected subgraph (V, E') such that the sum

$$w(E') = \sum_{e \in E'} w(e) \qquad (4.1)$$

has the minimum possible value.

It is easy to see that there is always at least one spanning tree of G among the solutions of this problem. If the weights are strictly positive, *each* solution must be a spanning tree. For instance, if all edges have weight 1, then the solutions of the problem are exactly the spanning trees of the graph, and the expression (4.1) has minimum value $|V| - 1$.

Hence, it suffices to deal with the following problem:

4.4.1 Problem (Minimum spanning tree problem). For a connected graph $G = (V, E)$ with a weight function w on the edges, find a spanning tree $T = (V, E')$ of the graph G with the smallest possible value of $w(E')$.

> An attentive reader might have observed that here we do not assume nonnegativity of the weights. Indeed, the algorithms we are going to discuss solve even this more general problem with arbitrary weights. There is more to this remark than meets the eye: many graph problems that are easy for nonnegative weights turn into algorithmically intractable problems if we admit weights with arbitrary signs. An example of such a problem is finding the shortest path in a network, where the length of a path is measured as the sum of the edge weights.

A given graph may have very many spanning trees (see Chapter 7) and it may seem difficult to find the best one. It is not really so difficult, and nowadays it can be done by an easy modification of the algorithms from the previous section. We present several algorithms. A simple and very popular one is the following:

4.4.2 Algorithm (Kruskal's or the "greedy" algorithm). The input is a connected graph $G = (V, E)$ with edge weight function w. Let us denote the edges by $e_1, e_2, \ldots, e_m$ in such a way that

$$w(e_1) \le w(e_2) \le \cdots \le w(e_n).$$

For this ordering of edges, execute Algorithm 4.3.2.

Before proving the correctness of Kruskal's algorithm, which is not completely easy, let us illustrate the algorithm with a small example.

Example. Let us apply Kruskal's algorithm for the following network:

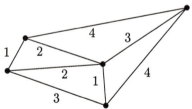

One possible execution of the algorithm is shown in the next diagram:

4.4.3 Proposition (Correctness of Kruskal's algorithm). *Algorithm 4.4.2 solves the minimum spanning tree problem.*

Proof. This proof is not really deep but it seems to require a concentrated attention, for otherwise it is very easy to make a mistake in it (both the authors have a rich experience of it by presenting it in lectures).

Let T be the spanning tree found by the algorithm, and let $\check{T}$ be any other spanning tree of the graph $G = (V, E)$. We need to show that $w(E(T)) \leq w(E(\check{T}))$. Let us denote the edges of T by $e'_1, e'_2, \ldots, e'_{n-1}$ in such a way that $w(e'_1) \leq w(e'_2) \leq \cdots \leq w(e'_{n-1})$ (the edge e'_i has been denoted by some e_j in the algorithm, so that it now has two names!). Similarly let $\check{e}_1, \ldots, \check{e}_{n-1}$ be the edges of $\check{T}$ ordered increasingly by weights.

We show that for $i = 1, \ldots, n-1$, we even have

$$w(e'_i) \leq w(\check{e}_i). \tag{4.2}$$

This of course shows that T is a minimum spanning tree. For contradiction, let us assume that (4.2) is not true, and let i be the smallest index for which it is violated, i.e. $w(e'_i) > w(\check{e}_i)$. We consider the sets

$$E' = \{e'_1, \ldots, e'_{i-1}\},$$
$$\check{E} = \{\check{e}_1, \ldots, \check{e}_i\}.$$

The graphs (V, E') and $(V, \check{E})$ contain no cycles and, moreover, $|E'| = i - 1$, $|\check{E}| = i$.

For the desired contradiction it suffices to show that there exists an edge $e \in \check{E}$ for which the graph $(V, E' \cup \{e\})$ contains no cycle. Then we obtain $w(e) \leq w(\check{e}_i) < w(e'_i)$ and this means that at the moment the edge e was considered in the algorithm we made a mistake. There was no reason to reject e, and we should have selected it instead of the edge e'_i or earlier.

Therefore it is enough to show the following: *If $E', \check{E} \subseteq \binom{V}{2}$ are two sets of edges, such that the graph $(V, \check{E})$ has no cycle and $|E'| < |\check{E}|$, then some edge $e \in \check{E}$ connects vertices of distinct components of the graph (V, E').* This can be done by a simple counting argument. Let $V_1, \ldots, V_s$ be the vertex sets of the components of the graph (V, E'). We have

$$\left| E' \cap \binom{V_j}{2} \right| \geq |V_j| - 1,$$

and by summing these inequalities over j we get $|E'| \geq n - s$. On the other hand, since $\check{E}$ has no cycles, we get

$$\left| \check{E} \cap \binom{V_j}{2} \right| \leq |V_j| - 1,$$

and therefore at most $n - s$ edges of $\check{E}$ are contained in some of the components V_j. But since we assumed $|\check{E}| > |E'|$, there is an edge $e \in \check{E}$ going between two distinct components. □

Kruskal's algorithm is a prototype of the so-called *greedy algorithm*. At every step, it selects the cheapest possible edge among those allowed by the restrictions of the problem (here "the graph should contain no cycle"). In general, "greedy algorithm" is a term for a strategy trying to gain as much as possible at any given moment, never contemplating the possible disadvantages such a choice may bring in the future.[3] For numerous problems, this short-sighted strategy may fail completely. For instance, a greedy strategy applied for a chess game would mean (in the simplest form) that a player would always take a piece whenever possible, and always the most valuable one. And by following such a naive play he would lose very soon.

---

[3]In the minimum spanning tree problem, a "parsimonious algorithm" would perhaps be a more appropriate name, since we always take the cheapest possible edge. But "greedy algorithm" is a universally accepted name. It is derived from situations where one tries to maximize something by grabbing as much as possible in each step. Since the minimizing and maximizing problems aren't conceptually very different, it seems better to stick to the single term "greedy algorithm" in both situations.

In this context, it seems somewhat surprising that the greedy algorithm finds a minimum spanning tree correctly. The greedy strategy
may be useful also for many other problems (especially if we have no
better idea). Often it at least yields a good approximate solution. Problems for which the greedy algorithm always finds an optimal solution
are studied in the so-called *matroid theory*. The reader can learn about
it in Oxley [24], for instance.

Exercises

1. Analogously to the minimum spanning tree problem, define the *maximum spanning tree problem*. Formulate a greedy algorithm for this
   problem and show that it always finds an optimal solution.

2. Prove that if $T = (V, E')$ is a spanning tree of a graph $G = (V, E)$
   then the graph $T + e$, where e is an arbitrary edge of $E \setminus E'$, contains
   exactly one cycle.

3. Prove that if T is a spanning tree of a graph G then for every $e \in
   E(G) \setminus E(T)$ there exists an $e' \in E(T)$ such that $(T - e') + e$ is a
   spanning tree of G again.

4. Let G be a connected graph with a weight function w on the edges,
   and assume that w is injective. Prove that the minimum spanning tree
   of G is determined uniquely.

5. *Let G be a connected graph with a weight function w on the edges.
   Prove that for each minimum spanning tree T of G, there is an initial
   ordering of the edges in Kruskal's algorithm such that the algorithm
   outputs the tree T.

6. Let w and w' be two weight functions on the edges of a graph $G =
   (V, E)$. Suppose that $w(e_1) < w(e_2)$ holds if and only if $w'(e_1) <
   w'(e_2)$, for any two edges $e_1, e_2 \in E$. Prove that (V, E') is a minimum
   spanning tree G for the weight function w if and only if (V, E') is a
   minimum spanning tree of G for the weight function w'. (This means:
   the solution to the minimum spanning tree problem only depends on
   the ordering of edge weights.)

7. ^{CS} Using the discussion of Algorithm 4.3.2 in the preceding section,
   design the details of Kruskal's algorithm in such a way that its time
   complexity is $O((n + m) \log n)$.

8. Consider an n-point set V in the plane. We define a weight function on
   the edge set of the complete graph on V: the weight of an edge $\{x, y\}$
   is the distance of the points x and y.

   (a) *Show that no minimum spanning tree for this network has a vertex
   of degree 7 or higher.

(b) *Show that there exists a minimum spanning tree whose edges do not cross.

9. *Let V be a set of $n > 1$ points in the unit square in the plane. Let T be a minimum spanning tree for V (i.e. for the complete graph with edge weights given by distances as in Exercise 8). Show that the total length of the edges of T is at most $10\sqrt{n}$. (The constant 10 can be improved significantly; the best known estimate is about $1.4\sqrt{n} + O(1)$.)

10. Let $G = (V, E)$ be a graph and let w be a nonnegative weight function on its edges.

(a) *Each set $E' \subseteq E$ of pairwise disjoint edges (i.e. sharing no vertices) is called a *matching* in the graph G. Let $\nu_w(G)$ denote the maximum possible value of $w(E')$ for a matching $E' \subseteq E$. A greedy algorithm for finding a maximum matching works similar to Kruskal's algorithm for a maximum spanning tree, i.e. it considers edges one by one in the order of decreasing weights, and it selects an edge if it has no common vertex with the previously selected edges. Show that this algorithm always finds a matching with weight at least $\frac{1}{2}\nu_w(G)$.

(b) Show that the bound in (a) cannot be improved; that is, for any constant $\alpha > \frac{1}{2}$ there exists an input for which the greedy algorithm finds a matching with weight smaller than $\alpha\,\nu_w(G)$.

11. A set $C \subseteq E$ in a graph $G = (V, E)$ is called an *edge cover* if each vertex $v \in V$ is contained in at least one edge $e \in C$. Let us look for a small edge cover by a greedy algorithm: if there is an edge containing 2 uncovered vertices take an arbitrary such edge, otherwise take any edge covering some yet uncovered vertex, and repeat until all is covered. Show that the number of edges in a cover thus found is

(a) at most twice the size of the smallest possible cover,

(b) **and (even) at most $\frac{3}{2}$ of the size of the smallest possible cover.

12. *A set $D \subseteq V$ in a graph $G = (V, E)$ is called a *dominating set* if $\bigcup_{e \in E:\ e \cap D \neq \emptyset} e = V$. Let us look for a small dominating set by a greedy algorithm: we always select a vertex connected to the maximum possible number of yet uncovered vertices. Show that for any number C there exists a graph for which $|D_G| \geq C|D_M|$, where D_G is a dominating set selected by the greedy algorithm and D_M is a dominating set of the smallest possible size. (Start by finding examples for small specific values of C.)

4.5 Jarník's algorithm and Borůvka's algorithm

What we call "Jarník's algorithm" is mostly known under the name "Prim's algorithm". However, since Prim's paper is dated 1957 while

Jarník[4] already described the same algorithm in an elegant and precise way in 1930 (continuing the work of Borůvka who published the first documented algorithm for the minimum spanning tree problem in 1928), we believe it is appropriate to use the name of the first inventor.

Nowadays, Jarník's algorithm can be viewed as a simple extension of Algorithm 4.3.5.

4.5.1 Algorithm (Jarník's algorithm). Proceed according to Algorithm 4.3.5, and always choose the newly added edge e_i as an edge of the smallest possible weight from the set $\{\{x,y\} \in E(G) \colon x \in V_{i-1}, y \notin V_{i-1}\}$.

4.5.2 Proposition (Correctness of Jarník's algorithm). *Jarník's algorithm finds a minimum spanning tree for every connected network.*

Proof. Let $T = (V, E')$ be the spanning tree resulting from Jarník's algorithm, and suppose that the edges of E' are numbered e_1 through e_{n-1} in the order they were added to T. For contradiction, suppose that T is not a minimum spanning tree.

Let T' be some minimum spanning tree. Let $k(T')$ denote the index k for which all the edges $e_1, e_2, \ldots, e_k$ belong to $E(T')$ but $e_{k+1} \notin E(T')$. Among all minimum spanning trees, select one which has the maximum possible value of k and denote it by $\check{T} = (V, \check{E})$. Write $k = k(\check{T})$.

Now consider the moment in the algorithm's execution when the edge e_{k+1} has been added to T. Let $T_k = (V_k, E_k)$ be the tree formed by the edges $e_1, \ldots, e_k$. Then e_{k+1} has the form $\{x,y\}$, where $x \in V(T_k)$ and $y \notin V(T_k)$. Consider the graph $\check{T} + e_{k+1}$. This graph contains some cycle C (since it is connected and has more than $n-1$ edges), and such a C necessarily contains the edge e_{k+1} (see also Exercise 4.4.2).

The cycle C consists of the edge $e_{k+1} = \{x,y\}$ plus a path P connecting the vertices x and y in the spanning tree $\check{T}$. At least one edge of the path P has one vertex in the set V_k and the other vertex outside V_k. Let e be some such edge. Obviously $e \neq e_{k+1}$, and further we know that $e \in \check{E}$ and $e_{k+1} \notin \check{E}$; see Fig. 4.1. Both the edges e and e_{k+1} connect a vertex of V_k with a vertex not lying in V_k, and by the edge selection rule in the algorithm we get $w(e_{k+1}) \leq w(e)$.

---

[4]An approximate pronunciation is YAR-neekh, and for Borůvka it is BOH-roof-kah.

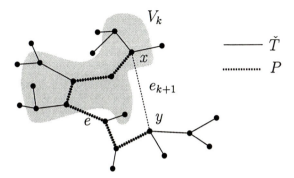

Fig. 4.1   Illustration for the correctness proof of Jarník's algorithm.

Now consider the graph $T' = (\check{T} + e_{k+1}) - e$. This graph has $n - 1$ edges and, as is easy to check, it is connected; hence it is a spanning tree. We have $w(E(T')) = w(\check{E}) - w(e) + w(e_{k+1}) \leq w(\check{E})$, and thus T' is a minimum spanning tree as well, but with $k(T') > k(\check{T})$. This contradiction to the choice of $\check{T}$ proves Proposition 4.5.2.

Warning. This is another proof of a slippery nature: make one step slightly differently and the whole thing falls apart.   □

Borůvka's algorithm. In conclusion, let us mention the historically first algorithm for minimum spanning tree computation due to Borůvka. As is usual in science, the first method discovered was not the simplest— both Kruskal's and (in particular) Jarník's algorithm are conceptually simpler. But yet it was Borůvka's algorithm that recently became a starting point for the theoretically fastest known algorithm for the minimum spanning tree problem (Karger, Klein, and Tarjan [37]). This latter algorithm is fairly complicated and uses a number of other ideas (which we will not pursue here) to make the computation fast.

4.5.3 Algorithm (Borůvka's algorithm). The input is a graph $G = (V, E)$ with edge weight function w. We need to assume, moreover, that *distinct edges get distinct weights,* i.e. that the weight function is one-to-one. This assumption is not particularly restrictive. Each weight function can be converted into a one-to-one function by arbitrarily small changes of the weights, which changes the weight of a minimum spanning tree by an arbitrarily small amount. (Alternatively, the algorithm can be modified to work with arbitrary weight functions, by adding a simple tie-breaking rule outlined in Exercise 6 below.)

   The algorithm successively constructs sets $E_0, E_1, \ldots \subseteq E$ of edges, beginning with $E_0 = \emptyset$.

Suppose that the set E_{i-1} has already been computed, and let $(V_1, \ldots, V_t)$ be the partition of the vertex set according to the components of the graph (V, E_{i-1}). Strictly speaking, this partition should also have the index i since it is different in each step, but we omit this index in order to make the notation simpler. For each set V_j of this partition we find the edge $e_j = \{x_j, y_j\}$ (where $x_j \in V_j$, $y_j \notin V_j$) whose weight is minimum among all edges of the form $\{x, y\}$, $x \in V_j$, $y \in V \setminus V_j$ (it may happen that $e_j = e_{j'}$ for $j \neq j'$). We put $E_i = E_{i-1} \cup \{e_1, \ldots, e_t\}$. The algorithm finishes when the graph (V, E_i) has a single component.

The algorithm could also be called a "bubbles algorithm". The graph G is covered by a collection of "bubbles". In each step, we merge each bubble with its nearest neighboring bubble.

We will not prove the correctness of this algorithm. We only show that the constructed graph has no cycle (which, unlike the previous algorithms, is not quite obvious). So suppose for contradiction that a cycle arose in some step i. This means that there are pairwise distinct indices $j(1), j(2), \ldots, j(k)$ for which

$$x_{j(1)} \in V_{j(1)}, \qquad\qquad y_{j(1)} \in V_{j(2)}$$
$$x_{j(2)} \in V_{j(2)}, \qquad\qquad y_{j(2)} \in V_{j(3)}$$
$$\vdots$$
$$x_{j(k-1)} \in V_{j(k-1)}, \qquad\qquad y_{j(k-1)} \in V_{j(k)}$$
$$x_{j(k)} \in V_{j(k)}, \qquad\qquad y_{j(k)} \in V_{j(1)}.$$

Here is an illustration:

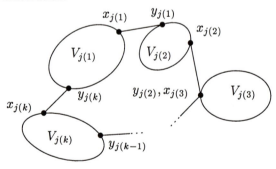

Since distinct edges have distinct weights, the edge $e_{j(\ell)}$ always has the smallest weight among the edges leaving the component $V_{j(\ell)}$, and in particular, we get

$$w(e_{j(1)}) < w(e_{j(2)}) < \cdots < w(e_{j(k)}) < w(e_{j(1)}).$$

This chain of strict inequalities cannot hold and so Borůvka's algorithm finds some spanning tree of G. With some more effort, it can be proved that it finds the minimum spanning tree. $\qquad\square$

Example. Consider the following network (the weights are given as labels of the edges):

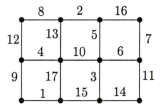

Jarník's algorithm started in the upper left corner proceeds as follows:

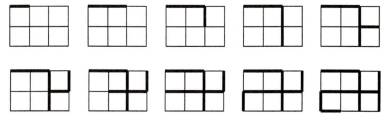

Kruskal's algorithm needs 17 steps (but only in 10 of them is a new edge added). Borůvka's algorithm, on the other hand, is quite short:

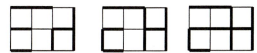

But in each step we have to do much more work.

Exercises

1. (General spanning tree algorithm) Consider the following algorithm for the minimum spanning tree problem. The input is a connected graph $G = (V, E)$ with weight function w. We put $E_0 = \emptyset$. Suppose that E_{i-1} has already been defined. Choose an arbitrary component V_i of the graph (V, E_{i-1}), select an edge e_i of the minimum weight among the edges with one vertex in V_i and the other vertex not in V_i, and set $E_i = E_{i-1} \cup \{e_i\}$. Prove that (V, E_{n-1}) is a minimum spanning tree. (Imitate the correctness proof for Jarník's algorithm.)

   Check that this proves the correctness of both Kruskal's and Jarník's algorithm.

2. ("Inverse" greedy algorithm) Consider the following algorithm for the minimum spanning tree problem. The input is a connected graph $G = (V, E)$ with weight function w. Label the edges $e_1, \ldots, e_m$ in such a way that $w(e_1) \geq \cdots \geq w(e_m)$. Put $E_0 = E$, and

$$E_i = \begin{cases} E_{i-1} \setminus \{e_i\} & \text{if the graph } (V, E_{i-1} \setminus \{e_i\}) \text{ is connected} \\ E_i & \text{otherwise.} \end{cases}$$

Prove that (V, E_m) is a minimum spanning tree of G.

3. *Prove the correctness of Borůvka's algorithm.

4. ^{CS} Design the details of Jarník's algorithm in such a way that its time complexity is $O((m+n)\log n)$ (this probably requires some basic knowledge of data structures).

5. (a) Prove that Borůvka's algorithm has at most $O(\log n)$ phases, i.e. the graph (V, E_i) is connected already for some $i = O(\log n)$.

    (b) ^{CS} Design the details of Borůvka's algorithm in such a way that its time complexity is at most $O((m+n)\log n)$.

6. (Tie-breaking in Borůvka's algorithm) Given an arbitrary weight function on edges of a graph $G = (V, E)$, we choose an arbitrary ordering $e_1, e_2, \ldots, e_m$ of the edges once and for all, and we define $e_i \lhd e_j$ if either $w(e_i) < w(e_j)$ or both $w(e_i) = w(e_j)$ and $i \le j$.

    (a) Formulate Borůvka's algorithm with edges ordered by the linear ordering $\lhd$ (instead of the usual ordering $\le$ on the weights).

    (b) Check that the algorithm in (a) computes a spanning tree of G.

    (c) Prove that the algorithm in (a) computes a minimum spanning tree of G with respect to the weight function w (modify the proof in Exercise 3).

5

Drawing graphs in the plane

5.1 Drawing in the plane and on other surfaces

Often it is advantageous to draw graphs. As you can see, most of the graphs in this book are specified by a picture (instead of a list of vertices and edges, say). But so far we have been studying properties of graphs not related to their drawings, and the role of drawings was purely auxiliary. In this chapter the subject of analysis will be the drawing of graphs itself and we will mainly investigate graphs that can be drawn in the plane without edge crossings. Such graphs are called planar.

From the numerous pictures shown so far and from the informal definition given in Section 3.1, the reader might have gained a quite good intuition about what is meant by a drawing of a graph. Such an intuition is usually sufficient if we want to show, say, that some graph is planar—we can simply draw a suitable picture of the graph with no edge crossings. However, if we want to prove, in a strictly logical way, that some graph is *not* planar, then we cannot do without a mathematical definition of the notion of a drawing, based on other exact mathematical notions. Today's mathematics is completely built from a few primitive notions and axioms of set theory—or at least the majority of mathematicians try to ensure it is. So, for instance, the notion of a "plane" is being modeled as the Cartesian product $\mathbf{R} \times \mathbf{R}$. Each real number is defined as a certain subset of the rationals, the rational numbers are created from natural numbers, and finally the natural numbers are defined as certain special sets produced from the empty set. (This is seldom apparent in everyday mathematics, but if you look at a book on the foundations of mathematics you can find it in there.)

In order to introduce the notion of a drawing formally, we define an *arc* first: this is a subset α of the plane of the form $\alpha = \gamma([0,1]) = \{\gamma(x) : x \in [0,1]\}$, where $\gamma : [0,1] \to \mathbf{R}^2$ is an injective continuous map of the closed interval $[0,1]$ into the plane. The points $\gamma(0)$ and $\gamma(1)$ are called the *endpoints* of the arc α.

This definition, frightening as it may look, is very close to the intuitive notion of drawing. The interval $[0, 1]$ can be thought of as a time interval during which we draw a line from the point $\gamma(0)$ to the point $\gamma(1)$. Then $\gamma(t)$ is the position of the pencil's tip at time t. The continuity of the mapping γ means a continuous motion on the paper's surface in time, and the injectivity says that the line being drawn never intersects itself.

5.1.1 Definition. *By a drawing of a graph $G = (V, E)$ we mean an assignment as follows: to every vertex v of the graph G, assign a point $b(v)$ of the plane, and to every edge $e = \{v, v'\} \in E$, assign an arc $\alpha(e)$ in the plane with endpoints $b(v)$ and $b(v')$. We assume that the mapping b is injective (different vertices are assigned distinct points in the plane), and no point of the form $b(v)$ lies on any of the arcs $\alpha(e)$ unless it is an endpoint of that arc. A graph together with some drawing is called a topological graph[1].*

A drawing of a graph G in which any two arcs corresponding to distinct edges either have no intersection or only share an endpoint is called a planar drawing. A graph G is planar if it has at least one planar drawing.

We have given the above formal definition of a graph drawing a bit "for show", in order to illustrate that the notion of a drawing can be included in the logical construction of mathematics. We will not continue building the subsequent theory of planar graphs in a strictly logical way, though. We would have to use notions and results concerning planar curves. These belong to a branch of mathematics called *topology*. Only very little from topology is usually covered in introductory mathematical courses, and we would have to introduce quite complicated machinery in order to do everything rigorously. Moreover, proofs of certain "intuitively obvious" statements are surprisingly difficult. For these reasons, we will sometimes rely on the reader's intuition in the subsequent text, and we will ask the reader to believe in some (valid!) statements without a proof. A rigorous treatment can be found, for example, in the recent book by Mohar and Thomassen [22]. Fortunately, in the theory of graph drawing, the basic intuition about drawing seldom leads one astray.

A planar drawing is advantageous for a visualization of a graph (edge crossings in a nonplanar drawing could be mistaken for vertices), and in some applications where the drawing has a physical

---

[1] A planar graph with a given planar drawing, i.e. a topological planar graph, is sometimes called a *plane graph*.

meaning, edge crossings can be inadmissible (for instance, in the design of single-layer integrated circuits).

Faces of a graph drawing. Let $G = (V, E)$ be a topological planar graph, i.e. a planar graph together with a given planar drawing. Consider the set of all points in the plane that lie on none of the arcs of the drawing. This set consists of finitely many connected regions (imagine that we cut the plane along the edges of the drawing):

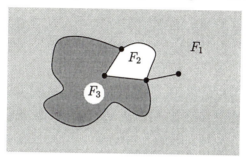

(We say that a set $A \subseteq \mathbf{R}^2$ is *connected* if for any two points $x, y \in A$ there exists an arc $\alpha \subseteq A$ with endpoints x and y. "Being connected"[2] is an example of a topological notion.) These regions will be called the *faces* of the considered topological planar graph. The region spreading out to infinity, such as F_1 in the picture, is called the *outer face* (or the *unbounded face*) of the drawing, and all the other faces are called *inner faces* (or *bounded faces*).

Let us stress that faces are defined *for a given planar drawing.* Faces are usually not defined for a nonplanar drawing, and also we should not speak about faces for a planar graph without having a specific drawing in mind.

Drawing on other surfaces. A graph can also be drawn on other surfaces than the plane. Let us list some examples of interesting surfaces.

Everyone knows the sphere (i.e. the surface of a ball). The surface of a tire-tube is scientifically called the *torus*:[3]

---

[2]What we called a "connected set" is usually called an *arc-connected set* in topology. A connected set is defined as follows. A set $A \subseteq \mathbf{R}^2$ is connected if no two disjoint open sets $A_1, A_2 \subseteq \mathbf{R}^2$ exist such that $A \subseteq A_1 \cup A_2$ and $A_1 \cap A \neq \emptyset \neq A_2 \cap A$. For sets considered in this chapter, such as faces of a graph, both these notions of being connected coincide, and so we use the shorter name.

[3]People with an US-centered worldview might want to speak about the surface of a doughnut, but since doughnuts in other countries (Australia, United Kingdom) are often spherical, we don't consider this name 100% politically correct.

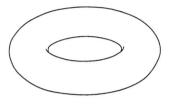

If we take a long strip of paper, turn one of its ends by 180 degrees, and glue it to the other end, we obtain an interesting surface called the *Möbius band*:

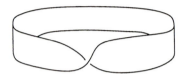

Other examples are a sphere with two handles:

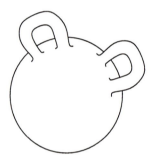

(after a suitable deformation, this is also the surface of a "fat figure 8"), or the so-called *Klein bottle*:

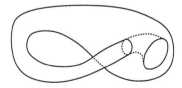

Each of these surfaces can be created from a planar polygon by "gluing" and a suitable deformation. In the above examples, with the exception of the sphere with two handles, we would always start from a planar rectangle, and we would identify (glue) some its edges in a suitable way. We have already introduced the Möbius band using such a procedure, and this method is also a basis for a rigorous definition of such surfaces (which we do not present here). For example, the torus can be made as follows:

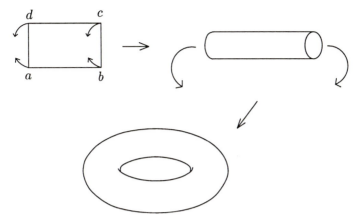

That is, we identify the opposite edges of the rectangle *abcd*, in such a way that the edge *ab* is glued to the edge *dc* and the edge *ad* is glued to the edge *bc*. The orientation of edges for gluing is usually marked by arrows. The arrows in the following picture mean that when gluing the edge *ad* to the edge *bc*, the point *a* goes to the point *b* and the point *d* goes to *c*.

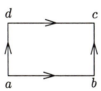

The following picture once again indicates how to manufacture the Möbius band:

(we only identify the two edges marked by arrows in such a way that the arrow directions are the same on the glued edge). The Klein bottle is produced by following the instructions of the next picture:

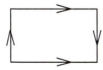

In fact, these are not quite honest directions for producing your own Klein bottle, since the Klein bottle cannot be realized in the 3-dimensional Euclidean space. The indicated gluing cannot be done in **R**^{3} unless the rectangle intersects itself (as it does in the picture showing the Klein bottle). Nevertheless, the definition of the Klein bottle by gluing makes good sense mathematically, and the surface can be realized in $\mathbf{R}^4$, say.

There is a general theorem saying that any closed surface (having no "boundary points" and not "running to infinity" anywhere; a scholarly term is "a compact 2-manifold without boundary") can be created by a suitable gluing and deformation from a regular convex polygon. Moreover, if the resulting surface is two-sided (note that both the Möbius band and the Klein bottle only have one side!) then it can be continuously deformed into a sphere with finitely many handles. The basics of this theory and a number of related topics are beautifully explained in the book by Stillwell [27].

Graphs can be classified according to the surfaces they can be drawn on. As will be shown in the next section, neither the graph K_5, the complete graph on 5 vertices, nor $K_{3,3}$, the complete bipartite graph on $3+3$ vertices, is planar. But K_5 can be drawn on the torus, for instance:

 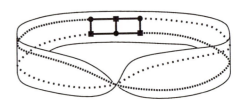

and $K_{3,3}$ on the Möbius band:

 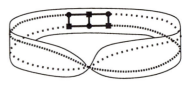

As we said above, these surfaces can be obtained by a suitable gluing of the edges of a rectangle. In order that our spatial imagination is not overstrained, we can convert drawing on surfaces into a modified planar drawing, where edges can "jump" among the glued rectangle edges. The drawings just shown can thus be recast as follows:

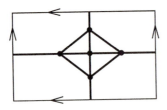 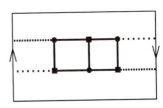

Let us remark that even the graph $K_{4,4}$ can be drawn on the torus:

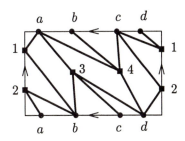

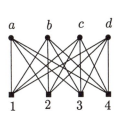

(K_7 can be so drawn as well; see Exercise 2). In general, we have

5.1.2 Proposition. *Any graph can be drawn without edge crossings on a sphere with sufficiently many handles.*

(This proposition must be taken informally, since we gave no exact definition of a sphere with handles.)

Informal proof. Let us draw the given graph $G = (V, E)$ on the sphere, possibly with edge crossings. Let $e_1, e_2, \ldots, e_n$ be all edges having a crossing with another edge. For each edge e_i, add a handle serving as a "bridge" for that edge to avoid the other edges, in such a way that the handles are disjoint and the edges drawn on handles do not cross anymore:

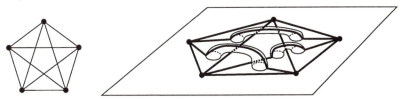

Since we deal with finitely many edges only, it is easy to find such handles. □

Hence the following definition makes sense.

5.1.3 Definition. *The smallest number of handles that must be added to the sphere so that a graph G can be drawn on the resulting surface without edge crossings is called the genus[4] of the graph G.*

In conclusion of this section, let us show that planar graphs are exactly graphs of genus 0, i.e. ones that can be drawn on the sphere. This becomes quite obvious if we use the *stereographic projection*. We place

---

[4]The notion of genus is primarily used for surfaces. For example, the genus of a sphere with handles is the number of handles. Interestingly, 2-dimensional surfaces were first systematically studied in connection with algebraic equations! The set of all complex solutions of a polynomial equation in 2 variables is typically a 2-dimensional surface, and its genus is crucial for various properties of the equation. The (difficult) area of mathematics studying algebraic equations in this spirit is called *algebraic geometry* (Cox, Little, and O'Shea [17] is a nice introduction to the subject).

the sphere in the 3-dimensional space in such a way that it touches the considered plane ρ. Let o denote the point of the sphere lying farthest from ρ (the "north pole"):

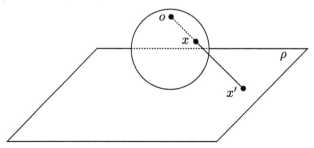

Then the stereographic projection maps each point $x \neq o$ of the sphere to a point x' in the plane, where x' is the intersection of the line ox with the plane ρ. (For the point o, the projection is undefined.) This defines a bijection between the plane and the sphere without the point o. Given a drawing of a graph G on the sphere without edge crossings, where the point o lies on no arc of the drawing (which we may assume by a suitable choice of o), the stereographic projection yields a planar drawing of G. Conversely, from a planar drawing we get a drawing on the sphere by the inverse projection.

Exercises

1. Find
   (a) a planar graph all of whose vertices have degree 5,
   (b) *connected graphs as in (a) with arbitrarily many vertices.

2. (a) Check that the picture shown in the text indeed gives a drawing of $K_{4,4}$ on the torus.
   (b) Find a drawing of K_6 on the torus.
   (c) *Draw K_7 on the torus.

3. *Let G be a planar Eulerian graph. Consider some planar drawing of G. Show that there exists a closed Eulerian tour that never crosses itself in the considered drawing (it may touch itself at vertices but it never "crosses over to the other side").

4. Try to invent some suitable definition of a closed 2-dimensional surface. Then look up what the usual definition is in a book on topology.

5.2 Cycles in planar graphs

We will investigate various combinatorial properties of planar graphs. Among others, it turns out that the notion of a planar graph itself can be equivalently defined by purely combinatorial means, without using topological properties of the plane or intuition about graph drawing.

Should the geometric definition of planar graphs be converted into a combinatorial definition, we have to use some property of the plane connecting geometry to combinatorics. Such a property is expressed by the Jordan curve theorem below. First, a definition: a *Jordan curve*[5] is a closed curve without self-intersections. More formally, a Jordan curve is defined as an arc whose endpoints coincide, i.e. a continuous image of the interval $[0, 1]$ under a mapping f that is one-to-one except for the equality $f(0) = f(1)$.

5.2.1 Theorem (Jordan curve theorem). *Any Jordan curve k divides the plane into exactly two connected parts, the "interior" and the "exterior" of k, and k is the boundary of both the interior and the exterior. (Both the interior and exterior will be called the regions of k.) This means that if we define a relation $\approx$ on the set $\mathbf{R}^2 \setminus \gamma$ by setting $x \approx y$ if and only if x and y can be connected by an arc disjoint from k, then $\approx$ is an equivalence with 2 classes, one of them being a bounded set and the other one an unbounded set.*

This theorem is intuitively obvious, but its proof is by no means simple, although significant simplifications have been found recently by Thomassen [44]. For some Jordan curves in the plane, the statement is very evident,

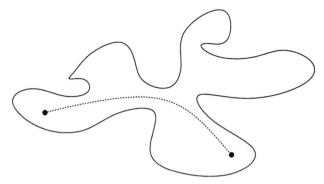

but for others it is perhaps less obvious (try finding an arc connecting the points ○ and ● and not intersecting the curve):

---

[5]Another common name for this object is a *simple closed curve*.

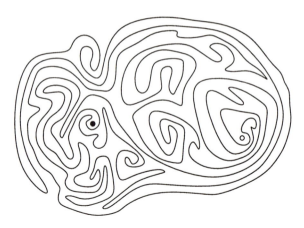

In order to illustrate that intuition is not always reliable for such "obvious" statements, let us mention a related theorem. An extension of the Jordan curve theorem, the *Jordan–Schönflies theorem*, tells us that for any Jordan curve, the interior can be continuously deformed onto the interior of the (usual geometric) circle. More precisely, there exists a continuous mapping whose inverse mapping is continuous as well, a so-called *homeomorphism*, between the (closed) region bounded by any Jordan curve and the ordinary circular disk. Similarly, one would expect that if we define a "topological sphere" as the image of the usual geometric sphere by an injective continuous map, such a thing will bound a region that can be continuously deformed onto the ordinary ball. But this is false—a counterexample is known under the name "Alexander's horned sphere" (see e.g. the excellent but somewhat more advanced book by Bredon [16]).

Let us remark that the difficulties with proving the Jordan curve theorem mainly stem from the considerable generality of the notion of an arc. We admit an arbitrary injective continuous mapping of the unit interval in the definition of an arc, and such mappings can be quite "wild". A simpler way to build a logically precise theory of planar graphs is to permit only arcs consisting of a finite number of straight segments—let us call them *polygonal arcs*. We can thus call a graph polygonally planar if it can be drawn without edge crossings using polygonal arcs. To prove the Jordan curve theorem for polygonal arcs only is reasonably easy (see Exercise 7). And it is not too difficult to verify that any planar graph is polygonally planar too;[6] for this one needs some topology but very little. Therefore, allowing for general nonpolygonal arcs achieves nothing new in graph drawing—only complications with the Jordan curve theorem.

---

[6]Even a much stronger statement holds: any planar graph can be drawn without edge crossings in such a way that every edge is a straight segment! But this is not an easy theorem.

As was announced earlier, we prefer to rely on intuition at some points in deriving results about planar graphs. At the price of longer and more complicated proofs, such imperfections could be removed, and everything could be derived from the Jordan curve theorem and its variations.

Let us begin with a proof of nonplanarity of the graph K_5. Later on, we will prove it again by other means.

5.2.2 Proposition. *K_5 is not planar.*

Proof. Proceed by contradiction. Let b_1, b_2, b_3, b_4, b_5 be the points corresponding to the vertices of K_5 in some planar drawing. The arc connecting the points b_i and b_j will be denoted by $\alpha(i, j)$.

Since b_1, b_2, and b_3 are vertices of a cycle in the graph K_5, the arcs $\alpha(1, 2)$, $\alpha(2, 3)$, and $\alpha(3, 1)$ form a Jordan curve k, and hence the points b_4 and b_5 lie either both inside or both outside k, for otherwise the arc $\alpha(4, 5)$ would cross k. First suppose that b_4 lies inside k, as in the following picture:

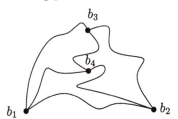

Then b_5 lies inside the Jordan curve formed by the arcs $\alpha(1, 4)$, $\alpha(2, 4)$, and $\alpha(1, 2)$, or in the Jordan curve made up by $\alpha(2, 3)$, $\alpha(3, 4)$, and $\alpha(2, 4)$, or inside the Jordan curve consisting of $\alpha(1, 3)$, $\alpha(3, 4)$, and $\alpha(1, 4)$.[7] In the first of these cases, the arc $\alpha(3, 5)$ has to intersect the Jordan curve formed by the arcs $\alpha(1, 4)$, $\alpha(2, 4)$, and $\alpha(1, 2)$, however, and similarly in the remaining two cases.

If the points b_4 and b_5 lie both outside k, we proceed quite analogously. □

Faces and cycles in 2-connected graphs. If $e_1, \ldots, e_n$ are the edges of a cycle in a topological planar graph G, then the arcs $\alpha(e_1), \ldots, \alpha(e_n)$ form a Jordan curve. By the Jordan curve theorem, we get that each face of G lies either inside or outside this Jordan

---

[7]We haven't proved that the interiors of these Jordan curves together cover the interior of k, and so this is one of the points where we rely on intuition somewhat.

curve. For brevity, let us call this Jordan curve a *cycle of G* too (so a cycle of G may now mean either a cycle in the graph-theoretic sense, i.e. a subgraph of G, or the Jordan curve corresponding to a graph-theoretic cycle of G in some drawing of G).

For some topological planar graphs, each face is the interior or the exterior of some cycle of G. But it need not always be so. For instance, a planar drawing of a tree has only one face. Another example might look as follows:

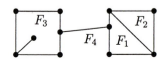

It turns out that the bad examples are exactly the graphs that are not 2-connected.

5.2.3 Proposition. *Let G be a 2-vertex-connected planar graph. Then every face in any planar drawing of G is a region of some cycle of G.*

Proof. We proceed by induction, making use of Proposition 3.7.5 (characterization of 2-connected graphs). If the graph G is a triangle, the statement we are proving follows from the Jordan curve theorem.

Let $G = (V, E)$ be a connected topological planar graph with at least 4 vertices. By Proposition 3.7.5, either there exists an edge $e \in E$ such that the graph $G' = G - e$ is 2-connected, or there are a 2-connected graph $G' = (V', E')$ and an edge $e \in E'$ such that $G = G' \% e$, where $\%$ denotes the operation of edge subdivision.

Since G is a topological planar graph, G' is a topological planar graph as well, in both cases. Since G' is 2-connected, we can use the inductive hypothesis. Each face of the topological graph G' is thus a region of some cycle of G'.

Let us consider the first case, where $G' = G - e$, $e = \{v, v'\}$. The vertices v and v' are connected by the arc $\alpha(e)$ corresponding to the edge e, and hence they both lie on the boundary of a face F of G'. Let k_F be the cycle bounding the face F. As the following picture indicates, the arc $\alpha(e)$ divides F into two new faces F' and F'':

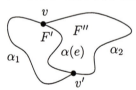

(This is one of the points where we rely on pictorial intuition and omit a rigorous proof, but see Exercise 7.) The faces F' and F'' are regions of the cycles $\alpha_1 \cup \alpha(e)$ and $\alpha_2 \cup \alpha(e)$, where α_1 and α_2 are the two arcs connecting v and v' and together forming the cycle k_F. Hence the faces of G are all bounded by cycles as claimed. This finishes the inductive step in the first case, where $G = G' + e$.

The remaining second case is easier: if $G = G'\%e$ and each face of G' is a region of some cycle of G', then G has the same property, as follows immediately from the definition of edge subdivision. This concludes the proof of the proposition. □

Proposition 5.2.3 shows that 2-connected planar graphs behave, in some sense, more nicely than arbitrary planar graphs. But it turns out that a still much nicer behavior results by requiring that the considered planar graphs be 3-vertex-connected. Such graphs have an essentially unique planar drawing (up to a continuous deformation of the plane and mirror reflection), and in many ways they are easier to work with (see Section 5.3). When proving theorems or designing algorithms concerning planar graphs, it is usually advisable to deal with 3-connected planar graphs first, and then try to handle the general case by decomposing a given graph into 3-connected pieces. Here we will not pursue this matter any further.

A combinatorial characterization of planar graphs. Let us remark that the following clearly holds: *A graph G is planar if and only if each subdivision of G is planar.* This property can be used for a combinatorial characterization of planar graphs—a characterization purely in graph-theoretic notions, using no geometric notions at all. It is the following celebrated result:

5.2.4 Theorem (Kuratowski's theorem). *A graph G is planar if and only if it has no subgraph isomorphic to a subdivision of $K_{3,3}$ or to a subdivision of K_5.*

It is easy to prove one of the implications in this theorem (if G is planar then it cannot contain a subdivision of a nonplanar graph), but the reverse implication is more demanding and we will not prove it in this book.

This theorem shows that the nonplanarity of any nonplanar graph can be certified by finding a subdivision of $K_{3,3}$ or K_5 in it. From a computational point of view, i.e. if we want to really test graph planarity by a computer and perhaps also look for a planar drawing, this method is not very efficient. Algorithms are known for testing whether a given graph contains a subdivision of a fixed (small) graph, but these algorithms are fairly complicated and impractical. For planarity testing and finding "nice" planar drawings, a number of fast (although also complicated) methods have been invented. Such methods can, for instance, test planarity of a given graph on n vertices in time $O(n)$. Concerning "nice" drawings, it is known that, for example, any planar graph on n vertices can be drawn in such a way that the vertices have integer coordinates between 1 and n and the edges are straight segments. Recent results in this direction, as well as further references, can be found in Kant [36]. There are also many interesting problems related to graph drawing which remain open. For instance, it is not known (at the time of writing this book) whether every planar graph has a planar drawing where all edges are straight segments with integer lengths. (This is an example of how easily one can sometimes formulate even difficult problems in discrete mathematics.)

Kuratowski's theorem characterizes planar graphs by specifying 2 "obstacles to planarity", namely the presence of a subdivision of K_5 or of a subdivision of $K_{3,3}$. Recently, many theorems in a similar spirit have been found, characterizing various classes of graphs by finite sets of "obstacles". The obstacles are usually not forbidden subdivisions of certain graphs but rather so-called *forbidden minors* (see Exercise 5.4.11). Many outstanding problems in graph theory, including numerous questions about efficient algorithms, have been solved by this approach and related ideas, and currently this area (called the *structural graph theory*) constitutes one of the most dynamic and successful parts of modern graph theory. A sample of recent progress in this area is Robertson, Seymour, and Thomas [41], where a long-standing open problem has been resolved by related methods.

Exercises

1. Show that the graph $K_{3,3}$ is not planar, in a manner similar to the proof of nonplanarity of K_5 given in the text.

2. (a) Find a subdivision of either $K_{3,3}$ or K_5 in the graph on the left in Fig. 3.3.

   (b) Is the graph in the middle in Fig. 3.3 planar?

   (c) Is the graph in Fig. 8.3 planar?

3. The *complete k-partite graph* $K_{n_1,n_2,\ldots,n_k}$ has vertex set $V = V_1 \dot\cup V_2 \dot\cup \cdots \dot\cup V_k$, where $V_1,\ldots,V_k$ are disjoint sets with $|V_i| = n_i$, and each vertex $v \in V_i$ is connected to all vertices of $V \setminus V_i$, $i = 1,2,\ldots,k$. Describe all k-tuples $(n_1,n_2,\ldots,n_k)$ of natural numbers, $k = 1,2,\ldots$, such that $K_{n_1,n_2,\ldots,n_k}$ is a planar graph.

4. In the proof of Proposition 5.2.3, we proceeded by induction, but induction on what? (There was no explicitly mentioned natural number as an induction parameter.)

5. Prove that if each face of a topological planar graph G is a region of some cycle of G then G is 2-connected.

6. *Consider an arbitrary drawing (not necessarily planar) of the complete graph K_n. Prove that at least $\frac{1}{5}\binom{n}{4}$ pairs of edges have to cross. (Use the nonplanarity of K_5.)

7. The goal of this exercise is to give a rigorous proof, without relying on geometric intuition.

    (a) *Let k be a Jordan curve consisting of finitely many segments (i.e. a polygon). Define two points of $\mathbf{R}^2 \setminus k$ to be equivalent if they can be connected by a polygonal arc not intersecting k. Prove that this equivalence has at most two classes.

    (b) *Show that in the situation as in (a) there are at least two classes. Hint: define an "interior point" as one for which a vertical semiline emanating upwards from it has an odd number of intersections with k.

    (c) Let k be a polygonal Jordan curve as in (a), let p, q be two distinct points of k, let k_1, k_2 be the two polygonal arcs into which k is divided by the points p and q, and let $r \in k_1$, $s \in k_2$ be points inside these arcs. Let ℓ be a polygonal arc connecting p and q and lying completely in the interior of k (except for its endpoints). Prove that any polygonal arc connecting r to s and lying in the interior of k (except for the endpoints) must intersect ℓ.

5.3 Euler's formula

There exists essentially only one basic quantitative formula for planar graphs. One can say that all other results use this formula to some extent. At the same time, it is the oldest formula. It was known to Euler in 1752, and sometimes it is asserted that it was known to Descartes in 1640 as well; the original statement was about convex polytopes rather than about planar graphs.

5.3.1 Proposition (Euler's formula). *Let $G = (V, E)$ be a connected planar graph, and let f be the number of faces of some planar drawing of G. Then we have*

$$|V| - |E| + f = 2.$$

In particular, the number of faces does not depend on the particular way of drawing.

Proof. We proceed by induction on the number of edges of the graph G. If $E = \emptyset$ then $|V| = 1$ and $f = 1$, and the formula holds. So let $|E| \geq 1$. We distinguish two cases:

1. The graph G contains no cycle. Then G is a tree and hence $|V| = |E| + 1$; at the same time we have $f = 1$ since a planar drawing of a tree has only one (unbounded) face.

2. Some edge $e \in E$ is contained in a cycle. In this case the graph $G - e$ is connected. Hence by the inductive hypothesis, Euler's formula holds for it (we consider the drawing arising from the given drawing of G by removing the edge e). The edge e in the considered drawing of G is adjacent to two distinct faces F and F', by the Jordan curve theorem. These faces become a single face after deleting e. Hence both the number of faces and edges increases by 1 by adding e back to the drawing, and the number of vertices is unchanged; hence Euler's formula is true for G too.

□

Application: Platonic solids. A Greek school of thinkers associated with Plato's name used to attribute a particular significance to highly regular geometric solids, the so-called *regular polytopes*, looking for them even in the foundations of the structure of the universe. (Besides, Kepler also regarded as one of his most important discoveries a theory, most likely a mistaken one, according to which the spacing among the planets' orbits is determined by the geometry of the regular polytopes.) A regular polytope is a 3-dimensional convex body[8] bounded by a finite number of faces. All faces should be congruent copies of the same regular convex polygon, and the same number of faces should meet at each vertex of the body. One reason for the above-mentioned great interest in these objects is most likely their exceptionality. There are only 5 types of regular polytopes: the regular tetrahedron, the cube, the regular octahedron, the regular dodecahedron, and the regular icosahedron (surely the reader will know which is which):

---

[8] *Convexity* means that whenever x and y are two points of the considered body, then the whole segment xy is a part of the body, i.e. the surface has no "dips" in it.

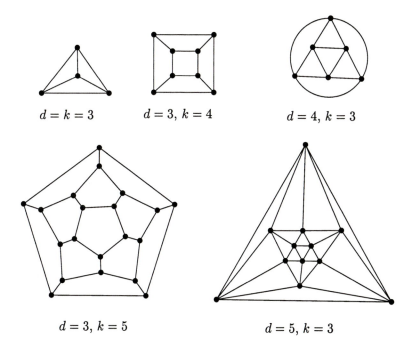

$d = k = 3$ $d = 3,\ k = 4$ $d = 4,\ k = 3$

$d = 3,\ k = 5$ $d = 5,\ k = 3$

Fig. 5.1 Graphs of the Platonic solids.

This fact was already known to the ancient Greeks. Let us remark that if we relax the conditions on regularity a little (we do not insist on convexity, or we allow for two types of faces, etc.), or if we go to higher dimensions, we can meet many more interesting and beautiful geometric shapes. Research in this area is still quite active.

Using Euler's formula, we will show that no other regular polytopes than the 5 Platonic ones exist. (The actual existence of these regular polytopes must be checked geometrically, which we do not consider here.) The first step in proving the non-existence of other regular polytopes is converting a convex polytope into a planar graph. We place the considered polytope inside a sphere, in such a way that the center of the sphere lies inside the polytope. Then we project the polytope onto the sphere (imagine that the edges of the polytope are made from wire and we place a tiny lamp in the center). This yields a graph drawn on the sphere without edge crossings, and as we know from Section 5.1, such a drawing can be metamorphosed into a planar drawing using the stereographic projection. The vertices, edges, and faces of the polytope become the vertices, edges, and faces of this planar drawing, respectively. This is where the terms "vertex" and "edge" for graphs and "face" for planar graphs come from. For the 5 regular solids, we thus obtain the graphs in Fig. 5.1.

For a regular polytope, the resulting topological planar graph has the same degree, d, of each vertex (where $d \geq 3$), and each face has the same number, $k \geq 3$, of vertices on its boundary. The nonexistence of any other regular polytopes is thus a consequence of the following:

5.3.2 Proposition. *Let G be a topological planar graph in which each vertex has degree d and each face is adjacent to k vertices, for some integers $d \geq 3$ and $k \geq 3$. Then G is isomorphic to one of the graphs in Fig. 5.1.*

Proof. Let us denote the number of vertices of the considered graph $G = (V, E)$ by n, the number of edges by m, and the number of faces by f. First we use the equation $\sum_{v \in V} \deg_G(v) = 2|E|$ (Proposition 3.3.1), which in our case specializes to

$$dn = 2m.$$

Similarly we obtain the equality

$$2m = kf.$$

We double-count the number of ordered pairs (e, F), where F is a face of G and $e \in E$ is an edge lying on the boundary of F. Each edge contributes 2 such pairs (as each face is bounded by a cycle), and each face k pairs.

Next, we express both n and f in terms of m using the just derived equations, and we substitute the results into Euler's formula:

$$2 = n - m + f = \frac{2m}{d} - m + \frac{2m}{k}.$$

By adding m and dividing by $2m$, we obtain

$$\frac{1}{d} + \frac{1}{k} = \frac{1}{2} + \frac{1}{m}.$$

Hence if both d and k are known, the other parameters n, m, and f are already determined uniquely. Obviously $\min(d, k) = 3$, for otherwise $\frac{1}{d} + \frac{1}{k} \leq \frac{1}{2}$. For $d = 3$ we get $\frac{1}{k} - \frac{1}{6} = \frac{1}{m} > 0$, and therefore $k \in \{3, 4, 5\}$. Similarly for $k = 3$ we derive $d \in \{3, 4, 5\}$. Hence one of the following possibilities must occur:

d	k	n	m	f
3	3	4	6	4
3	4	8	12	6
3	5	20	30	12
4	3	6	12	8
5	3	12	30	20

Now it is easy to check that in each of these cases the graph is completely determined by the values d, k, n, m, f, and it is isomorphic to one of the graphs in Fig. 5.1. □

Let us remark that the connection between planar graphs and 3-dimensional convex polytopes is closer than it might seem. As we have seen, we obtain a planar graph from each convex polytope. A quite difficult theorem, due to Steinitz, asserts that any vertex 3-connected planar graph (i.e. a planar graph that remains connected even after deleting any 2 vertices) is the graph of some 3-dimensional convex polytope. A delightful account of the theory of convex polytopes is Ziegler [29].

A very important property of planar graphs is that they can only have relatively few edges: a planar graph on n vertices has $O(n)$ edges. Here is a precise formulation of this property:

5.3.3 Proposition (Planar graph has $O(n)$ edges).

(i) *Let $G = (V, E)$ be a planar graph with at least 3 vertices. Then $|E| \leq 3|V| - 6$. Moreover, equality holds for any maximal planar graph; that is, a planar graph such that adding any new edge (while preserving the same vertex set) makes it nonplanar.*

(ii) *If, moreover, the considered planar graph contains no triangle (i.e. K_3 as a subgraph) and has at least 3 vertices, then $|E| \leq 2|V| - 4$.*

(We do not admit graphs with multiple edges in this proposition, of course!)

Proof of (i). If the graph G is not maximal planar we keep adding edges until it becomes maximal planar. Hence part (i) will be proved as soon as we show that $|E| = 3|V| - 6$ is true for any maximal planar graph with at least 3 vertices.

As a first step, we want to show that each face (including the outer one) of a maximal planar graph with at least 3 vertices is a triangle,[9] i.e. it is bounded by a cycle of length 3.

If G is disconnected, we can clearly connect its two distinct components by a new edge. If G is connected but not 2-connected, it has some vertex v whose removal disconnects the graph, creating components $V_1, V_2, \ldots, V_k$, $k \geq 2$ (note that we're using the assumption of G having at least 3 vertices at this moment!). Choose two edges e and e' connecting v to two distinct components V_i, V_j such that e and e' are drawn next to each other. Their endpoints can be connected by a new edge $\bar{e}$ without destroying planarity:

---

[9]For this reason, a maximal topological planar graph is also called a *triangulation*.

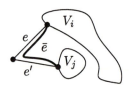

Hence a maximal planar graph with at least 3 vertices is necessarily 2-connected, and by Proposition 5.2.3, each face is bounded by a cycle of the graph. For contradiction, assume that the bounding cycle of some face F contains $t \geq 4$ vertices $v_1, \ldots, v_t$. If the vertex v_1 is not connected by an edge with the vertex v_3 then we can draw the edge $\{v_1, v_3\}$ inside the face F. On the other hand, if $\{v_1, v_3\} \in E(G)$, this edge must be drawn outside the face F, and therefore $\{v_2, v_4\}$ cannot be an edge, for otherwise $\{v_1, v_3\}$ and $\{v_2, v_4\}$ would have to cross:

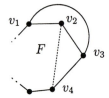

Hence we can safely draw the edge $\{v_2, v_4\}$ inside the face F.

Thus, each face of a maximal planar graph is a triangle as was asserted above. From this we get, similar to the proof of Proposition 5.3.2, the equality $3f = 2|E|$, where f is the number of faces. Expressing f from Euler's formula and substituting into the just derived equation, we obtain

$$|V| - |E| + \frac{2}{3}|E| = 2.$$

The desired equality $|E| = 3|V| - 6$ follows by a simple algebraic manipulation. This proves part (i).

Proof of (ii). We proceed likewise. After adding some edges if necessary, we may suppose that our graph is an edge-maximal triangle-free planar graph, meaning that by adding any new edge we create a triangle or make the graph nonplanar (or both). We can again assume that the graph is connected.

If G is not (vertex) 2-connected then it has a vertex v whose deletion splits G into components $V_1, \ldots, V_k$, $k \geq 2$. Certainly we can add some edge going between distinct components in such a way that G remains planar, but for some edges we could introduce a triangle

(this happens in case we connect two vertices both adjacent to v), and so we have to proceed more carefully. If each of the components V_i consists of a single vertex then G is a tree and the formula being proved holds for it. So let us suppose $|V_1| \geq 2$, and consider a face F having both a vertex of V_1 and a vertex of some other V_i on its boundary, as in the picture:

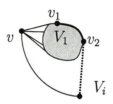

The component V_1 must have at least one edge, $\{v_1, v_2\}$, on the boundary of F, and since G has no triangle, it is not possible that both v_1 and v_2 are adjacent to v. Hence v_1 or v_2 can be connected to a vertex of V_i within F without possibly creating a triangle.

Therefore, we may assume that G is 2-connected. In this case, each face is bounded by a cycle in G. Each such cycle has length at least 4, and by double-counting we get $2|E| \geq 4f$ this time. Using Euler's formula we finally arrive at $|E| \leq 2|V| - 4$. □

A typical wrong proof. Suppose we want to prove that any topological planar graph with $n \geq 3$ vertices such that each face is bounded by 3 edges has $3n - 6$ edges. Students sometimes give an (incorrect) answer of the following type. We proceed by induction on n. For $n = 3$ we have a single triangle with 3 edges and 3 vertices, so the claim holds. Next, let us assume the statement holds for any topological planar graph G with n vertices. For any such G, we add a vertex into some face and connect it to the 3 vertices of that face as in the following picture:

This yields a graph G' with $n + 1$ vertices. The number of edges of G is $3n - 6$ by the inductive hypothesis, and we have added 3 new edges. Hence G' has $3(n + 1) - 6$ edges and the statement holds for graphs with $n + 1$ vertices as well. Well, the problem is that not all possible topological planar graphs G' on $n + 1$ vertices with triangular faces can arise from some G by the operation described. For instance, the graph of the regular octahedron (drawn in Fig. 5.1) has all degrees 4, while

any graph produced from a smaller graph by the operation has at least one vertex of degree 3.

Part (i) of Proposition 5.3.3 has an important and often applied consequence, namely that every planar graph has some vertex of degree at most 5. Similarly part (ii) guarantees that a planar graph without triangles contains a vertex of degree no more than 3.

Part (i) also shows that K_5 is not planar, because it has 10 edges, while a planar graph on 5 vertices has at most 9 edges. Similarly (ii) implies the nonplanarity of $K_{3,3}$, since a triangle-free graph on 9 vertices has at most 8 edges.

We prove yet another proposition, giving us more information about possible scores of planar graphs.

5.3.4 Proposition. *Let $G = (V, E)$ be a 2-connected planar graph with at least 3 vertices. Let n_i be the number of its vertices of degree i, and let f_i be the number of faces (in some fixed planar drawing) bounded by cycles of length i. Then we have*

$$\sum_{i \geq 1}(6 - i)n_i = 12 + 2\sum_{j \geq 3}(j - 3)f_j,$$

or, rewritten differently,

$$5n_1 + 4n_2 + 3n_3 + 2n_4 + n_5 - n_7 - 2n_8 - \cdots = 12 + 2f_4 + 4f_5 + 6f_6 + \cdots.$$

Hence $5n_1 + 4n_2 + 3n_3 + 2n_4 + n_5 \geq 12$, and so every planar graph with at least 3 vertices contains at least 3 vertices of degree no larger than 5.

Proof. Clearly $|V| = \sum_i n_i$, $f = \sum_i f_i$. By substituting for $|V|$ and f from these equations into Euler's formula we have

$$2|E| = 2(|V| + f - 2) = \sum_i 2n_i + \sum_j 2f_j - 4. \qquad (5.1)$$

By a double-counting similar to that in previous proofs we obtain further relations: $\sum_i in_i = 2|E| = \sum_j jf_j$. By expressing $2|E|$ using (5.1), these equalities are transformed into

$$\sum_j (j - 2)f_j + 4 = \sum_i 2n_i \qquad \sum_j 2f_j = \sum_i (i - 2)n_i + 4.$$

We multiply the first of these equalities by 2 and we subtract the second one from it. The result is

$$\sum_i (6 - i)n_i - 4 = 2\sum_j (j - 3)f_j + 8.$$

This already gives the proposition. $\square$

Exercises

1. Prove that the bound $|E| \leq 2|V| - 4$ for triangle-free planar graphs is the best possible in general. That is, for infinitely many n construct examples of triangle-free planar graphs with n vertices and $2n - 4$ edges.

2. (a) Show that a topological planar graph with $n \geq 3$ vertices has at most $2n - 4$ faces.

   (b) Show that a topological planar graph without triangles has at most $n - 2$ faces.

3. Prove that a planar graph in which each vertex has degree at least 5 must have at least 12 vertices.

4. For which values of k can you prove the following statement? *There exists an n_0 such that any planar graph on at least n_0 vertices contains at least k vertices of degree at most 5.* Could it hold for *every* k?

5. *Consider a maximal triangle-free planar graph $G = (V, E)$, i.e. a triangle-free planar graph such that any graph of the form $G + e$, where $e \in \binom{V}{2} \setminus E$, contains a triangle or is nonplanar. Prove that each face in any drawing of such a graph is a quadrilateral or a pentagon.

6. Find an example, other than the Platonic solids, of a convex polytope in the 3-dimensional space such that all faces are congruent copies of the same regular convex polygon. *Can you list all possible examples?

7. (Game "Sprouts") The following game has been invented by J. H. Conway and M. S. Paterson. Initially, n dots are drawn on a sheet of paper (the game is already interesting for small n, say for $n = 5$). The players alternate their moves, and the player with no legal move left loses. In each move, a player connects two dots with an arc and draws a new dot somewhere on the newly drawn arc. A dot can be used as an endpoint of a new arc only if there are at most 2 other arc ends leading into that dot, and a new arc must not cross any other arcs already drawn. (So at each moment we have a planar drawing of a graph with maximum degree at most 3; the dot on the newly added arc initially has degree 2.) An example:

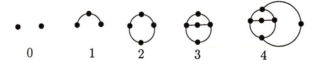

   (a) Prove that a game with n initial dots lasts no more than $3n - 1$ moves (for any strategy of the players).

   (b) *Prove that a game with n initial dots lasts at least $2n$ moves (for any strategy of the players).

(c) *("Brussels sprouts") We modify the game as follows. Instead of dots we draw little crosses, and the ends of new arcs are connected to the arms of the crosses (thus, the vertices can have maximum degree 4 this time). On each new arc, a new cross is drawn by crossing the arc with a short segment, as in the following example:

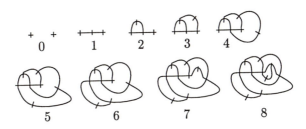

Prove that this game always has exactly $5n - 2$ moves (and so it is easy to determine who wins).

8. Consider a set L consisting of n lines in the plane. No two of them are parallel but many can pass through a single point. By drawing these lines we create vertices (intersection points of the lines), edges (parts of the lines between intersections and semiinfinite rays extending from the first and last intersections on a line to infinity), and faces (connected parts of the plane after removing the lines from it).

(a) *Express the number of edges in terms of the number of vertices and faces.

(b) Prove that unless all the lines pass through a single point, there exist at most n faces bounded by 2 (semiinfinite) edges.

(c) *Prove that unless all lines pass through a single point, there exists at least one intersection with only 2 lines passing through it. (This is the famous *Sylvester's problem.*)

9. **Consider an arbitrary topological planar graph. Suppose that every edge has one of the two colors red, blue. Show that there exists a vertex of the following type:

blue edges   red edges

(red edges form a contiguous segment when going around the vertex, and similarly for blue edges; one of the two groups can be empty).

Fig. 5.2 An example of a map for the four-color problem.

5.4 Coloring maps: the four-color problem

Consider a political map showing the boundaries of states as in Fig. 5.2. Suppose that each state is a connected region (that's why we haven't drawn islands like Britain, Ireland, Sardinia, Sicily, Corsica, etc., on our schematic map, and we also had to leave out Russia—a disconnected state in AD 1997!). We consider two regions neighbors if they have at least a small piece of border in common, so it is not sufficient if the boundaries touch in one or several points (but such situations are very rare on maps anyway). We want to color each state in such a map by some color, and no two neighboring states should get the same color, as is usual on political maps. What is the minimum necessary number of colors? For the map shown, 4 colors are sufficient (try finding such a coloring!).

Here is one of most celebrated combinatorial problems:

5.4.1 Problem (Four-color problem). Can each planar map be colored by at most 4 colors?

Four is certainly the minimum number of colors coming into consideration for coloring all planar maps. This is illustrated by the map shown above (consider Austria or Luxembourg, say) or by the following examples:

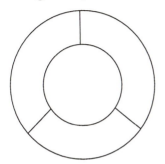 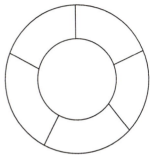

Here we prove that any planar map can be colored by 5 colors. Although this result has been known for more than 100 years, the four-color problem was only solved (positively) in the late 1970s. Known proofs of the fact that any planar map can be colored by 4 colors are difficult and they substantially depend on analyzing a large number of cases by computer. So far no one has managed to do the proof by hand, although the original proof was greatly simplified in 1995 (Robertson et al. [40]). But some basic ideas of these complicated proofs appear in two proofs of the "five-color theorem" given below.

The four-color problem looks like a geometric question, but it can be reformulated in a purely combinatorial manner. Coloring regions of a map can be translated to coloring vertices of a planar graph. First we define the notion of coloring for an arbitrary graph.

5.4.2 Definition (Chromatic number of a graph). *Let* $G = (V, E)$ *be a graph, and let* k *be a natural number. A mapping* $c\colon V \to \{1, 2, \ldots, k\}$ *is called a coloring of the graph[10]* G *if* $c(x) \neq c(y)$ *holds for every edge* $\{x, y\} \in E$. *The chromatic number of* G, *denoted by* $\chi(G)$, *is the minimum* k *such that there exists a coloring* $c\colon V(G) \to \{1, 2, \ldots, k\}$.

The chromatic number of a graph belongs among the most important combinatorial notions. However, in this book we mention it only in this section.

Mathematically, a map can be regarded as a drawing of a planar graph. The states on the map are faces of the graph, the vertices of

---

[10]Sometimes this is called a *proper coloring* of G.

the graph are the points lying on borders of 3 or more states, and the edges of the graph are the portions of the borders between vertices:

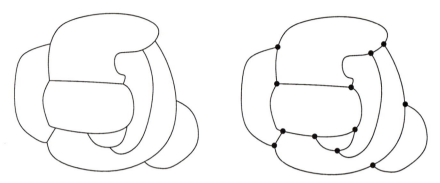

Note that multiple edges may arise in the graph corresponding to a map.

In order to convert the map-coloring problem (i.e. coloring the faces of a topological planar graph) to a problem of coloring graph vertices in the sense of the definition just given, we introduce the notion of the so-called dual graph. Imagine that we mark the capital city of each state in the considered map by a dot, and that the capitals of each two neighboring states are connected by a highway lying in the territory of these two states and crossing their common boundary. The vertices of the dual graph are the capitals and the edges are the highways.

For defining the dual graph formally, we need also multiple edges and loops, so let us recall one possible way of introducing them. A graph with multiple edges and loops can be represented as a triple (V, E, ε), where V and E are disjoint sets and $\varepsilon \colon E \to \binom{V}{2} \cup V$ is a mapping assigning to each edge its two endpoints and to each loop its single endpoint (see Section 3.4 for a more detailed discussion). Now we can give a mathematical definition of the dual graph:

5.4.3 Definition (Dual graph). *Let G be a topological planar graph, i.e. a planar graph (V, E) with a fixed planar drawing. Let $\mathcal{F}$ denote the set of faces of G. We define a graph, possibly with loops and multiple edges, of the form $(\mathcal{F}, E, \varepsilon)$, where ε is defined by $\varepsilon(e) = \{F_i, F_j\}$ whenever the edge e is a common boundary of the faces F_i and F_j (we also permit $F_i = F_j$, in the case when the same face lies on both sides of a given edge). This graph $(\mathcal{F}, E, \varepsilon)$ is called the (geometric) dual of G, and it is denoted by G^*.*

An example:

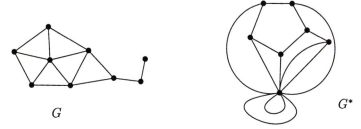

G G^*

The dual graph G^* can be drawn together with the drawing of the graph G, as was already suggested in the above informal explanation with capitals and highways. We choose a point b_F inside each face F of G, and for each edge e of G we draw an arc crossing e and connecting the points b_F and $b_{F'}$, where F and F' are the faces adjacent to the edge e. This arc lies completely in the faces F and F'. In this way, we obtain a planar drawing of G^*:

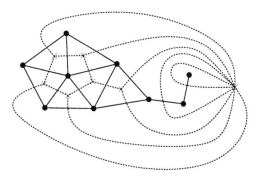

This way of drawing the dual graph witnesses its planarity. Further examples of dual graphs can be found in Fig. 5.1. The graphs of the cube and of the regular octahedron are dual to each other, similarly the graphs of the regular dodecahedron and icosahedron are mutually dual, and finally the graph of the tetrahedron (i.e. K_4) is dual to itself.

Consider a planar map, and regard it as a drawing of a planar graph G in the manner indicated above. The colorability of this map by k colors is then equivalent to the colorability of the vertices of the dual graph G^* by k colors.

On the other hand, any planar graph can be obtained as a subgraph of a suitable dual graph. We indicate a proof by a picture only: for example, the graph

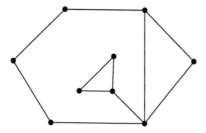

is contained in the dual of

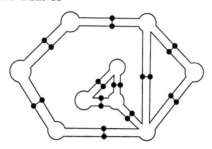

(Alternatively, one can argue that any topological planar graph is isomorphic to the dual of its dual.) Hence problems of map colorability can be reformulated as problems concerning the colorability of planar graphs. In particular, we can rephrase:

5.4.4 Problem (Four-color problem again). Does $\chi(G) \leq 4$ hold for every planar graph G?

We prove a weaker result:

5.4.5 Proposition (Five-color theorem). *Any planar graph G satisfies $\chi(G) \leq 5$.*

First proof. We proceed by induction on the number of vertices of the graph $G = (V, E)$. For $|V| \leq 5$, the statement holds trivially.

By the results of Section 5.3 we know that any planar graph has a vertex v of degree at most 5. If we even have $\deg_G(v) < 5$ then consider the graph $G - v$, and apply the inductive hypothesis on it. Assuming that the graph $G - v$ is colored by colors $1,2,\ldots,5$, then we color the vertex v by some color $i \in \{1, 2, \ldots, 5\}$ not occurring among the (at most 4) colors used on the neighbors of v. In this way, we get a coloring of G by 5 colors.

A very similar argument already shows that the chromatic number of every planar graph is at most 6, so in the rest of the proof we work on improving this 6 to 5.

Also, it may be instructive to formulate an algorithm for coloring every planar graph by at most 6 colors (or, with the improvement below, by at most 5 colors) based on this proof. Note that such an algorithm colors the low-degree vertex v last, after all other vertices have been colored!

It remains to investigate the case when $\deg_G(v) = 5$. Let us consider the graph G with some fixed planar drawing, and let t, u, x, z, y be the vertices connected to v by an edge, listed in the order the corresponding edges emanate from the vertex v (in the clockwise direction, say).

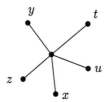

Let us again consider a coloring $c\colon V(G - v) \to \{1, 2, \ldots, 5\}$ of the graph $G - v$ by 5 colors guaranteed by the inductive assumption. If at most 4 colors occur at the neighbors of v then the vertex v can be assigned a color distinct from the colors of all its neighbors. Thus, we suppose that the neighbors of v have all the 5 distinct colors. Consider the vertices x and y, and define $V_{x,y}$ as the set of all vertices of the graph $G - v$ having color $c(x)$ or $c(y)$. Clearly x, $y \in V_{x,y}$. It may be that there exists a path from x to y in the graph $G - v$ using only the vertices of the set $V_{x,y}$, or such a path need not exist. We distinguish these two cases.

1. No such path exists. Let $V'_{x,y}$ be the set of all the vertices $s \in V(G - v)$ that can be reached from x by a path using only the vertices from $V_{x,y}$. In particular, we have $y \notin V'_{x,y}$. We define a new coloring c' of the graph $G - v$:

$$c'(s) = \begin{cases} c(s) & \text{if } s \notin V'_{x,y} \\ c(y) & \text{if } s \in V'_{x,y} \text{ and } c(s) = c(x) \\ c(x) & \text{if } s \in V'_{x,y} \text{ and } c(s) = c(y) \end{cases}$$

(this means that we exchange the colors on the set $V'_{x,y}$). It is easy to see that c' is a coloring of $G - v$ again, and since $c'(x) = c'(y) = c(y)$, we can set $c'(v) = c(x)$, obtaining a valid coloring of G by 5 colors in this way.

2. Next, suppose that there exists a path P from x to y with all vertices in $V_{x,y}$. In this case, we consider the pair of vertices t and z and we define a set $V_{t,z}$ as the set of vertices of $G - v$ colored by the colors $c(t)$ and $c(z)$. The sets $V_{x,y}$ and $V_{t,z}$ are disjoint. The drawing of the path P forms, together with the edges $\{v, x\}$ and $\{v, y\}$, a cycle:

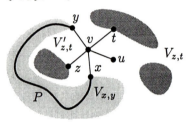

One of the points z, t lies inside this cycle and the other one outside, and hence any path from z to t has to use some vertex of the cycle. Hence there is no path from z to t only using vertices of the set $V_{z,t}$, and so we can construct a coloring of G by 5 colors in the same way as in case 1, only with z and t in the role of x and y.

This finishes the first proof of the five-color theorem. $\qquad\square$

Edge contraction. Before commencing a second proof, let us introduce one more important graph operation. Let $G = (V, E)$ be a graph (not necessarily planar at this moment) and let $e \in E$ be one of its edges. The *contraction* of e means that we "glue together" both vertices of e into a single new vertex, and then we remove multiple edges that may have arisen by this gluing. The resulting graph is denoted by $G.e$, and formally it is defined as follows:

$$G.e = (V', E'),$$

where $e = \{x, y\}$ and

$$V' = (V \setminus \{x, y\}) \cup \{z\}$$
$$E' = \{e \in E : e \cap \{x, y\} = \emptyset\}$$
$$\cup \{\{z, t\} : t \in V \setminus \{x, y\}, \{t, x\} \in E \text{ or } \{t, y\} \in E\};$$

here $z \notin V$ denotes a new vertex.

Lemma. *If G is a planar graph and $e \in E(G)$ is an edge then the graph $G.e$ is planar as well.*

Informal proof. By a picture:

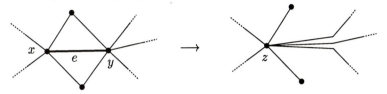

$\square$

Second proof of Proposition 5.4.5. We again work by induction
on the number of vertices of the considered planar graph $G = (V, E)$.
We begin as in the first proof. So we may assume that $G = (V, E)$ is a
planar graph with at least 6 vertices in which each vertex has degree
at least 5. Let us choose a vertex v of degree 5. Since the graph G
is planar, it contains no K_5 as a subgraph, and hence there exists
a pair of neighbors of v not connected by an edge. Let us denote
the vertices in some such pair by x and y, and let t, u, and z be
the remaining 3 neighbors. We look at the graph G' produced from
G by contracting the edges $\{x, v\}$ and $\{y, v\}$ (i.e. the triple x, y, v
of vertices is replaced by a single new vertex w; this generalizes the
definition of contraction of a single edge in an obvious manner). This
graph is planar and has fewer vertices than G. Hence some coloring c'
of G' by 5 colors exists by the inductive hypothesis. In this situation,
we define a coloring c of the graph G as follows:

$$c(s) = \begin{cases} c'(s) & \text{if } s \notin \{x, y, v\} \\ c'(w) & \text{if } s = x \text{ or } s = y \\ i \in \{1, \ldots, 5\} \setminus \{c'(w), c'(u), c'(t), c'(z)\} & \text{if } s = v. \end{cases}$$

The picture below illustrates this definition:

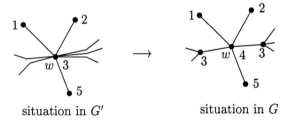

situation in G' situation in G

It is not difficult to check that the c thus defined is a valid coloring
of the graph G. $\square$

Yet another proof is outlined in Exercise 13.

Let us remark that the question of colorability can also be studied for graphs that can be drawn on other types of surfaces. This question is completely solved by now: the maximum possible chromatic number of a graph of genus k equals

$$\left\lfloor \frac{7 + \sqrt{1 + 48k}}{2} \right\rfloor .$$

It is remarkable that the case of genus 0, i.e. of planar graphs, is by far the most difficult one. For larger genus, this result (Heawood's formula) was proved much earlier than the four-color theorem.

Exercises

1. Prove that $\chi(G) \le 1 + \max\{\deg_G(x) : x \in V\}$ holds for every (finite) graph $G = (V, E)$.

2. For a graph G, put $\delta(G) = \min\{\deg_G(v) : v \in V\}$ (the *minimum degree* of G). Prove $\chi(G) \le 1 + \max\{\delta(G') : G' \subseteq G\}$, where $G' \subseteq G$ means that G' is a subgraph of G.

3. *Call a graph G *outerplanar* if a drawing of G exists in which the boundary of one of the faces contains all the vertices of G (we can always assume the outer face has this property). Prove that every outerplanar graph has chromatic number at most 3.

4. Let G be a planar graph containing no K_3 as a subgraph. Prove $\chi(G) \le 4$. (A difficult theorem due to Grötsch asserts that, actually, $\chi(G) \le 3$ holds for all planar triangle-free graphs.)

5. ^{CS} Based on one of the proofs of the five-color theorem, design an algorithm for coloring a given planar graph by at most 5 colors. Assume that a planar drawing of the graph is given in the following form: for each vertex of G, we have a circular list of the neighbors of v, listed in the order of the corresponding outgoing edges in the drawing.

6. ^{CS}

    (a) Consider a greedy algorithm for graph coloring. Pick one of the yet uncolored vertices arbitrarily and color it with the smallest color (the colors are natural numbers $1, 2, \ldots$) not used on any of its already colored neighbors. For each number K, find a graph G having a coloring by 2 colors (i.e. bipartite) but such that the algorithm just described sometimes colors G by at least K colors.

    (b) Can you find a planar G as the example in (a)?

    (c) The algorithm can be made more sophisticated as follows: among the yet uncolored vertices, pick one with the largest degree, and color it as in (a). Show that graphs G as in (a) still exist.

(d) This can be continued as a game. Participants propose versions of the greedy algorithm for graph coloring, and others try to find graphs for which the algorithm performs badly.

Remark. It is known that no polynomial-time algorithm can do approximate coloring very well, for instance to color all 3-colorable graphs on n vertices by at most K colors for some constant K, unless polynomial-time algorithms for exact coloring and many other difficult problems exist, which is considered very unlikely.

7. Consider a map M where each state has at most k connected regions (a more realistic model of the world situation). Using Exercise 2, show that the chromatic number of every such map is no larger than $6k$ (for each state, we insist that all its regions be colored by the same color).

8. The following picture is an example of a map where each state has 2 regions:

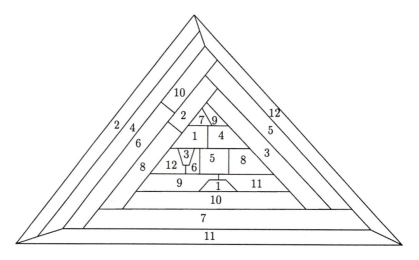

Prove that the chromatic number of this map is 12.

9. Prove that each graph G has at least $\binom{\chi(G)}{2}$ edges.

10. (a) *Consider a planar graph G with all degrees even. Prove that the map arising from (any) planar drawing of G can be colored by 2 colors.

(b) *Using (a), prove that there exists no topological planar graph with all degrees even such that all the inner faces are triangles (i.e. bounded by cycles of length 3) and the outer face is a pentagon.

11. *Let us say that a graph H is a *minor* of a graph G if a graph isomorphic to H can be obtained from G by a repeated application of the following operations: deleting an edge, deleting an isolated vertex, contracting an edge. Deduce the following result from Kuratowski's

theorem: *A graph G is planar if and only if neither K_5 nor $K_{3,3}$ is a minor of G.*

12. (List chromatic number)   Let $G = (V, E)$ be a graph, and suppose that a finite list $L(v)$ of natural numbers is given for each vertex $v \in V$. A mapping $c: V \to \mathbb{N}$ is called a *list coloring* (with respect to the given lists $L(v)$, $v \in V$) if $c(v) \in L(v)$ holds for all $v \in V$ and $c(v) \neq c(v')$ whenever $\{v, v'\} \in E$. The *list chromatic number* of G, denoted by $\chi_\ell(G)$, is the smallest number k such that G has a list coloring for any collection of lists $L(v)$ with $|L(v)| \leq k$ for all $v \in V$.

    (a) Find an example of a graph with $\chi_\ell(G) > \chi(G)$.

    (b) *For each integer k, construct a bipartite graph G with $\chi_\ell(G) > k$.

    Remark. There exists a planar graph G with $\chi_\ell(G) = 5$.

13. *The notion of list coloring from the previous exercise is used in a short and remarkable proof of the five-color theorem 5.4.5 due to Thomassen. To reconstruct that proof, establish the following statement by induction on the number of vertices:

    Let $G = (V, E)$ be a topological planar graph and let $L(v)$, $v \in V$, be lists with the following properties:

    - The boundary of the unbounded face is a cycle C with vertices $v_1, v_2, \ldots, v_k$ (numbered along the circumference).
    - All bounded faces are triangles.
    - $L(v_1) = \{1\}$, $L(v_2) = \{2\}$.
    - $|L(v_i)| = 3$ for $i = 3, 4, \ldots, k$.
    - $|L(v)| = 5$ for all vertices distinct from the v_i (i.e. lying in the interior of the bounding cycle C).

    Then G has a list coloring with respect to the lists $L(v)$.

    This proof uses neither Euler's formula nor the existence of a low-degree vertex, and it is an example of mathematical induction *par excellence*.

6
Double-counting

In pre-computer ages, double-counting was used by accountants. When adding up the numbers in a table, they first found the sum of the row totals and then they compared it to the sum of the column totals. If their computation was correct, both results were the same. Put mathematically, if A is an $n \times m$ matrix then

$$\sum_{i=1}^{n}\sum_{j=1}^{m} a_{ij} = \sum_{j=1}^{m}\sum_{i=1}^{n} a_{ij}.$$

Pictorially,

$$\boxed{\equiv} \; = \; \boxed{|||||}$$

In other words, the order of summation in a double sum like this may be changed. This simple idea underlies many mathematical tricks and proofs; in more complex cases it is accompanied by other ideas, of course. The main difficulty is usually in figuring out what exactly should be double-counted.

6.1 Parity arguments

In Section 3.3, we encountered the handshake lemma: *Any graph has an even number of odd-degree vertices.* The proof was in fact a typical double-counting (we double-counted "ends of edges"). Using this claim, we were able to exclude some vectors as possible graph scores—for instance, the vector $(3, 3, 3, 3, 3)$. But there are much more interesting ways to use the handshake lemma. For instance, we can sometimes prove the existence of a certain object. To this end, we reformulate the claim slightly: *If we know that a graph G has at least one vertex of an odd degree, then it must have at least two such vertices.* Next, we demonstrate a nice application.

Let us draw a big triangle in the plane with vertices A_1, A_2, A_3. We divide it arbitrarily into a finite number of smaller triangles, as in the following picture:

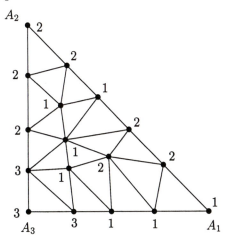

No triangle may have a vertex inside an edge of any other small triangle, so that if we consider the resulting picture as a drawing of a planar graph, all inner faces are triangular. Let us assign the labels 1, 2, 3 to the vertices of the big and the small triangles, under the following rules: the vertex A_i gets the label i, $i = 1, 2, 3$, and all vertices lying on the edge $A_i A_j$ of the big triangle may be assigned only the label i or j. Otherwise the assignment is completely arbitrary.

6.1.1 Proposition (Sperner's lemma—a planar version). *In the situation described above, a small triangle always exists whose vertices are assigned all the three labels 1, 2, 3.*

Proof. We define an auxiliary graph G; see Fig. 6.1. Its vertices are the faces of our triangulation, i.e. all small triangles plus the outer face. In the figure, the vertices are depicted as little black triangles inside the corresponding faces. The vertex for the outer face is denoted by v. Two vertices, i.e. faces of the original drawing, are joined by an edge in G if they are neighboring faces and the endpoints of their common edge have labels 1 and 2. This also concerns the outer face vertex v: it is connected to all small triangles adjacent to the circumference of the big triangle by a side labeled 12.

A small triangle can be connected to some of its neighbors in this graph G only if one of its vertices is labeled by 1 and another by 2. If the remaining vertex is labeled 1 or 2, the considered small

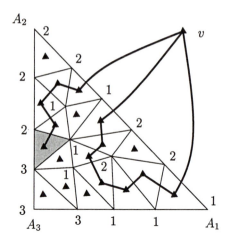

Fig. 6.1 Illustration for the proof of Sperner's lemma.

triangle is adjacent to exactly two of its neighbors. If the remaining vertex has label 3, then the considered triangle is adjacent to exactly one neighbor, and *this is the only case where the degree of a small triangle in the graph G is odd.* We now show that the vertex v (the outer face) has an odd degree in G. Then, by the handshake lemma, there exists at least one other vertex of odd degree in G, and this is the desired small triangle labeled 1, 2, 3.

The edges of the graph G incident to v can obviously only cross the side A_1A_2 of the big triangle. By the rules of the labeling, this side only contains vertices labeled by 1 or 2. Let us write down the sequence of these labels, starting at A_1 and ending at A_2. The number of neighbors of v is just the number of alterations between 1 and 2 in this sequence (the number of times a segment of 1s ends and a segment of 2s starts or the other way round). Since the sequence begins with 1 and ends with 2, the number of such alterations must be odd. Hence v has an odd degree in G. ◻

Sperner's lemma is not a toy problem only; it is a crucial step in the proof of a famous theorem. Before we state this theorem, let us mention a simpler theorem of a similar kind as a warm-up.

6.1.2 Proposition (One-dimensional fixed point theorem). *For any continuous function $f\colon [0,1] \to [0,1]$, there exists a point $x \in [0,1]$ such that $f(x) = x$.*

Such an x is called a *fixed point* of the function f. The proposition can be proved by considering the function $g(x) = f(x) - x$. This is a continuous function with $g(0) \geq 0$ and $g(1) \leq 0$. Intuitively it is quite clear that the graph of such a continuous function cannot jump across the x-axis and therefore it has to intersect it, and hence g is 0 at some point of $[0, 1]$. Proving the existence of such a point rigorously requires quite some work. In analysis, this result appears under the heading "Darboux theorem".

Fixed point theorems generally state that, under certain circumstances, some function f must have a fixed point, i.e. there exists an x such that $f(x) = x$. Such theorems belong to the key results in many areas of mathematics. They often serve as a tool for proving the existence of solutions to equations of various types (differential equations, integral equations, etc.). They even play a role in the theory of the meaning of computer programs, the so-called program semantics.

In Brouwer's fixed point theorem, the 1-dimensional interval from Proposition 6.1.2 is replaced by a triangle in the plane, or by a tetrahedron in the 3-dimensional space, or by their analogs in higher dimensions (simplices). Here we prove only the 2-dimensional version since we have only proved Sperner's lemma in 2 dimensions (but see Exercise 5). The proof belongs more to mathematical analysis. However, we will try to present it using a minimum of facts and notions from analysis, and we will recall the necessary facts as we go along.

Let Δ denote a triangle in the plane. For simplicity, let us take the triangle with vertices $A_1 = (1, 0)$, $A_2 = (0, 1)$, and $A_3 = (0, 0)$:

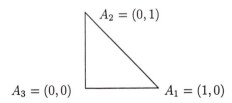

A function $f \colon \Delta \to \Delta$ is called *continuous* if for every $a \in \Delta$ and for every $\varepsilon > 0$ there exists $\delta > 0$ such that if $b \in \Delta$ is a point at distance at most δ from a then the distance of $f(a)$ and $f(b)$ is at most ε. Briefly, f maps close points to close points.

6.1.3 Theorem (Planar Brouwer's fixed point theorem). *Every continuous function $f \colon \Delta \to \Delta$ has a fixed point.*

Proof. We define three auxiliary real-valued functions β_1, β_2, and β_3 on the triangle Δ. For a point $a \in \Delta$ with coordinates (x, y), we set

$$\beta_1(a) = x, \quad \beta_2(a) = y, \quad \beta_3(a) = 1 - x - y.$$

Geometrically, the β_i are as in the following illustration:

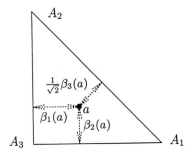

The key properties of these functions are: $\beta_i(a) \geq 0$ and $\beta_1(a) + \beta_2(a) + \beta_3(a) = 1$ for all $a \in \Delta$.

Further we define sets $M_1, M_2, M_3 \subseteq \Delta$:

$$M_i = \{a \in \Delta: \beta_i(a) \geq \beta_i(f(a))\},$$

$i = 1, 2, 3$. Thus, M_i consists of the points that are not moved farther apart from the side opposite to A_i by the function f.

Note that every point $p \in M_1 \cap M_2 \cap M_3$ is a fixed point of the function f, for if p is not fixed then f must move it away from some of the sides. In more detail, if $p \in M_1 \cap M_2 \cap M_3$ then we have $\beta_i(p) \geq \beta_i(f(p))$ for all $i = 1, 2, 3$, and since $\sum_i \beta_i(p) = \sum_i \beta_i(f(p)) = 1$ we get $\beta_i(p) = \beta_i(f(p))$ for all i, which implies $p = f(p)$. Our goal now is to find a point in the intersection $M_1 \cap M_2 \cap M_3$.

Consider a sequence of successively refining triangulations of the triangle Δ:

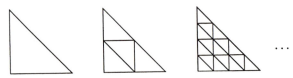

In each of these triangulations, we label all vertices of the triangles by 1, 2, or 3. We require that a vertex labeled i belongs to the set M_i, and, moreover, that the assignment satisfies the rules of Sperner's lemma. We have to make sure this can always be arranged.

The vertex A_1 has the largest possible distance from its opposite side; hence this distance cannot increase under f. Therefore $A_1 \in M_1$ and we can label A_1 by 1; similarly for A_2 and A_3. A point a lying on the side $A_1 A_2$ has $\beta_1(a) + \beta_2(a) = 1$, which implies that $f(a)$ cannot satisfy both $\beta_1(f(a)) > \beta_1(a)$ and $\beta_2(f(a)) > \beta_2(a)$. Thus $a \in M_1 \cup M_2$, and so we can label all vertices on the side $A_1 A_2$ only by 1s and 2s. A similar argument works for the other sides. Finally, each point of Δ belongs to at least one of the sets M_i since it cannot be moved farther from all three sides at once. (This time we leave a detailed verification to the reader.)

Now Sperner's lemma 6.1.1 implies that each of the successively refining triangulations has a triangle labeled 1, 2, 3. Let us denote the vertices of some such triangle in the jth triangulation by $a_{j,1}$, $a_{j,2}$, and $a_{j,3}$ in such a way that $a_{j,i}$ belongs to M_i for $i = 1, 2, 3$.

Consider the infinite sequence of points $(a_{1,1}, a_{2,1}, a_{3,1}, \ldots)$. We need to choose an infinite convergent subsequence from it. This is always possible; in fact, any infinite sequence of points inside the triangle contains a convergent infinite subsequence. (This property of the triangle—shared, for instance, by any closed and bounded subset of the plane—is called the *compactness*.) So suppose that we have chosen a convergent subsequence $(a_{j_1,1}, a_{j_2,1}, a_{j_3,1}, \ldots)$, $j_1 < j_2 < j_3 < \ldots$, and let us denote its limit point by p.

We claim that $p \in M_1$. Indeed, by the definition of M_1, we have $\beta_1(a_{j_k,1}) \geq \beta_1(f(a_{j_k,1}))$ for all j_k, and taking a limit on both sides yields $\beta_1(p) \geq \beta_1(f(p))$, because taking a limit preserves nonstrict inequalities between continuous functions.

Since the diameter of triangles in the successive triangulations tends to 0, the sequences of the other vertices, i.e. $(a_{j_1,2}, a_{j_2,2}, a_{j_3,2}, \ldots)$ and $(a_{j_1,3}, a_{j_2,3}, a_{j_3,3}, \ldots)$, also converge to the point p. This implies that $p \in M_2$ and $p \in M_3$ as well. Thus p is the desired fixed point of the function f. □

A number of more complicated results similar to Brouwer's fixed point theorem are known. For example, under any continuous mapping of the surface of a 3-dimensional ball to the plane, some two points at opposite ends of a diameter of the ball are mapped to the same point (the so-called Borsuk–Ulam theorem). Such theorems are proved in a branch of mathematics called *algebraic topology*. The proofs typically employ fairly complex techniques, but deep down they are often based on parity arguments similar to Sperner's lemma.

Let us present another example illustrating the use of the handshake lemma. We will analyze a game similar to the game called HEX. It takes place on a board like the one shown in Fig. 6.2 (but the triangulation inside the outer square may be arbitrary). The players take turns. Each player in turn marks an unmarked node with her symbol. For example, the first player (Alice) paints nodes gray and the second player (Betty) black (boring colors but we can illustrate them in a black-and-white book). In the starting position, Alice has the nodes a and c marked, and Betty has b and d. Alice wins if she manages to mark all nodes of a path from a to c, and Betty's goal is a path from b to d. If a player is supposed to make a move and has no more nodes to mark, the game ends in a draw.

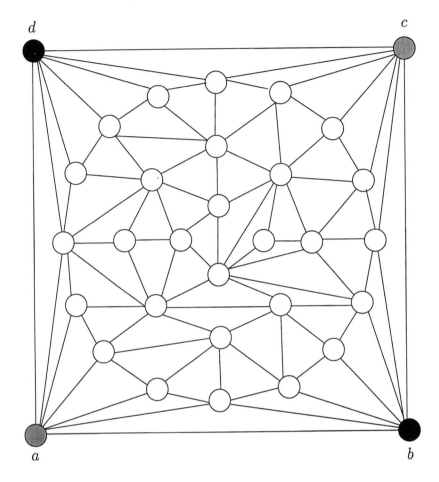

Fig. 6.2 A game board.

6.1.4 Proposition. *On a board of the given type (the outer face is a square, all inner faces are triangles), a draw is impossible.*

Proof. Assume for contradiction that a draw has occurred. Let A be the set of nodes marked by Alice and let B be the set of nodes marked by Betty.

Let us assign labels 1, 2, 3 to nodes according to the following rules. A node in A is labeled by 1 if it can be connected to a by a path with all vertices belonging to A. Similarly, nodes in B connected to b by a path lying entirely in B are labeled by 2. The remaining nodes get label 3. By the hypothesis, both c and d are labeled by 3, otherwise one of the players would have won.

We will show that there is an inner triangular face T labeled 1, 2, 3. This leads to a contradiction, since the node of T labeled 3 (call it x) can belong neither to A nor to B. Indeed, if x belongs to A, we consider the node y of the triangle T labeled by 1. By definition of the labeling, there is a path from a to y using only nodes of A, and this path could be extended to x, since y is adjacent to x. For a similar reason x does not belong to B, which is a contradiction. This reasoning is illustrated by the following picture:

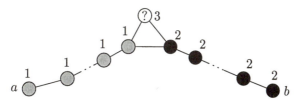

How do we prove the existence of a triangle labeled 1, 2, 3? Exactly like Sperner's lemma. We leave it to the reader. Another very similar proposition is presented in Exercise 1. □

Let us remark that this proof makes substantial use of the fact that all inner faces of the board are triangular and the outer face has just the 4 vertices a, b, c, d. If we allow, for instance, quadrilateral inner faces, the game might possibly end in a draw.

Various games from a mathematical point of view are investigated in the interesting book by Berlekamp, Conway, and Guy [13].

Exercises

1. Consider a drawing of a planar graph all of whose faces, including the outer one, are triangular (i.e. have 3 vertices). To each vertex we assign, quite arbitrarily, one of the labels 1, 2, 3. Prove that there are an even number of faces whose vertices get all 3 labels.

2. A well-known problem about a tourist climbing a mountain also relates to fixed points. A tourist starts climbing a mountain at 6 in the morning. He reaches the summit at 6pm and spends the night there (in a shelter built there especially for that purpose). At 6 the next morning he starts descending along the same trail. He often pauses to contemplate the view, and so he reaches the starting point at 6 in the evening again. Prove that there is a place on the trail that he passed through at the same time on both days.

3. A building engineer is standing in a freshly finished apartment and holding a floor plan of the same apartment. Prove that some point in

the plan is positioned exactly above the point on the apartment's floor it corresponds to.

4. Decide for which of the various sets X listed below is an analog of Brouwer's fixed point theorem 6.1.3 valid (i.e. every continuous function $f: X \to X$ has a fixed point). If it is valid, derive it from Theorem 6.1.3, and if not, describe a function f witnessing it.

   (a) X is a circle in the plane (we mean the curve, not the disk bounded by it);

   (b) X is a circular disk in the plane;

   (c) X is a triangle in the plane with one interior point removed;

   (d) X is a sphere in the 3-dimensional space (a surface);

   (e) X is a sphere in the 3-dimensional space with a small circular hole punctured in it;

   (f) X is the torus (see Section 5.1);

   (g) X is the Klein bottle (see Section 5.1).

5. (Sperner's lemma in dimension 3)

   (a) *Consider a tetrahedron $T = A_1 A_2 A_3 A_4$ in the 3-dimensional space and some subdivision of T into small tetrahedra, such that each face of each small tetrahedron either lies on a face of the big tetrahedron or is also a face of another small tetrahedron. Let us label the vertices of the small tetrahedra by labels 1, 2, 3, 4, in such a way that the vertex A_i gets i, the edge $A_i A_j$ only contains vertices labeled i and j, and the face $A_i A_j A_k$ has only labels i, j, and k. Prove that there exists a small tetrahedron labeled 1, 2, 3, 4.

   (b) Formulate and prove a 3-dimensional version of Brouwer's fixed point theorem (about continuous mappings of a tetrahedron into itself).

6. Consider a game as in Proposition 6.1.4.

   (a) Prove that on any board meeting the condition of Proposition 6.1.4, either Alice or Betty has a winning strategy (i.e. if she does not make a mistake she wins, regardless of how the other player plays).

   (b) Find an example of a game board (meeting the condition again) such that Betty has a winning strategy.

   (c) *Show that if the board is symmetric with respect to rotation by 90 degrees about the center (i.e. if it looks the same from both players' points of view), then Alice always has a winning strategy.

7. The real game HEX, as the reader might know, was invented by Piet Hein and it is played on a board like this:

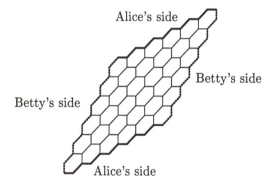

Alice's side

Betty's side

Betty's side

Alice's side

Alice's goal is to connect her two sides by a contiguous chain of fields occupied by her pieces, and Betty wants to connect her sides.

(a) Discover and explain the connection of this game to the version discussed in the text.

(b) Having no HEX game board at hand, Alice and Betty started playing HEX on an ordinary chessboard. Alice tried to connect two opposite sides and Betty the other two sides, with two squares considered adjacent if they share a side. Soon they found the game pretty boring. Can you guess why?

6.2 Sperner's theorem on independent systems

Yes, this is the second time that we meet the name Sperner in this chapter, but Sperner's theorem, which we consider next, deals with something quite different than Sperner's lemma of the previous section. It is about an n-element set X and a system $\mathcal{M}$ of its subsets. We call the system $\mathcal{M}$ *independent* if it contains no two different sets A, B such that $A \subset B$. Before reading further, you may want to try finding an independent set system on a 4-element set with as many sets as possible. How many sets can you get?

6.2.1 Theorem (Sperner's theorem). *Any independent system of subsets of an n-element set contains at most $\binom{n}{\lfloor n/2 \rfloor}$ sets.*

This is in fact a theorem on partially ordered sets (posets). Consider the set system 2^X consisting of all subsets of the set X. The relation $\subseteq$, "to be a subset of", is a partial order on 2^X (it is even one of the most important examples of a partial order; see Section 1.7). An independent system of sets is exactly a set of pairwise incomparable elements in the poset $(2^X, \subseteq)$. A set of pairwise incomparable elements in a poset is commonly called an *antichain*, and so Sperner's theorem gives an upper bound on the size of any antichain in $(2^X, \subseteq)$.

Before we start proving Sperner's theorem, we should remark that the upper bound in that theorem is certainly the best possible, because all subsets of X of size $\lfloor n/2 \rfloor$ constitute an independent system of size exactly $\binom{n}{\lfloor n/2 \rfloor}$.

Proof of Theorem 6.2.1. We first say what a *chain* of subsets of X is: it is any set $\{A_1, A_2, \ldots, A_k\}$ of subsets of X such that $A_1 \subset A_2 \subset \cdots \subset A_k$. In the language of ordered sets, it is simply a linearly ordered subset of the poset $(2^X, \subseteq)$.

The key observation is that any chain has at most one element in common with any antichain. For example, if we succeeded in proving that the whole poset in question can be expressed as a union of at most r chains, then no antichain would have more than r elements. Our proof uses this simple observation in a more sophisticated way, however.

We consider the *maximal chains* in $(2^X, \subseteq)$, where a maximal chain is a chain such that if we add to it any other set of 2^X, the result is no longer a chain. It is easy to see what the maximal chains look like: they contain one subset of X of each of the possible sizes; that is, they have the form

$$\emptyset \subset \{x_1\} \subset \{x_1, x_2\} \subset \{x_1, x_2, x_3\} \subset \cdots \subset \{x_1, x_2, \ldots, x_n\}, \quad (6.1)$$

where $x_1, x_2, \ldots, x_n$ are all elements of X written out in some arbitrary order. Every maximal chain therefore induces a linear ordering of elements of X, and, on the other hand, every linear ordering yields exactly one maximal chain. As a result, the number of maximal chains equals the number of permutations of X, i.e. $n!$.

Let $\mathcal{M}$ be an antichain (an independent system of subsets). Form all ordered pairs $(\mathcal{R}, M)$, where $M \in \mathcal{M}$ is a set and $\mathcal{R}$ is a maximal chain containing M. We count such pairs in two ways.

First, by the observation mentioned above, every chain contains at most one $M \in \mathcal{M}$ (because $\mathcal{M}$ is an antichain), so the number of pairs $(\mathcal{R}, M)$ is less than or equal to the number of maximal chains, which is $n!$.

On the other hand, we can take a set $M \in \mathcal{M}$ and ask how many maximal chains contain it. A maximal chain of the form (6.1) contains M if and only if $\{x_1, x_2, \ldots, x_k\} = M$, where $k = |M|$. Hence we ask how many linear orderings of X there are such that the first k elements are just the elements of M. We can still order the elements of M in $k!$ ways, thus determining the first k sets of the chain, and the elements outside M can be ordered in $(n - k)!$ ways, which determines the rest of the chain. Altogether M is contained in $k!(n - k)!$ maximal chains. So the number of ordered pairs $(\mathcal{R}, M)$ is equal to

$$\sum_{M \in \mathcal{M}} |M|!(n - |M|)!,$$

while according to the first way of counting, it is at most $n!$. Dividing the resulting inequality by $n!$, we obtain

$$\sum_{M \in \mathcal{M}} \frac{|M|!(n - |M|)!}{n!} = \sum_{M \in \mathcal{M}} \frac{1}{\binom{n}{|M|}} \leq 1. \tag{6.2}$$

We use the fact that $\binom{n}{\lfloor n/2 \rfloor}$ is at least as large as any binomial coefficient of the form $\binom{n}{k}$, $k = 0, 1, \ldots, n$. Therefore

$$1 \geq \sum_{M \in \mathcal{M}} \frac{1}{\binom{n}{|M|}} \geq |\mathcal{M}| \frac{1}{\binom{n}{\lfloor n/2 \rfloor}},$$

and hence $|\mathcal{M}| \leq \binom{n}{\lfloor n/2 \rfloor}$. $\qquad \square$

The remarkable inequality (6.2) is called the LYM inequality, after its (independent) discoverers Lubell, Meshalkin, and Yamamoto.

Another proof of Sperner's theorem. As with many other important theorems, Sperner's theorem can be proved in several essentially different ways. From each of the various proofs one can learn something new, or just enjoy the beautiful ideas. We will describe two more proof methods. The first one cleverly covers 2^X with chains of a special type.

Let us consider a chain in the poset $(2^X, \subseteq)$, i.e. a sequence of sets in inclusion: $M_1 \subset M_2 \subset \cdots \subset M_t$. Call such a chain *symmetric* if it contains one set of size k, one set of size $k + 1$, one set of size $k + 2, \ldots$, one set of size $n - k$, and no other sets (for some number k). For example, for $n = 3$, the chain consisting of the sets $\{2\}$ and $\{2, 3\}$ is symmetric, as well as the chain $\{\emptyset, \{3\}, \{2, 3\}, \{1, 2, 3\}\}$, but the chains $\{\{1\}\}$ and $\{\emptyset, \{1, 2, 3\}\}$ are not symmetric in this sense. A *partition into symmetric chains* is a way of expressing 2^X as a union of several disjoint symmetric chains.

Any partition into symmetric chains (if it exists at all) has to consist of exactly $\binom{n}{\lfloor n/2 \rfloor}$ symmetric chains, because each symmetric chain contains exactly one set of size $\lfloor n/2 \rfloor$. Each chain has at most one set in common with any independent set system (this was the basic observation in the first proof of Sperner's theorem). Hence Sperner's theorem is a consequence of the following:

Claim. *For any finite set X, the system 2^X has a partition into symmetric chains.*

Proof of the claim. We may assume that $X = \{1, 2, \ldots, n\}$. The proof is based on the following construction:

To each set $M \subseteq X$, we assign a sequence "$m_1 m_2 \ldots m_n$" consisting of left and right parentheses by the rule

$$m_i = \begin{cases} \text{"("} & \text{if } i \in M \\ \text{")"} & \text{if } i \notin M. \end{cases}$$

For example, for $n = 7$ and the set $M = \{2, 6\}$, we get the sequence "$m_1 m_2 \ldots m_7$"$=$ ")()))()". The resulting sequence of parentheses is quite general, and certainly it need not be a "correct" parenthesizing, i.e. the left and right parentheses need not match. We can get a "partial pairing of parentheses" from it, however. First, we pair up all pairs "()" of adjacent parentheses. Then we ignore these already paired parentheses, and we continue pairing up the remaining parentheses according to the same rule. Here are two examples:

$$) \underbrace{(\,)})\,) \underbrace{(\,)}$$

$$)\,)\,)\,(\,(\underbrace{(\,)\underbrace{(\,)}}\,)\,($$

After finishing this pairing procedure, some parentheses may remain unmatched. But the rule of the pairing implies that the sequence of the remaining unmatched parentheses has only closing parentheses at the beginning and then, from some position on, only opening parentheses.

We say that two sequences of parentheses have the *same partial pairing* if the paired parentheses are the same in both sequences (also in the same positions). This is the case for the sequences corresponding to the 3 sets below (the sets are regarded as subsets of $\{1, 2, \ldots, 11\}$):

$$
\begin{array}{ll}
M_1 = \{4, 5, 6, 8, 11\} & \ldots \quad)\,)\,)\,(\,(\,(\,)\,(\,)\,)\,(\\
M_2 = \{5, 6, 8, 11\} & \ldots \quad)\,)\,)\,)\,(\,(\,)\,(\,)\,)\,(\\
M_3 = \{5, 6, 8\} & \ldots \quad)\,)\,)\,)\,(\,(\,)\,(\,)\,)\,)
\end{array}
$$

The only way two sequences with the same partial pairing may differ is that one has more unmatched right parentheses on the left than the other. From this it is easy to see that two sets with the same partial pairing of the corresponding sequences of parentheses have to be in inclusion (one is a subset of the other).

We now define an equivalence $\sim$ on the set 2^X, by letting $M \sim M'$ hold if and only if both M and M' have the same partial pairing of their sequences. We claim that each class of this equivalence is a symmetric chain. We leave the verification as an easy exercise. So we have proved Sperner's theorem once more. □

Finally we demonstrate one more proof of Sperner's theorem. It is remarkably different from the previous two proofs, and it uses the highly symmetric structure of the poset $(2^X, \subseteq)$.

We begin with a general definition (which has already been mentioned in Exercise 1.7.9). Let $(X, \leq)$ and $(Y, \preceq)$ be some posets. A mapping $f: X \to Y$ is called an *isomorphism* of posets if f is a bijection and for any two elements $x, y \in X$, $x \leq y$ holds if and only if $f(x) \leq f(y)$. An isomorphism of a poset $(X, \leq)$ onto itself is called an *automorphism* of $(X, \leq)$. An automorphism preserves all properties which can be defined in terms of the ordering relation $\leq$. For example, x is the largest element[1] of some subset $A \subseteq X$ if and only if $f(x)$ is the largest element of the set $f(A)$, and so on.

Third proof of Sperner's theorem. Let X be a given n-element set. Each permutation $f: X \to X$ induces a mapping $f^{\#}: 2^X \to 2^X$ (i.e. sending subsets of X to subsets of X) given by $f^{\#}(A) = \{f(x): x \in A\}$. It is clear that $f^{\#}$ is a bijection[2] $2^X \to 2^X$, and even an automorphism of the poset $(2^X, \subseteq)$.

Let us now consider a system $\mathcal{M}$ of subsets of the set X. For each permutation f of the set X, we get the set system $\{f^{\#}(M): M \in \mathcal{M}\}$; that is, the system of images of the sets from $\mathcal{M}$ under the mapping $f^{\#}$. In this way, we have defined a new mapping

$$f^{\#\#}: 2^{2^X} \to 2^{2^X}$$

(assigning set systems to set systems) by the formula

$$f^{\#\#}(\mathcal{M}) = \{f^{\#}(M): M \in \mathcal{M}\}.$$

The mapping $f^{\#\#}$ is again a bijection.

We introduce a relation $\lhd$ on the set of all set systems on X (i.e. on the set 2^{2^X}):

$$\mathcal{M} \lhd \mathcal{N} \iff \text{for each } M \in \mathcal{M} \text{ there exists an } N \in \mathcal{N} \text{ with } M \subseteq N.$$

Note that the relation $\lhd$ is something different than the inclusion between set systems. It is a possibly larger relation than inclusion (hence

---

[1] The largest element was defined in Exercise 1.7.7.

[2] In Section 1.4, we adopted a convention according to which one can write the image of a set under some mapping in the same way as for the image of an element. That is, in our case we could write the set $\{f(x): x \in A\}$ simply as $f(A)$. In the present proof, however, it is better to distinguish more exactly between the image of an element and the image of a set. That is why we have introduced a different symbol, $f^{\#}$, for the mapping of sets.

$M \subseteq \mathcal{N}$ implies $\mathcal{M} \lhd \mathcal{N}$). The reader is invited to check that the relation $\lhd$ is reflexive and transitive, but that it need not be antisymmetric (find an example on a 3-element set X).

Let the letter Ξ stand for the set of all independent set systems on the set X (so $\Xi \subset 2^{2^X}$). We claim that the relation $\lhd$ restricted to Ξ is already antisymmetric, and consequently it is a partial ordering on Ξ. Indeed, if $\mathcal{M}$ and $\mathcal{N}$ are independent systems of sets such that both $\mathcal{M} \lhd \mathcal{N}$ and $\mathcal{N} \lhd \mathcal{M}$, we consider an arbitrary set $M \in \mathcal{M}$. The system $\mathcal{N}$ has to contain some set $M' \supseteq M$, and then $\mathcal{M}$ contains some $M'' \supseteq M'$ too. Thus we get $M, M'' \in \mathcal{M}$ with $M \subseteq M''$, and by the independence of $\mathcal{M}$ it follows that $M = M'' = M'$, and hence $M \in \mathcal{N}$. This shows that $\mathcal{M} \subseteq \mathcal{N}$, and symmetrically we obtain $\mathcal{N} \subseteq \mathcal{M}$, whence $\mathcal{M} = \mathcal{N}$. Thus $(\Xi, \lhd)$ is an ordered set.

Further we claim that for any permutation f, the mapping $f^{\#\#}$ is an automorphism of the poset $(\Xi, \lhd)$—we leave the verification to the reader (this is a good way to a real understanding of the notions like Ξ and $f^{\#\#}$).

The proof of Sperner's theorem is based on the following lemma:

Lemma. *Let $\Xi_0 \subseteq \Xi$ denote the set of the independent set systems with the largest possible number of sets. The set Ξ_0 has a largest element $\mathcal{N}_0$ with respect to the ordering by $\lhd$. This means $\mathcal{M} \lhd \mathcal{N}_0$ for all $\mathcal{M} \in \Xi_0$.*

Proof of the lemma. Since there are only finitely many set systems on X, it suffices to prove that for any two set systems $\mathcal{M}, \mathcal{M}' \in \Xi_0$, there exists a set system $\mathcal{N} \in \Xi_0$ that is larger than both $\mathcal{M}$ and $\mathcal{M}'$, i.e. $\mathcal{M} \lhd \mathcal{N}$ and $\mathcal{M}' \lhd \mathcal{N}$.

So we consider some $\mathcal{M}, \mathcal{M}' \in \Xi_0$, and we form a new set system $\bar{\mathcal{M}} = \mathcal{M} \cup \mathcal{M}'$. Since both $\mathcal{M}$ and $\mathcal{M}'$ are independent, the longest chain in $\bar{\mathcal{M}}$, with respect to the ordering of $\bar{\mathcal{M}}$ by inclusion, has at most two sets. Next, let $\bar{\mathcal{M}}_{min}$ be the system of all sets from $\bar{\mathcal{M}}$ for which $\bar{\mathcal{M}}$ contains no proper subset, or in other words the system of all sets of $\bar{\mathcal{M}}$ minimal with respect to inclusion. Similarly we introduce the system $\bar{\mathcal{M}}_{max}$ as the system of all inclusion maximal sets of $\bar{\mathcal{M}}$. We want to check that the system $\mathcal{N} = \bar{\mathcal{M}}_{max}$ belongs to Ξ_0 and satisfies both $\mathcal{M} \lhd \bar{\mathcal{M}}_{max}$ and $\mathcal{M}' \lhd \bar{\mathcal{M}}_{max}$.

Both the systems $\bar{\mathcal{M}}_{min}$ and $\bar{\mathcal{M}}_{max}$ are independent, and we have $\bar{\mathcal{M}} = \bar{\mathcal{M}}_{min} \cup \bar{\mathcal{M}}_{max}$. Clearly also $\bar{\mathcal{M}}_{max} \rhd \mathcal{M}$ and $\bar{\mathcal{M}}_{max} \rhd \mathcal{M}'$. It remains to verify that $\bar{\mathcal{M}}_{max}$ has the largest possible number of sets, i.e. $|\bar{\mathcal{M}}_{max}| = |\mathcal{M}|$. Let us note that by the independence of $\mathcal{M}$ and $\mathcal{M}'$, we get $\mathcal{M} \cap \mathcal{M}' \subseteq \bar{\mathcal{M}}_{min} \cap \bar{\mathcal{M}}_{max}$ (check!). Hence $|\bar{\mathcal{M}}_{min}| + |\bar{\mathcal{M}}_{max}| = |\bar{\mathcal{M}}_{min} \cup \bar{\mathcal{M}}_{max}| + |\bar{\mathcal{M}}_{min} \cap \bar{\mathcal{M}}_{max}| \geq |\mathcal{M} \cup \mathcal{M}'| + |\mathcal{M} \cap \mathcal{M}'| = |\mathcal{M}| + |\mathcal{M}'|$, and so if we had $|\bar{\mathcal{M}}_{max}| < |\mathcal{M}| = |\mathcal{M}'|$ then we would get $|\bar{\mathcal{M}}_{min}| > |\mathcal{M}|$ and the systems $\mathcal{M}$ and $\mathcal{M}'$ would not have the maximum possible size. This proves the lemma.

It remains to finish the third proof of Sperner's theorem. Let us consider the largest element $\mathcal{N}_0$ of the set $(\Xi_0, \lhd)$. For each permutation f of the set X, the corresponding induced automorphism $f^{\#\#}$ maps the set Ξ_0 (independent systems of maximum size) onto itself, and so it must map its unique largest element $\mathcal{N}_0$ onto itself: $f^{\#\#}(\mathcal{N}_0) = \mathcal{N}_0$. This implies, however, that if $\mathcal{N}_0$ contains at least one k-element set then it already contains all k-element sets! In other words, $\binom{X}{k} \subseteq \mathcal{N}_0$. It is impossible to add any set to the system $\binom{X}{k}$ so that it remains independent, and thus $\mathcal{N}_0 = \binom{X}{k}$. The maximality of the binomial coefficients $\binom{n}{\lfloor n/2 \rfloor}$ and $\binom{n}{\lceil n/2 \rceil}$ then implies that $\mathcal{N}_0 = \binom{X}{\lceil n/2 \rceil}$. $\qquad\square$

Exercises

1. Let us call a system $\mathcal{N}$ of subsets of X *semiindependent* if it contains no three sets A, B, C such that $A \subset B \subset C$.

   (a) Show by a method similar to the first proof of Sperner's theorem that $|\mathcal{N}| \leq 2\binom{n}{\lfloor n/2 \rfloor}$, where $n = |X|$.

   (b) Show that for odd n, the estimate from (a) cannot be improved.

2. (a) Determine the number of maximal chains in the set $\{1, 2, \ldots, 10!\}$ ordered by the divisibility relation.

   (b) Count the number of maximal antichains in the set $\{1, 2, \ldots, 5!\}$ ordered by the divisibility relation.

3. *Show that the set systems $\binom{X}{\lfloor n/2 \rfloor}$ and $\binom{X}{\lceil n/2 \rceil}$ are the only independent set systems on an n-element set X with the largest possible number of sets.

4. Determine the number of automorphisms of the poset $(2^X, \subseteq)$.

5. By modifying the third proof of Sperner's theorem, show that for any finite poset $(P, \leq)$ there exists an antichain of maximum possible size that is mapped to itself by all automorphisms of $(P, \leq)$ (i.e. it is a "fixed point" of all automorphisms).

6. Let $a_1, a_2, \ldots, a_n$ be real numbers with $|a_i| \geq 1$. Let $p(a_1, \ldots, a_n)$ be the number of vectors $(\varepsilon_1, \varepsilon_2, \ldots, \varepsilon_n)$, where $\varepsilon_i = \pm 1$, such that

$$-1 < \sum_{i=1}^{n} \varepsilon_i a_i < 1.$$

   (a) *Prove that for any $a_1, a_2, \ldots, a_n$ we have $p(a_1, \ldots, a_n) \leq \binom{n}{\lfloor n/2 \rfloor}$. (This is one of the first applications of Sperner's theorem—the so-called Littlewood–Offord problem.)

   (b) Find $a_1, a_2, \ldots, a_n$ with $p(a_1, \ldots, a_n) = \binom{n}{\lfloor n/2 \rfloor}$.

7. Let n be a natural number that is not divisible by the square of any integer greater than 1. Determine the maximum possible size of a set of divisors of n such that no divisor in this set divides another (i.e. $\max |M|$, where $x \in M \Rightarrow x|n$ and $x, y \in M$, $x \neq y \Rightarrow x$ doesn't divide y).

8. Let $(X, \leq)$ be a poset with n elements. Let c stand for the length of the longest chain in $(X, \leq)$ and a for the length of the longest antichain.

   (a) *Show that X can be expressed as a (disjoint) union of at most c antichains (this implies that there exists an antichain with at least $\lceil n/c \rceil$ elements).

   (b) **(Dilworth's theorem) Show that X can be expressed as a (disjoint) union of at most a chains.

9. (Erdős–Szekeres lemma)

   (a) *Show that for an arbitrary sequence $(a_1, a_2, \ldots, a_n)$ of pairwise distinct real numbers there exist indices $i_1, i_2, \ldots, i_k$, where $1 \leq i_1 < i_2 < \cdots < i_k \leq n$ and $k = \lceil \sqrt{n} \rceil$, such that either $a_{i_1} < a_{i_2} < \cdots < a_{i_k}$ (the terms a_{i_j} form an *increasing subsequence*), or $a_{i_1} > a_{i_2} > \cdots > a_{i_k}$ (the terms a_{i_j} form a *decreasing subsequence*). You can use Exercise 8(a).

   (b) Show that the bound on k in (a) cannot be improved in general, that is, there exists an n-element sequence containing no increasing or decreasing subsequence with more than $\lceil \sqrt{n} \rceil$ terms.

   (c) This time we consider two sequences $a = (a_1, \ldots, a_n)$ and $b = (b_1, \ldots, b_n)$ of pairwise distinct numbers. Show that indices $i_1, \ldots, i_k$, $1 \leq i_1 < \cdots < i_k \leq n$, always exist with $k = \lceil n^{1/4} \rceil$ such that the subsequences determined by them in both a and b are increasing or decreasing (all 4 combinations are allowed, e.g. "increasing in a, decreasing in b", "decreasing in a, decreasing in b", etc.).

   (d) *Show that the bound for k in (c) cannot be improved in general.

6.3 A result in extremal graph theory

In some situations, we know that a graph G, usually with many vertices, does not contain a certain graph or a graph of a certain type as a subgraph, and we are interested in finding the maximum possible number of edges G may have. For instance, if we know that a given graph on n vertices contains no cycle then it is a tree or a forest, and so it has at most $n - 1$ edges. We can also use the converse of this statement: if a graph on n vertices has at least n edges then it contains a cycle. This is a very simple example, and for other "forbidden" subgraphs the situation is more complicated; some of the corresponding problems have not yet been completely solved. The branch of mathematics concerned with questions of this type is called *extremal graph theory*. We present

Exercises

1. Prove that for any $t \geq 2$, the maximum number of edges of a graph on n vertices containing no $K_{2,t}$ as a subgraph is at most

$$\frac{1}{2}\left(\sqrt{t-1}\,n^{3/2} + n\right).$$

2. Let X be an n-element set, and let $S_1, S_2, \ldots, S_n$ be subsets of X such that $|S_i \cap S_j| \leq 1$ whenever $1 \leq i < j \leq n$. Prove that at least one of the sets S_i has size at most $C\sqrt{n}$, for some absolute constant C (independent of n).

3. Let G be a bipartite graph with vertex classes of size n and m. Suppose that G contains no $K_{2,2}$ as a subgraph. Prove that G has at most $O(m\sqrt{n}+n)$ edges. (Note that for n much larger than m, this is better than the bound from Theorem 6.3.1.)

4. (a) Prove the Cauchy–Schwarz inequality by induction on n (square both sides first).

   (b) Prove the Cauchy–Schwarz inequality directly, starting from the inequality $\sum_{i,j=1}^{n}(x_i y_j - x_j y_i)^2 \geq 0$.

5. (a) Let $f \colon \mathbf{R} \to \mathbf{R}$ be a *convex function*, i.e. for any $x, y \in \mathbf{R}$ and $\lambda \in [0,1]$ we have $f(\lambda x + (1-\lambda)y) \leq \lambda f(x) + (1-\lambda)f(y)$. Geometrically, this means that if we connect any two points on the graph of f by a segment, then no part of this segment reaches below the graph of f:

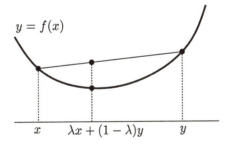

Prove (by induction) that for f convex, the inequality

$$f\left(\tfrac{1}{n}x_1 + \tfrac{1}{n}x_2 + \cdots + \tfrac{1}{n}x_n\right) \leq \tfrac{1}{n}f(x_1) + \tfrac{1}{n}f(x_2) + \cdots + \tfrac{1}{n}f(x_n) \quad (6.4)$$

holds for any real numbers $x_1, x_2, \ldots, x_n$ (this is sometimes called *Jensen's inequality*).

(b) Define a function f by the formula

$$f(x) = \begin{cases} 0 & \text{for } x \leq 1 \\ x(x-1)/2 & \text{for } x > 1. \end{cases}$$

Prove that f is convex.

(c) Prove Theorem 6.3.1 from (6.3) using (a) and (b). First derive $n \cdot f\left(\frac{m}{2n}\right) \leq \binom{n}{2}$, where $m = |E(G)|$.

6. *In a way similar to the method in Exercise 5, deduce that if a graph on n vertices does not contain $K_{3,3}$ as a subgraph then it has $O(n^{5/3})$ edges.

7. (a) *Let L be a set of n (distinct) lines in the plane and P a set of n (distinct) points in the plane. Prove that the number of pairs (p, ℓ), where $p \in P$, $\ell \in L$, and p lies on ℓ, is bounded by $O(n^{3/2})$.

Remark. It is known that the right bound is $O(n^{4/3})$; see Pach and Agarwal [25].

(b) *Show that the bound in (a) remains valid if lines are replaced by circles of unit radius. Use Exercise 1.

Remark. Here the best known upper bound is $O(n^{4/3})$ as well, but it is suspected that the number of incidences is actually bounded by $O(n^{1+\varepsilon})$ for an arbitrarily small constant $\varepsilon > 0$ (with the constant hidden in the $O(.)$ notation depending on ε). Proving ot disproving this is a very challenging open problem.

7

The number of spanning trees

7.1 The result

For a given graph G, let $T(G)$ denote the number of all spanning trees of G. For example, we have $T(K_3) = 3$ (so we really count all possible spanning trees, not just nonisomorphic ones). In this chapter, we present several proofs of the following result:

7.1.1 Theorem (Cayley's formula). *For each $n \geq 2$, the number $T(K_n)$, i.e. the number of trees on given n vertices, equals n^{n-2}.*

Although $T(K_n)$ can be expressed in such a nice and simple way, no completely straightforward method is known for deriving the formula. In the course of time, many proofs have been discovered. Their basic ideas differ substantially, but each of them involves a clever trick or follows from a nontrivial theory. We have collected several of these proofs into this chapter (by far not all known ones!), mainly as an illustration of the richness of mathematical thought.

The proof in Section 7.2 counts the number of spanning trees with a given score by a clever induction. Section 7.3 presents an elegant argument constructing a bijection between the set of all spanning trees of K_n with two marked vertices and the set of all mappings $\{1, 2, \ldots, n\} \rightarrow \{1, 2, \ldots, n\}$. Section 7.4 is a classical proof with encoding trees by sequences of length $n - 2$. Finally the method of Section 7.5 is the most advanced one and it gives perhaps the best insight into the problem. It expresses the number of spanning trees of an arbitrary graph as a certain determinant. Yet another proof is sketched in Exercise 7.2.2.

By no means do we want to claim that Theorem 7.1.1 belongs among the most fundamental results in mathematics. On the other hand, the number of spanning trees of a graph (and the related theory) has many theoretical and practical applications, and it was first studied in connection with electrical circuits.

As an illustration, let us mention without a proof an "electrotechnical" meaning of the number of spanning trees of a graph. Imagine that a given graph G is an electrical circuit, where each edge is a wire of unit resistance and the vertices are simply points where the wires are connected together. If x and y are vertices connected by an edge, then the resistance we would measure between x and y in this circuit equals the number of spanning trees of G containing the edge $\{x, y\}$ divided by the total number of spanning trees, $T(G)$. This result is not too useful directly for applications because the resistances are seldom all identical, but a generalization exists for graphs with weights on edges. More about the subject of this chapter can be found in Lovász [7] or Biggs [14].

In the whole chapter, we assume that the vertex set V of the considered graph G is the set $\{1, 2, \dots, n\}$.

Exercises

1. Prove that the number of nonisomorphic trees on n vertices is at least e^{n-1}/n^3 (see also Exercise 4.2.6!).

2. *Assume the validity of Theorem 7.1.1, and determine the number of spanning trees of the complete graph on n vertices minus one edge.

3. Put $T_n = T(K_n)$. Prove the recurrent formula

$$(n-1)T_n = \sum_{k=1}^{n-1} k(n-k)\binom{n-1}{k-1}T_k T_{n-k}.$$

   Remark. Theorem 7.1.1 can be derived from this recurrence too, but it's not so easy.

4. *Let G be a connected topological planar graph, and let G^* denote its dual (as in Definition 5.4.3). Prove that $T(G) = T(G^*)$. If convenient, you may assume that G is such that G^* has no loops and no multiple edges.

7.2   A proof via score

First we count the number of trees with a given score:

7.2.1 Proposition. *Let $d_1, d_2, \dots, d_n$ be positive integers summing up to $2n - 2$. Then the number of spanning trees of the graph K_n in which the vertex i has degree exactly d_i for all $i = 1, 2, \dots, n$ equals*

$$\frac{(n-2)!}{(d_1-1)!(d_2-1)!\cdots(d_n-1)!}.$$

Proof. By induction on n. For $n = 1, 2$, the proposition holds trivially, so let $n > 2$. Since the sum of the d_i is smaller than $2n$, there exists an i with $d_i = 1$. We now do the proof assuming that $d_n = 1$. This is just for notational convenience; exactly the same argument works for any other $d_i = 1$ (or, put differently, the number of the spanning trees of the required type obviously doesn't change by exchanging the values of d_n and d_i so we may assume $d_n = 1$ without loss of generality).

Let $\mathcal{T}$ be the set of all spanning trees of K_n with the given degrees (i.e. in which each vertex i has degree d_i). Classify the trees of $\mathcal{T}$ into $n-1$ groups $\mathcal{T}_1, \ldots, \mathcal{T}_{n-1}$: the set $\mathcal{T}_j$ consists of all trees of $\mathcal{T}$ in which the vertex n is connected to the vertex j. Next, we consider a tree from $\mathcal{T}_j$, and we delete the vertex n together with its (single) edge. We obtain a spanning tree of K_{n-1}, whose degree at the vertex i is d_i for $i \neq j$ and $d_j - 1$ for $i = j$. It is easy to see that in this way, we get a bijection between the set $\mathcal{T}_j$ and the set $\mathcal{T}'_j$ of all spanning trees of K_{n-1} with degrees $d_1, d_2, \ldots, d_{j-1}, d_j - 1, d_{j+1}, \ldots, d_{n-1}$ (since distinct trees of $\mathcal{T}_j$ give rise to distinct trees of $\mathcal{T}'_j$, and from each tree of $\mathcal{T}'_j$ we can get a tree of $\mathcal{T}_j$ by adding the vertex n back and connecting it to the vertex j).

By the inductive hypothesis, we have

$$|\mathcal{T}_j| = |\mathcal{T}'_j|$$
$$= \frac{(n-3)!}{(d_1-1)! \cdots (d_{j-1}-1)!(d_j-2)!(d_{j+1}-1)! \cdots (d_{n-1}-1)!}$$
$$= \frac{(n-3)!(d_j-1)}{(d_1-1)!(d_2-1)! \cdots (d_{n-1}-1)!}.$$

This formula also holds when $d_j = 1$—then it gives 0 which agrees with the fact that no spanning tree with degree $d_j - 1 = 0$ at the vertex j exists.

Therefore, the total number of spanning trees on n vertices with degrees $d_1, d_2, \ldots, d_n$, where $d_n = 1$, is

$$|\mathcal{T}| = \sum_{j=1}^{n} |\mathcal{T}_j| = \sum_{j=1}^{n-1} \frac{(n-3)!(d_j-1)}{(d_1-1)!(d_2-1)! \cdots (d_{n-1}-1)!}$$
$$= \left(\sum_{j=1}^{n-1} (d_j-1) \right) \frac{(n-3)!}{(d_1-1)!(d_2-1)! \cdots (d_{n-1}-1)!}$$

$$= \frac{(n-2)(n-3)!}{(d_1-1)!(d_2-1)!\cdots(d_{n-1}-1)!}.$$

Since $d_n = 1$, we can multiply the denominator by the factor $(d_n - 1)! = 0! = 1$ with no harm, and this finishes the inductive step. □

Next, we prove Theorem 7.1.1. We will sum over all possible scores of spanning trees, and we will use the multinomial theorem 2.3.5:

$$T(K_n) = \sum_{\substack{d_1,d_2,\ldots,d_n \geq 1 \\ d_1+d_2+\cdots+d_n=2n-2}} \frac{(n-2)!}{(d_1-1)!(d_2-1)!\cdots(d_n-1)!}$$

$$= \sum_{\substack{k_1+k_2+\cdots+k_n=n-2 \\ k_1,\ldots,k_n \geq 0}} \frac{(n-2)!}{k_1!k_2!\cdots k_n!} = \underbrace{(1+1+\cdots+1)}_{n\times}{}^{n-2} = n^{n-2}.$$

□

Exercises

1. (a) *Find the number of trees (on given n vertices) in which all vertices have degree 1 or 3.

   (b) *What if we allow degrees 1, 2, or 3?

2. (Yet another proof of Theorem 7.1.1) Let N_k denote the number of spanning trees of K_n in which the vertex n has degree k, $k = 1, 2, \ldots, n-1$ (recall that we assume $V(K_n) = \{1, 2, \ldots, n\}$).

   (a) *Prove that $(n-1-k)N_k = k(n-1)N_{k+1}$.

   (b) Using (a), derive $N_k = \binom{n-2}{k-1}(n-1)^{n-1-k}$.

   (c) Prove Theorem 7.1.1 from (b).

7.3 A proof with vertebrates

Consider a spanning tree of the complete graph K_n. Mark one of its vertices by a circle, and one vertex by a square, as in Fig. 7.1(a). We do not exclude the case that the same vertex is marked by both a circle and a square. Each object that can arise in this way from some spanning tree on K_n is called a *vertebrate*. Let $\mathcal{V}$ denote the set of all vertebrates (for the considered value of n).

From each given spanning tree, we can create n^2 vertebrates. Therefore the number of all spanning trees equals $|\mathcal{V}|/n^2$. We now show

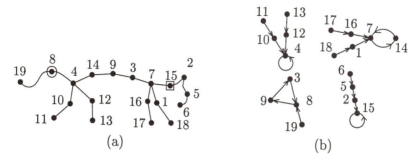

Fig. 7.1 (a) A vertebrate on 19 vertices; (b) the corresponding mapping.

Lemma. *There exists a bijection F between the set $\mathcal{V}$ of all vertebrates and the set of all mappings of the vertex set V to itself.*

Since the number of all mappings of an n-element set to itself is n^n, the number of vertebrates is the same by the lemma, and therefore the number of spanning trees is n^{n-2}.

Proof of the lemma. We demonstrate the definition of the bijection F on the example in Fig. 7.1. We start from the vertebrate W drawn in Fig. 7.1(a). The marked vertices $\square$ and $\bigcirc$ are connected by a unique path, which we call the *chord*. Let us write out the numbers of vertices of the chord ordered by magnitude. Then, we write these numbers again on the next line, in the order as they appear along the chord from $\bigcirc$ to $\square$:

$$
\begin{array}{ccccccc}
3 & 4 & 7 & 8 & 9 & 14 & 15 \\
8 & 4 & 14 & 9 & 3 & 7 & 15
\end{array}
$$

We define an auxiliary directed graph P: the vertex set consists of the vertices of the chord, and we make an arrow (directed edge) from each vertex written in the first line to the vertex below it in the second line. Since there is exactly one arrow going from each vertex and also exactly one arrow entering it, the graph P is a disjoint union of directed cycles (including possibly also isolated vertices with a directed loop). We can also say that the chord defines a permutation of its vertices, and P consists of the cycles of this permutation (see Section 2.2). In our example, these cycles are $(3, 8, 9)$, (4), $(7, 14)$, and (15).

We now look back at the whole vertebrate W. If we remove the edges of the chord from it, it splits into components (which are trees again). We direct the edges of the components so that they point to the (single) vertex of the chord contained in that component. This gives rise to one more set of directed edges on the set V. We now define a directed graph with vertex set V: its edges are all directed edges of the components, plus the edges of the graph P. In the figure, this is very intuitive. We draw the cycles of the graph P, and then we draw to each vertex (originally coming from the chord) the tree that was hanging by that vertex on the chord of the considered vertebrate; see Fig. 7.1(b).

We claim that the resulting directed graph, G, is a graph of a mapping, i.e. there is exactly one arrow going from each vertex. For the vertices of the chord, this has already been mentioned. For the other vertices, the reason is that there is a unique path to the chord from each vertex in the vertebrate W. Using the graph G, we can finally define a mapping $f: \{1, 2, \ldots, n\} \to \{1, 2, \ldots, n\}$: for each $i \in V$, we set $f(i) = j$, where j is the vertex of G that the arrow emanating from i ends in. In our specific example, we get the mapping $1 \mapsto 7$, $2 \mapsto 15$, $3 \mapsto 8$, $4 \mapsto 4$, $5 \mapsto 2$, $6 \mapsto 5$, $7 \mapsto 14$, $8 \mapsto 9$, $9 \mapsto 3$, $10 \mapsto 4$, $11 \mapsto 10$, $12 \mapsto 4$, $13 \mapsto 12$, $14 \mapsto 7$, $15 \mapsto 15$, $16 \mapsto 7$, $17 \mapsto 16$, $18 \mapsto 1$, and $19 \mapsto 8$. In this way, each vertebrate W determines a mapping $F(W)$.

It remains to prove that the original vertebrate W can be reconstructed from the mapping f produced as above, and that every mapping can be obtained from some vertebrate. This is left as an exercise. $\qquad\square$

Exercises

1. For a mapping $f: V \to V$, where V is a finite set, we define the *(directed) graph of f* as the directed graph with vertex set V and edge set $\{(i, f(i)): i \in V\}$ (such a graph was used in the proof above). Prove that each (weakly connected) component of such a graph is a directed cycle, possibly with some trees hanging at the vertices of the cycle, with edges directed towards the cycle.

2. Given a mapping $f: \{1, 2, \ldots, n\} \to \{1, 2, \ldots, n\}$ of the form $F(W)$ for some vertebrate W, describe how the vertebrate W can be reconstructed from the knowledge of f. Prove that any mapping f can be obtained as $F(W)$ for some vertebrate W (use Exercise 1).

3. *,CS Let $f: \{1, 2, \ldots, n\} \to \{1, 2, \ldots, n\}$ be a mapping. For each $i \in \{1, 2, \ldots, n\}$, the sequence $(i, f(i), f(f(i)), \ldots)$ must be eventually periodic. Design an algorithm that finds the shortest period of this sequence for a given i. That is, in the language of the directed graph

of the mapping f, we want to find the length of the directed cycle in the component containing the vertex i. The algorithm should use only an amount of memory bounded by some constant independent of n. It has a subroutine (black box) at its disposal for evaluating the function f for any given i.

4. ^{CS} (a) Design the details of an algorithm for producing a vertebrate from a mapping. How many steps does it need in the worst case? *Can you make it run in $O(n)$ time?

(b) *Program the algorithm from (a), and use it to generate random spanning trees of the complete graph from randomly generated mappings. Using this program, experimentally estimate the average (expectation) of the maximum degree and the diameter of a random spanning tree on a given number of vertices (10^4, say).

7.4 A proof using the Prüfer code

We show how each spanning tree of the complete graph K_n can be encoded by an $(n-2)$-term sequence such that each term is one of the numbers $1, 2, \ldots, n$. This coding will define a bijection between all spanning trees and all sequences of the type just described. Since the number of such sequences is obviously n^{n-2}, this will establish Theorem 7.1.1.

Consider a spanning tree T; our running example on 8 vertices is drawn in Fig. 7.2(a). We explain how to construct a sequence $p = P(T) = (p_1, p_2, \ldots, p_{n-2})$, the so-called Prüfer code of the tree T. The basic idea[1] is to tear off the leaves of T one by one until the tree is reduced to a single edge. We will thus construct an auxiliary sequence $T_0 = T, T_1, T_2, \ldots, T_{n-2} = K_2$ of trees, and produce the sequence p simultaneously. Suppose that the tree T_{i-1} has already been constructed for some i (initially we have $T_0 = T$). As we know, this T_{i-1} has at least one leaf (i.e. a vertex of degree 1). We take the smallest of the leaves of T_{i-1} (recall that the vertices of T are the numbers $1, 2, \ldots, n$), and we form T_i by removing this leaf from T_{i-1} together with the edge incident to it. At the same time, we also define the ith term, p_i, of the constructed sequence as the *neighbor* of the leaf just torn off from T_{i-1}. This is the main trick: we do not record the leaf but its neighbor! By doing this successively for $i = 1, 2, \ldots, n-2$, we have defined the whole sequence $p = P(T)$.

Now we derive how to reconstruct the original tree $T = P^{-1}(p)$ from the sequence $p = (p_1, p_2, \ldots, p_{n-2})$. More exactly, we give a rule

---

[1] A somewhat vandalic one.

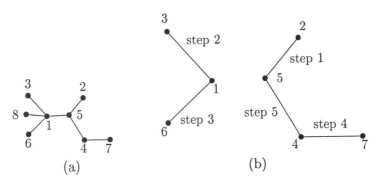

Fig. 7.2   (a) A spanning tree with code $(5, 1, 1, 4, 5, 1)$; (b) the procedure of its reconstruction from the code.

creating a spanning tree from each given sequence p, and we proceed in such a way that if p arose from some spanning tree T then our rule gives us back the tree T. This will show that the coding of spanning trees by sequences indeed defines a bijection as we claimed.

So let us suppose that a given sequence p arose by the above-described construction from some (as yet unknown) spanning tree T. Let ℓ_1 denote the leaf of T that was removed first. How can we tell ℓ_1 from the sequence p? This ℓ_1 cannot occur anywhere in the sequence p (since we wrote only vertices still present in the current tree into the sequence p). Further, any vertex not contained in the set $\{p_1, p_2, \ldots, p_{n-2}\}$ has to be a leaf of the tree T_0, for otherwise we would sooner or later tear off a leaf hanging from it, and it would appear in the sequence p at that moment. According to the rule for removing leaves, ℓ_1 must thus be the minimum element of the set $\{1, 2, \ldots, n\} \setminus \{p_1, p_2, \ldots, p_{n-2}\}$. Now we can draw the vertices ℓ_1 and p_1 and connect them by an edge (see Fig. 7.2(b)).

Further, we proceed similarly. Having found the leaves $\ell_1, \ell_2, \ldots,$ ℓ_{i-1} removed from T in the steps 1 through $i - 1$, we determine the leaf ℓ_i. It cannot be any of the vertices $p_i, p_{i+1}, \ldots, p_{n-2}$, and neither any one of $\ell_1, \ldots, \ell_{i-1}$, of course—therefore ℓ_i is the minimum of the set $\{1, 2, \ldots, n\} \setminus \{p_i, p_{i+1}, \ldots, p_{n-2}, \ell_1, \ell_2, \ldots, \ell_{i-1}\}$. We connect this ℓ_i by an edge to the vertex p_i; if ℓ_i has not been drawn yet, we draw it as well, and similarly for p_i. The first 5 steps of this construction are depicted in Fig. 7.2(b). In the 6th step, we would draw the edge $\{1, 5\}$.

After $n - 2$ steps, we have drawn $n - 2$ edges of the spanning tree T, namely all the edges that have been removed in the construction of p. It remains to deduce which edge was the last remaining one. One of its endpoints must be p_{n-2}, i.e. the neighbor of the last leaf torn off, and the other endpoint is the vertex not occurring among the removed leaves $\ell_1, \ldots, \ell_{n-2}$ and distinct from p_{n-2}. In Fig. 7.2, this edge is $\{1,8\}$. This finishes the description of the reconstruction procedure, and the considerations made along the way also confirm its correctness.

It remains to note that this algorithm executes properly for an arbitrary input sequence p, since the set from which ℓ_i is chosen is always nonempty, and thus we obtain a spanning tree of K_n for each input sequence p. So we have indeed defined a bijection. □

Exercises

1. Let T be a spanning tree of K_n, and let $p = P(T)$ be its Prüfer code. Let m_i be the number of times the vertex i appears in the sequence p, $i = 1, 2, \ldots, n$. Prove that $\deg_T(i) = m_i + 1$ holds for all i.

2. ^{CS} (a) Design the details of an algorithm for producing a spanning tree of K_n from its Prüfer code. How many steps does it need in the worst case? *Can you make it run in $O(n \log n)$ time, or even faster?

   (b) *Program the algorithm from (a). Use it as in part (b) of Exercise 7.3.4. Which of the two algorithms runs faster?

7.5 A proof working with determinants

The proof of Theorem 7.1.1 given in this section is based on linear algebra, and it illustrates a nice combinatorial meaning of the determinant (the definition and facts about determinants we need can be found in the Appendix). It is somewhat more difficult than the previous proofs and it requires basic results concerning determinants, but it provides a formula for the number of spanning trees of an arbitrary graph.

Let G be an arbitrary graph with vertices $1, 2, \ldots, n$, $n \geq 2$, and with edges $e_1, e_2, \ldots, e_m$. We introduce an $n \times n$ matrix Q, called the *Laplace matrix* of the graph G, whose elements q_{ij} are determined by the following formula:

$$
\begin{aligned}
q_{ii} &= \deg_G(i) & i = 1, 2, \ldots, n \\
q_{ij} &= \begin{cases} -1 & \text{for } \{i,j\} \in E(G) \\ 0 & \text{otherwise} \end{cases} & i, j = 1, 2, \ldots, n, \ i \neq j.
\end{aligned}
$$

Further, let Q_{ij} denote the $(n-1) \times (n-1)$ matrix arising from the matrix Q by deleting the ith row and the jth column.

The following remarkable theorem holds:

7.5.1 Theorem. *For every graph G, we have $T(G) = \det Q_{11}$.*

Let us remark that also $T(G) = |\det Q_{ij}|$ holds for any two indices $i, j \in \{1, 2, \ldots, n\}$. We do not prove this here; a proof is indicated in Exercise 1.

Before we start proving Theorem 7.5.1, we calculate the number of spanning trees of a complete graph (Theorem 7.1.1) from it. For $G = K_n$, the Laplace matrix has the number $n - 1$ everywhere on the diagonal, and -1 everywhere else. If we delete the first row and the first column, we obtain an $(n - 1) \times (n - 1)$ matrix of the form

$$\begin{pmatrix} n-1 & -1 & -1 & \cdots & -1 \\ -1 & n-1 & -1 & \cdots & -1 \\ \vdots & \vdots & \vdots & \cdots & \vdots \\ -1 & -1 & -1 & \cdots & n-1 \end{pmatrix}.$$

We calculate the determinant by suitable row and column operations. We subtract the first row from all rows except for the first one, and then we replace the first column by the sum of all columns. We get a matrix having the numbers $1, n, n, n, \ldots, n$ on the main diagonal and zeros everywhere below the main diagonal. The determinant is then the product of all diagonal elements, i.e. n^{n-2}.

We move on to the proof of Theorem 7.5.1. First we fix some arbitrarily chosen *orientation* $\vec{G}$ of the graph G, i.e. for each edge e_k, we select one of its endpoints as the *head* and the other one becomes the *tail* (see Section 3.6 for the terminology of directed graphs). This directed edge will be denoted by $\vec{e}_k$. Interestingly, we need some orientation for the proof, although the conclusion is independent of the particular orientation, and only depends on G! We define an auxiliary matrix $D = D_{\vec{G}}$, called the *incidence matrix* for the chosen orientation $\vec{G}$. This matrix has n rows, corresponding to the vertices of $\vec{G}$, and m columns, corresponding to the edges of $\vec{G}$, and it is defined as follows:

$$d_{ik} = \begin{cases} -1 & \text{if } i \text{ is the tail of } \vec{e}_k \\ 1 & \text{if } i \text{ is the head of } \vec{e}_k \\ 0 & \text{otherwise} \end{cases}$$

(recall that the edges are numbered $e_1, \ldots, e_m$). Note that the matrix D has exactly one entry 1 and one entry -1 in each column, the other entries are 0, and the sum of all rows is the zero vector.

Let us recall that if A is a matrix, the symbol A^T denotes the transposed matrix A; that is, A^T has the element a_{ji} at position (i, j). Next, let $\bar{D}$ denote the matrix arising from D by deleting the first row. Here is a connection between the matrix D and the Laplace matrix of G.

7.5.2 Lemma. *For any orientation $\vec{G}$ of the graph G, the equalities $DD^T = Q$ and $\bar{D}\bar{D}^T = Q_{11}$ hold, where $D = D_{\vec{G}}$.*

Proof. By the definition of matrix multiplication, the element at position (i, j) in the product DD^T equals $\sum_{k=1}^m d_{ik}d_{jk}$. For $i = j$, the product $d_{ik}d_{jk} = d_{ik}^2$ is 1 if i is the head or the tail of the edge $\vec{e}_k$, and it is 0 otherwise, and therefore the considered sum is just the degree of the vertex i in G. For $i \neq j$, the product $d_{ik}d_{jk}$ is nonzero only if $\vec{e}_k = (i, j)$ or $\vec{e}_k = (j, i)$, and in this case it equals -1. By comparing this with the definition of the Laplace matrix, we see that $DD^T = Q$. The second equality claimed in the lemma is a simple consequence of the definition of matrix multiplication. $\square$

The following key lemma connects spanning trees to determinants:

7.5.3 Lemma. *Let T be a graph on the vertex set $\{1, 2, \ldots, n\}$ with $n - 1$ edges ($n \geq 2$), and let $\vec{T}$ be an orientation of T. Let $C = D_{\vec{T}}$ be the incidence matrix of the directed graph $\vec{T}$, and let $\bar{C}$ denote the square matrix obtained from C by deleting its first row. Then $\det \bar{C}$ has one of the values 0, 1, -1, and it is nonzero if and only if T is a tree (this means it is a spanning tree of the complete graph on vertex set $\{1, 2, \ldots, n\}$).*

Proof. We proceed by induction on n. For $n = 2$, the situation is simple: T has one edge, so it is a spanning tree, and the single element of the matrix $\bar{C}$ is either 1 or -1.

Let us consider an arbitrary $n > 2$, and let us distinguish two cases, depending on whether any of the vertices $2, 3, \ldots, n$ has degree 1 in T.

First, suppose that such a vertex of degree 1 exists among the vertices $2, 3, \ldots, n$. Without loss of generality, we may assume that it is the vertex n (should it be another vertex, we simply renumber the vertices). The vertex n belongs to a single edge, $\vec{e}_k$. This means that the matrix $\bar{C}$ has a single nonzero element in the last row (equal to 1 or -1), namely in the kth column.

We expand the determinant of the matrix $\bar{C}$ according to the row corresponding to the vertex n (this is the $(n-1)$st row of $\bar{C}$):

$$\det \bar{C} = \sum_{j=1}^{n-1} (-1)^{n-1+j} \bar{c}_{n-1,j} \det \bar{C}_{n-1,j},$$

where $\bar{C}_{ij}$ denotes the matrix $\bar{C}$ after deleting the ith row and the jth column. Since the $(n-1)$st row has only one nonzero element, namely $\bar{c}_{n-1,k}$, we get $\det \bar{C} = (-1)^{n-1+k} \bar{c}_{n-1,k} \det \bar{C}_{n-1,k}$, and therefore $|\det \bar{C}| = |\det \bar{C}_{n-1,k}|$.

Let $\vec{T}'$ be the directed graph obtained from $\vec{T}$ by deleting the vertex n and the edge $\vec{e}_k$. The matrix $\bar{C}'$ arising from $C' = D_{\vec{T}'}$ by deleting the first row is just $\bar{C}_{n-1,k}$. By the inductive assumption, we thus know that $|\det \bar{C}'|$ is 1 or 0 depending on whether T' (the undirected version of $\vec{T}'$) is a spanning tree on its vertex set or not. Since we have removed a degree 1 vertex from T, T is a spanning tree if and only if T' is a spanning tree. This concludes the inductive step for the case when at least one of the vertices $2, 3, \ldots, n$ has degree 1 in T.

Let us discuss the second case, when none of the vertices $2, 3, \ldots, n$ has degree 1 in T. First we observe that T has an isolated vertex in this case (if it were not so, the vertex 1 would have degree at least 1 and the other vertices degrees at least 2, and hence the sum of degrees would be greater than $2|E(T)| = 2(n-1)$—a contradiction).

Because of an isolated vertex, T is disconnected, and hence it is not a spanning tree. To finish the proof we need to show that $\det \bar{C} = 0$. If there is an isolated vertex among $2, 3, \ldots, n$ then it corresponds to a zero row in the matrix $\bar{C}$. If the vertex 1 is isolated then the sum of all rows of $\bar{C}$ is the zero vector because the incidence matrix $D_{\vec{T}}$ has zero row sum. In both cases we have $\det \bar{C} = 0$. $\square$

By the lemma just proved, we know that the number of spanning trees of the graph G equals the number of square $(n-1) \times (n-1)$ submatrices with nonzero determinant of the matrix $\bar{D}$. For finishing the proof of Theorem 7.5.1, we use an algebraic result about determinants.

7.5.4 Theorem (Binet–Cauchy theorem). *Let A be an arbitrary matrix with n rows and m columns. Then*

$$\det(AA^T) = \sum_I \det(A_I)^2,$$

where the sum is over all n-element subsets $I \in \binom{\{1,2,\ldots,m\}}{n}$, and where A_I denotes the matrix obtained from A by deleting all columns whose indices do not lie in I.

We give a proof of this theorem for completeness, but first let us look at how the Binet–Cauchy theorem implies Theorem 7.5.1. With the preparation we have made, this is actually quite straightforward. By Lemma 7.5.2 and then by Theorem 7.5.4 we get

$$\det Q_{11} = \det(\bar{D}\bar{D}^T) = \sum_{I \in \binom{\{1,2,\ldots,m\}}{n-1}} \det(\bar{D}_I)^2,$$

and by Lemma 7.5.3, we see that the last expression is exactly the number of spanning trees of G. □

Proof of the Binet–Cauchy theorem 7.5.4. Let us denote $M = AA^T$. We expand the determinant of M according to the definition of a determinant, i.e.

$$\det M = \sum_{\pi \in S_n} \mathrm{sgn}(\pi) \prod_{i=1}^n m_{i,\pi(i)},$$

where the sum is over all permutations π of the set $\{1, 2, \ldots, n\}$, and where $\mathrm{sgn}(\pi)$ stands for the *sign* of the permutation π (for any permutation, the sign is $+1$ or -1). We will not need the definition of the sign of a permutation directly, and so we only recall the following:

7.5.5 Fact. *For any permutation π of the set $\{1, 2, \ldots, n\}$ and for indices i, j, $1 \le i < j \le n$, let the symbol $\pi_{i\leftrightarrow j}$ denote the permutation whose value at i is $\pi(j)$, the value at j is $\pi(i)$, and the values for all other numbers agree with those of π. Then we have $\mathrm{sgn}(\pi) = -\mathrm{sgn}(\pi_{i\leftrightarrow j})$.*

Now we substitute the values of the elements m_{ij} of the matrix M, namely $m_{ij} = \sum_{k=1}^m a_{ik}a_{jk}$, into the above expansion of the determinant of M. Then we multiply out each of the products. This leads to

$$\det M = \sum_{\pi} \mathrm{sgn}(\pi) \prod_{i=1}^n \left(\sum_{k=1}^m a_{ik}a_{\pi(i)k} \right)$$

$$= \sum_{\pi} \mathrm{sgn}(\pi) \sum_{k_1,k_2,\ldots,k_n=1}^m \prod_{i=1}^n a_{i,k_i}a_{\pi(i),k_i}.$$

Let us change the notation in this last formula to a more suitable one. The choice of the n-tuple $k_1, \ldots, k_n$ of the summation indices in the inner sum can be understood as a choice of a mapping $f: \{1, 2, \ldots, n\} \to \{1, 2, \ldots, m\}$ defined by $f(i) = k_i$. With this new fancy notation, we thus have

$$\det M = \sum_{\pi} \operatorname{sgn}(\pi) \sum_{f:\{1,2,\ldots,n\} \to \{1,2,\ldots,m\}} \prod_{i=1}^{n} a_{i,f(i)} a_{\pi(i),f(i)}.$$

In this sum, we exchange the order of summation: we sum according to the permutation π (inner sum) and then according to the function f (outer sum). For a yet more convenient notation, we introduce the symbols

$$P(f, \pi) = \prod_{i=1}^{n} a_{i,f(i)} a_{\pi(i),f(i)},$$

$$S(f) = \sum_{\pi} \operatorname{sgn}(\pi) P(f, \pi),$$

and then we have

$$\det M = \sum_{f:\{1,2,\ldots,n\} \to \{1,2,\ldots,m\}} S(f).$$

A key lemma for the whole proof is the following one:

7.5.6 Lemma. *If a function $f: \{1, 2, \ldots, n\} \to \{1, 2, \ldots, m\}$ is not injective, then we have $S(f) = 0$ (for any choice of the matrix A).*

Proof of the lemma. Let i, j be indices such that $f(i) = f(j)$. Then for any permutation π, the products $P(f, \pi)$ and $P(f, \pi_{i \leftrightarrow j})$ are the same (the notation is as in Fact 7.5.5). If π runs through all permutations, then also $\pi_{i \leftrightarrow j}$ runs through all permutations (although in a somewhat different order), and therefore we have

$$S(f) = \sum_{\pi} \operatorname{sgn}(\pi_{i \leftrightarrow j}) P(f, \pi_{i \leftrightarrow j}) = \sum_{\pi} -\operatorname{sgn}(\pi) P(f, \pi) = -S(f);$$

hence $S(f) = 0$ as the lemma claims.

By the lemma, we can write

$$\det(AA^T) = \sum_{f:\{1,2,\ldots,n\} \hookrightarrow \{1,2,\ldots,m\}} S(f),$$

where the summation is over all *injective* functions $\{1, 2, \ldots, n\} \hookrightarrow \{1, 2, \ldots, m\}$.

Our goal now is to show that the right-hand side of the last equation equals $\sum_I \det(A_I)^2$, where the sum runs over $I \in \binom{\{1,2,\ldots,m\}}{n}$. Let us choose some $I \in \binom{\{1,2,\ldots,m\}}{n}$, and calculate

$$\det(A_I)^2 = \det(A_I)\det(A_I^T) = \det(A_I A_I^T).$$

This determinant can be expanded in exactly the same manner as we did for the determinant of AA^T. This time we get

$$\det(A_I A_I^T) = \sum_{f:\{1,2,\ldots,n\} \hookrightarrow I} S(f).$$

We thus have

$$\sum_{I \in \binom{\{1,2,\ldots,m\}}{n}} \det(A_I)^2 = \sum_{I \in \binom{\{1,2,\ldots,m\}}{n}} \sum_{f:\{1,2,\ldots,n\} \hookrightarrow I} S(f)$$

$$= \sum_{f:\{1,2,\ldots,n\} \hookrightarrow \{1,2,\ldots,m\}} S(f) = \det(AA^T);$$

the second equality follows from the fact that any injective function $f: \{1, 2, \ldots, n\} \hookrightarrow \{1, 2, \ldots, m\}$ uniquely determines the n-element set I of its values. This proves the Binet–Cauchy theorem 7.5.4. □

Let us mention a geometric interpretation of the Binet–Cauchy theorem in terms of volumes. We omit all proofs, since these, although elementary, would lead us farther into geometry than we want to go in this book. Let us consider an $n \times m$ matrix A as in the theorem, and let us interpret its n rows as n vectors $a_1, a_2, \ldots, a_n$ in the m-dimensional space $\mathbf{R}^m$. These n vectors span an n-dimensional parallelotope P in $\mathbf{R}^m$; Fig. 7.3 illustrates this for $m = 3$ and $n = 2$ (about the only nontrivial case where a picture can be drawn). It can be shown that $|\det(AA^T)|$ is the square of the n-dimensional volume of P (in the figure, it is simply the squared area of P). Choosing an $n \times n$ submatrix B corresponds to projecting the vectors $a_1, a_2, \ldots, a_n$ to a n-dimensional subspace of $\mathbf{R}^m$ spanned by some n coordinate axes. In the figure, there are 3 possible 2×2 submatrices corresponding to the projections into the 3 coordinate planes. The quantity $|\det(BB^T)|$ is the squared volume of the corresponding projection of P, and the Binet–Cauchy theorem asserts that the squared volume of P equals the sum of the squared volumes of the n-dimensional projections of P. Actually, for $n = 1$ and $m = 2$, the reader may want to check that this is just the theorem of Pythagoras!

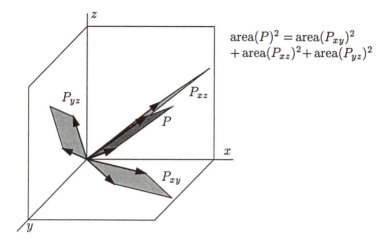

$$\text{area}(P)^2 = \text{area}(P_{xy})^2$$
$$+ \text{area}(P_{xz})^2 + \text{area}(P_{yz})^2$$

Fig. 7.3   Illustration for the geometric meaning of the Binet–Cauchy theorem.

Exercises

1. In this exercise, G is a graph on n vertices, Q is its Laplace matrix, and Q^* denotes the matrix whose element at position (i, j) equals $(-1)^{i+j} \det Q_{ij}$.

   (a) Prove that $\det Q = 0$.

   (b) Prove that if the graph G is connected then its Laplace matrix has rank $n - 1$.

   (c) *Prove that the rank of the Laplace matrix of a disconnected graph is at most $n - 2$. Derive that for a disconnected G, Q^* is the zero matrix.

   (d) Show that if G is connected and $x \in \mathbf{R}^n$ is an arbitrary vector, then $Qx = 0$ holds if and only if x is a multiple of the vector $(1, 1, \dots, 1)$.

   (e) *Prove that the product QQ^* is a zero matrix. Using (d), infer that all the elements of the matrix Q^* are identical.

2. Let G be a graph, which (exceptionally, for the purposes of this exercise) may also contain loops and multiple edges. Let e be an edge that is not a loop. Define $G - e$ as G with the edge e deleted, and $G : e$ as the graph arising from G by deleting the edge e and a subsequent gluing of the end-vertices of e into a single vertex. The other edges are preserved, and so this operation can produce new loops or multiple edges (note that this is different from the operation of edge contraction introduced in Section 5.4).

   (a) Prove that $T(G) = T(G - e) + T(G : e)$.

(b) Derive the number of spanning trees of the 3-dimensional cube (this graph is drawn in Fig. 5.1) by a calculation based on part (a).

(c) Find the number of spanning trees of the 3-dimensional cube using Theorem 7.5.1.

3. Solve Exercise 7.1.2 using Theorem 7.5.1.

4. *Calculate $T(K_{n,m})$ (the number of spanning trees of the complete bipartite graph).

5. *Let G be an (undirected) graph and M its incidence matrix; if G has n vertices $v_1, v_2, \ldots, v_n$ and m edges $e_1, e_2, \ldots, e_m$ then M is an $n \times m$ matrix with

$$m_{ik} = \begin{cases} 1 & \text{if } v_i \in e_k \\ 0 & \text{otherwise.} \end{cases}$$

Prove that the following two conditions are equivalent:

(i) G is bipartite.

(ii) Any square submatrix of M (arising by deleting some rows and columns) has determinant 0, 1, or -1. A matrix M with this property is called *totally unimodular*.

8

Finite projective planes

Mathematicians are often interested in objects that are highly regular in some sense. A good example is the regular Platonic solids we have seen in Section 5.3. They are scarce, beautiful, and also very useful. For instance, symmetry groups related to them play an important role in physics.

In this chapter, we will consider certain very regular families of finite sets, the so-called finite projective planes. As the name suggests, this notion has a geometric inspiration. Finite projective planes are highly symmetric and also somewhat rare. Appreciating their mathematical usefulness, and maybe also their beauty, requires learning something about them first. Let us remark that the study of regular configurations of sets with a flavor somewhat similar to finite projective planes is an extensive branch of combinatorial mathematics (we will say a little more about such objects in Chapter 11). Suitable further reading for the present chapter is Van Lint and Wilson [6].

8.1 Definition and basic properties

A finite projective plane is a system of subsets of a finite set with certain properties.

8.1.1 Definition (Finite projective plane). *Let X be a finite set, and let $\mathcal{L}$ be a system of subsets of X. The pair $(X, \mathcal{L})$ is called a finite projective plane if it satisfies the following axioms:*

- *(P0) There exists a 4-element set $F \subseteq X$ such that $|L \cap F| \leq 2$ holds for each set $L \in \mathcal{L}$.*
- *(P1) Any two distinct sets $L_1, L_2 \in \mathcal{L}$ intersect in exactly one element, i.e. $|L_1 \cap L_2| = 1$.*
- *(P2) For any two distinct elements $x_1, x_2 \in X$, there exists exactly one set $L \in \mathcal{L}$ such that $x_1 \in L$ and $x_2 \in L$.*

If $(X, \mathcal{L})$ is a finite projective plane, we call the elements of X *points* and the sets of $\mathcal{L}$ *lines*. If $x \in X$ is a point and $L \in \mathcal{L}$ a line and $x \in L$, we say that "the point x lies on the line L" or also that "the line L passes through the point x" and similarly.

If we express the axioms (P0)–(P2) in this new language, they start resembling familiar geometric statements. Axiom (P1) says that any two distinct lines meet at exactly one point (of course, in the usual planar geometry, this is not quite true—there is the exception of parallel lines!). Axiom (P2) tells us that there is exactly one line passing through any two distinct points. Axiom (P0) then requires the existence of 4 points such that no 3 of them are collinear. This axiom is of a somewhat auxiliary nature and it serves just for excluding a few "degenerate" types of set systems that satisfy (P1) and (P2) but are rather uninteresting.

If $a, b \in X$ are two distinct points of a finite projective plane, the unique line $L \in \mathcal{L}$ containing both a and b will be denoted by the symbol $\overline{ab}$. If $L, L' \in \mathcal{L}$ are distinct lines, the unique point of $L_1 \cap L_2$ is called their intersection (although, strictly speaking, the intersection of L_1 and L_2 in the usual sense is a one-point set).

Finite projective planes are a finite analogy of the so-called projective plane (more exactly, *real projective plane*) studied in geometry. The terminology introduced above ("points", "lines", etc.) follows this analogy. Thus, we make a short detour and mention what a real projective plane is. First, we should perhaps remark that the adjective "real" indicates that the real projective plane is constructed from the set of real numbers, and not that the other kinds of projective planes would be somehow faked.

In the usual (Euclidean) plane, any two lines intersect at a single point, but with an exception: parallel lines do not intersect at all. In many geometric considerations, such an exception is unpleasant, since it may require treating many special cases, both in proofs and in analytic calculations. The real projective plane is a suitable extension of the Euclidean plane by a set of additional points called the *points at infinity*.

Roughly speaking, each direction of lines in the plane corresponds to one point at infinity, and all the lines parallel to that direction are defined to intersect at this point. All the points at infinity lie on a single *line at infinity*. In this way, one achieves that now every two distinct lines intersect at a single point (which can lie at infinity, of course), and hence all the axioms (P1), (P2), and (P0) hold in the real projective plane.

The points at infinity are no philosophical mystery here. The projective plane is a mathematical construction of a similar kind as, for example, producing rational numbers from the integers or real numbers from the rationals, i.e. a kind of completion. Readers interested in a detailed construction of the real projective plane will find it in Section 8.2.

We still owe the reader an explanation of the adjective "projective" in the term projective plane. First, we should indicate what is a *projective transform*. Consider two planes ρ and σ in the 3-dimensional Euclidean space and a point c lying neither on ρ nor on σ, and project each point of ρ from the point c into the plane σ. This defines a mapping, called a projective transform, of ρ into σ. Or does it? Well, not quite: if $x \in \rho$ is a point such that the segment cx is parallel to the plane σ then, in the usual Euclidean geometry, the image of x is undefined. But if we complement both ρ and σ by lines at infinity so that they become projective planes, then this projective transform is a bijection between these two projective planes. If both ρ and σ are considered as copies of the same plane, we can regard this mapping as a bijective mapping of the projective plane onto itself. The projective plane is thus an appropriate domain for doing *projective geometry*, a branch of geometry concerned with properties of geometric objects and configurations that are preserved by projective transforms. For example, projective transforms map conic sections (circles, ellipses, hyperbolas, and parabolas) to conic sections, but an ellipse can be transformed to a hyperbola etc., and an elegant unified theory of conic sections can be built in the projective geometry.

The analogy of the finite projective planes with the real projective plane is useful as a motivation for various notions, and often also for intuition (we can draw geometric pictures). Geometric considerations made in the real projective plane and using axioms (P0), (P1), and (P2) only can be carried out in finite projective planes as well. It should not be forgotten, though, that a finite projective plane is only a system of finite sets with properties (P0)–(P2) and nothing else, and hence other geometric notions cannot be transferred to it automatically. For instance, there is no good notion of distance in a finite projective plane, and hence it is not clear what a "circle" should mean. Another important difference is that in the "usual" geometric plane, the points of each line are naturally ordered "along" that line, but no such ordering can be reasonably introduced in finite projective planes.

As was remarked above, finite projective planes are rare creatures, and, given only the definition, it is not easy to discover any example at all (try it if you don't believe). Even the smallest example is interesting.

8.1.2 Example. The smallest possible example of a finite projective plane has 7 points and 7 lines, each line with 3 points, and it is called the *Fano plane*. It is shown in Fig. 8.1; the points are marked by

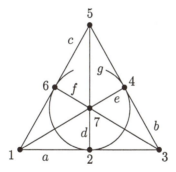

Fig. 8.1 The Fano plane.

dots labeled 1–7, and the three points of each line are connected by
a segment, in one case by a circle arc.[1] These connecting lines are
labeled by a–g in the figure.

The Fano plane, although small, is a useful mathematical object (see
Section 8.4 for one application), and it can also appear as a solution to
various puzzles or even totally serious problems, such as the following
one. Seven policemen were transferred to the 87th precinct from various
districts. A good opportunity for them to get acquainted with each
other is to watch the Plano Bar on Southwest C Drive, which otherwise
is a simple and somewhat boring duty as the bar's patrons are mostly
computer criminals, digital money counterfeiters, and the like. A shift
of three men is required for that service, seven days a week. How can a
weekly schedule of shifts be arranged in such a way that every two of
the seven men have a common shift? The Fano plane provides a good
solution (points correspond to policemen and the shifts are the lines,
arranged in some order); see Fig. 8.2. Everyone has the same number
of shifts, no one has more than two consecutive days, but on the other
hand, every shift has one man who was also present the previous day
and knows what was going on, etc.[2] We are not aware of such a schedule
actually being used by police forces, but, for instance, some motorcycle
races are organized according to a scheme based on an affine plane of
order 4 (an affine plane is a concept closely related to projective planes;
see Exercise 10 for a definition).

---

[1]It can be shown that 7 points cannot be drawn in the Euclidean plane in
such a way that each triple corresponding to a line in the Fano plane lies on a
Euclidean straight line—see Exercise 11.

[2]A quiz for aficionado of classical detective stories: can you recall where the
seven policemen in Fig. 8.2 come from, and the names of their more famous rivals
or partners, the great detectives?

Weekly schedule for location: Plano Bar		
Mon	Cramer Hoong	Japp
Tue	Cramer Holcomb	Lestrade
Wed	Holcomb Hoong	Janvier
Thu	Cramer Janvier	Parker
Fri	Holcomb Japp	Parker
Sat	Janvier Japp	Lestrade
Sun	Hoong Lestrade	Parker
Officer's signature: *Camlla*		

Fig. 8.2   Allocating 7 persons to 7 shifts by 3.

We now prove several propositions showing that in a construction of a finite projective plane, our freedom is much more restricted than might appear at first sight.

8.1.3 Proposition. *Let $(X, \mathcal{L})$ be a finite projective plane. Then all its lines have the same number of points; that is, $|L| = |L'|$ for any two lines $L, L' \in \mathcal{L}$.*

Proof.   Choose two lines $L, L' \in \mathcal{L}$ arbitrarily. First we prove an auxiliary claim: *There exists a point $z \in X$ lying neither on L nor on L'.*

Proof of the auxiliary claim.   Consider a set $F \subseteq X$ as in the axiom (P0). We have $|L \cap F| \leq 2$ and $|L' \cap F| \leq 2$. If F is not contained in $L \cup L'$ we are done. The only remaining possibility is that L intersects F at 2 points (call them a, b) and L' intersects F at the 2 remaining points (denote them by c, d). Then we consider the lines $L_1 = \overline{ac}$ and $L_2 = \overline{bd}$. Let z be the intersection of L_1 and L_2.

The following geometric picture illustrates the situation:

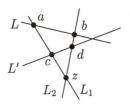

Of course, we have to be very careful to use only conditions (P0)–(P2), rather than some extra geometric information from the picture. In many respects, finite projective planes "look" quite different to the geometric (Euclidean) plane.

We assert that $z \notin L \cup L'$. The lines L and L_1 intersect at a single point, namely at a, and so if $z \in L$, we would get $z = a$. But this is impossible since then the line L_2 would contain the points $z = a$, b, and d, which are 3 points of F. This is forbidden by the condition (P0). Therefore $z \notin L$, and analogously one can show that $z \notin L'$. This finishes the proof of the auxiliary claim.

Now we show that the lines L and L' have the same size. To this end, we define a mapping $\varphi \colon L \to L'$; it will turn out that it is a bijection. We fix a point $z \notin L \cup L'$ and define the image $\varphi(x)$ of a point $x \in L$ as the intersection of the lines $\overline{zx}$ and L', as in the following picture:

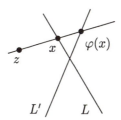

By axioms (P1) and (P2), the point $\varphi(x)$ is well defined. Next, we check that φ is a bijection. If $y \in L'$ is an arbitrary point, we consider the line $\overline{zy}$, and let x be its intersection with the line L. Then the lines $\overline{zy}$ and $\overline{zx}$ coincide, and hence we have $y = \varphi(x)$. The mapping φ is a bijection as claimed and so $|L| = |L'|$. $\qquad\square$

8.1.4 Definition (Order of a projective plane). *The order of a finite projective plane* $(X, \mathcal{L})$ *is the number* $|L| - 1$, *where* $L \in \mathcal{L}$ *is a line (according to the proposition just proved, the order doesn't depend on the choice of the line* L).

For example, the Fano plane has order 2 (the lines have 3 points), and it can be shown that it is the only projective plane of order 2 (up to renaming of the points, i.e. up to an isomorphism). Subtracting 1 from the line size in the definition of the order may seem odd, but this definition of order is very natural in other connections, e.g. for affine planes (Exercise 10) or for Latin squares (Section 8.3).

We continue proving properties of finite projective planes.

8.1.5 Proposition. *Let $(X, \mathcal{L})$ be a projective plane of order n. Then we have*

(i) *Exactly $n+1$ lines pass through each point of X.*
(ii) $|X| = n^2 + n + 1$.
(iii) $|\mathcal{L}| = n^2 + n + 1$.

Proof of (i). Consider an arbitrary point $x \in X$. First we observe that there exists a line L that doesn't contain x. That is, if F is the 4-point configuration as in (P0) and $a, b, c \in F$ are points distinct from x, then at least one of the lines $\overline{ab}$ and $\overline{ac}$ doesn't contain x, as is very easy to check.

Fix such a line L, $x \notin L$. For each point $y \in L$, we consider the line $\overline{xy}$; these are $n+1$ lines passing through x. On the other hand, any line containing x intersects L at some point $y \in L$ and hence it is counted among the above-mentioned $n+1$ lines. Therefore, exactly $n+1$ lines pass through x.

Proof of (ii). We choose some $L = \{x_0, x_1, x_2, \ldots, x_n\} \in \mathcal{L}$ and a point $a \notin L$, as in the following picture:

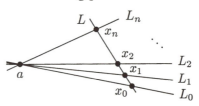

Let L_i denote the line $\overline{ax_i}$, $i = 0, 1, \ldots, n$. According to (P1), any two of these lines, L_i and L_j, intersect at a single point, and this is the point a. The lines $L_0, L_1, \ldots, L_n$ each have n more points besides a, and hence they together contain $(n+1)n + 1 = n^2 + n + 1$ distinct points. It remains to show that any point $x \in X \setminus \{a\}$ already lies on some of the lines L_i. By (P1), the line $\overline{ax}$ intersects the line L at some point, x_i, and by (P2), the line $\overline{ax}$ must thus coincide with L. This proves part (ii).

We omit the proof of part (iii) for now. In the sequel, we will learn an important principle, and from it we will see that (iii) follows immediately from what we have already proved. □

Duality. The meaning of duality in projective planes is "exchanging the roles of points and lines". In order to formulate this exactly, we first introduce the so-called *incidence graph* of a finite projective

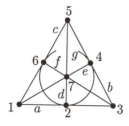 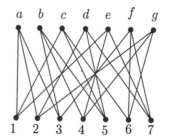

Fig. 8.3   The Fano plane and its incidence graph.

plane. In fact, the incidence graph can be defined for an arbitrary system S of subsets of a set X. The incidence graph is a bipartite graph with vertex set $X \cup S$, where each set $S \in S$ is connected by an edge to all points $x \in S$. Consequently, each point $x \in X$ is connected to all the sets containing it. Briefly, we can say that the edges of the incidence graph correspond to the membership relation "$\in$". The incidence graph of the Fano plane is shown in Fig. 8.3; the vertices are labeled by the labels of the corresponding points and lines. (By the way, the resulting graph is a pretty and important graph, although the drawing in our figure is rather ugly, and it even has a name: the *Heawood graph*.)

Given a finite projective plane $(X, \mathcal{L})$, the *dual* of $(X, \mathcal{L})$ is obtained by taking the incidence graph of $(X, \mathcal{L})$ and interpreting as lines the vertices that were understood as points, and conversely, vertices that used to be sets start playing the role of points. In Fig. 8.3, we could just flip the graph upside down. Hence $\mathcal{L}$ is now thought of as a point set, and for each point $x \in X$, the set of lines $\{L \in \mathcal{L}: x \in L\}$ is interpreted as a line. For the Fano plane example, the points of the dual are thus $\{a, b, \ldots, g\}$, and the lines of the dual are $\{a, c, e\}$ (for the point 1 in the Fano plane), $\{a, d, g\}$ (for the point 2), and so on.

Proposition.   *The dual of a finite projective plane is itself a finite projective plane.*

Proof.   Let $(X, \mathcal{L})$ be a finite projective plane. The dual of $(X, \mathcal{L})$ is a pair $(\mathcal{L}, \Lambda)$, where Λ is a system of subsets of $\mathcal{L}$, each of these subsets corresponding to some point of X. (Note that distinct points always yield distinct subsets of $\mathcal{L}$, since two points share only one line.)

We have to verify conditions (P0)–(P2) for $(\mathcal{L}, \Lambda)$. We begin with the condition (P0). If this condition is translated into the language of the original set system $(X, \mathcal{L})$, it means that we should find 4 lines $L_1, L_2, L_3, L_4 \in \mathcal{L}$ such that no 3 of them have a point in common. To this end, let us consider a 4-point configuration $F = \{a, b, c, d\} \subseteq X$ as in the condition (P0), and define $L_1 = \overline{ab}$, $L_2 = \overline{cd}$, $L_3 = \overline{ad}$, $L_4 = \overline{bc}$. If we look at any 3 of these 4 lines, any two of them share one point of F, and this point is not contained in the third one. Hence any 3 of the lines $L_1, \ldots, L_4$ have an empty intersection, and we have confirmed the validity of the condition (P0) for the dual set system.

The condition (P1) formulated for the dual $(\mathcal{L}, \Lambda)$ requires the following: if $x, x' \in X$ are two distinct points, then there exists exactly one line $L \in \mathcal{L}$ containing both x and x'. This is exactly condition (P2) for $(X, \mathcal{L})$! Similarly, we find that (P2) for the dual is a consequence of (P1) for the original projective plane $(X, \mathcal{L})$. $\qquad \square$

Now we can call the dual of a finite projective plane the *dual projective plane*. Proposition 8.1.5(i) implies that the dual projective plane has the same order as the projective plane we started with. Also, one can see that parts (ii) and (iii) of Proposition 8.1.5 are dual to each other, and if we prove one of them, the other one must be valid too.

In general, if we have some valid statement about finite projective planes of order n, and if we interchange the words "point" and "line" everywhere in it, we get a valid statement again. To get a meaningful sentence, we may have to rephrase other parts as well. For instance, if the original statement said "lines L_1, L_2 intersect at the point x", we should say "points x_1, x_2 are connected by the line L" in the dual statement etc. Hence, we have a "recipe for producing new theorems" which gives us about half of the theorems in projective geometry for free! It is sometimes called the *duality principle*, and it was noted by geometers studying the real projective plane a long time ago.

Exercises

1. Prove that the Fano plane is the only projective plane of order 2 (i.e. any projective plane of order 2 is isomorphic to it—define an isomorphism of set systems first).

2. *Construct a projective plane of order 3 (before reading the next section!).

3. (a) Find an example of a set system $(X, \mathcal{L})$ on a nonempty finite set X that satisfies conditions (P1) and (P2) but doesn't satisfy (P0).

   (b) Find an X and $\mathcal{L}$ as in (a) such that $|X| \geq 10$, $|\mathcal{L}| \geq 10$, and each $L \in \mathcal{L}$ has at least 2 points.

   (c) *Describe all set systems $(X, \mathcal{L})$ as in (a).

4. Let X be a finite set and let $\mathcal{L}$ be a system of subsets of X satisfying conditions (P1), (P2), and the following condition (P0'):

   There exist at least two distinct lines $L_1, L_2 \in \mathcal{L}$ having at least 3 points each.

   Prove that any such $(X, \mathcal{L})$ is a finite projective plane.

5. Prove part (iii) of Proposition 8.1.5 directly, without using duality.

6. Show that the number of sets in a set system consisting of 3-element sets on a 9-point set, such that no two sets share more than one point, is at most 12. *Find an example with 12 sets.

7. *Is it possible to arrange 8 bus routes in a city so that

   (i) if any single route is removed (doesn't operate, say) then any stop can still be reached from any other stop, with at most one change, and

   (ii) if any two routes are removed, then the network becomes disconnected?

8. *Let X be a set with $n^2 + n + 1$ elements, $n \geq 2$, and let $\mathcal{L}$ be a system consisting of $n^2 + n + 1$ subsets of X of size $n + 1$ each. Suppose that any two distinct sets of $\mathcal{L}$ intersect in at most one point. The goal is to prove that $(X, \mathcal{L})$ is a finite projective plane of order n. The following sequence of auxiliary statements give one possible way of arranging the proof.

   (a) Prove that each pair or points of X is contained in exactly one set of $\mathcal{L}$ (use double-counting).

   (b) Prove that at most $n + 1$ sets contain any given point.

   (c) Prove that each point is contained in exactly $n + 1$ sets.

   (d) Prove that any two sets of $\mathcal{L}$ intersect.

   (e) Verify that $(X, \mathcal{L})$ is a projective plane of order n.

9. *Let $(X, \mathcal{L})$ be a projective plane of order n, and let $A \subseteq X$ be a set with no 3 points lying on a common line. Prove that $|A| \leq n + 2$ (for n odd, it can even be shown that $|A| \leq n + 1$).

10. (Affine planes) Let us define an *affine plane* as a pair $(X, \mathcal{A})$, where X is a set and $\mathcal{A}$ is a collection of subsets of X (called *lines*) satisfying the following axioms: there exist 3 points not contained in a common line, any two distinct points are contained in exactly one line, and for

any point p and line $A \in \mathcal{A}$ with $p \notin A$, there exists exactly one line A' with $A' \cap A = \emptyset$ and $p \in A'$.

(a) Check that the axioms of an affine plane are valid for the usual Euclidean plane. Find out how to construct a finite affine plane from a finite projective plane.

(b) Define relation $\parallel$ on $\mathcal{A}$ by letting $A_1 \parallel A_2$ if and only if $A_1 \cap A_2 = \emptyset$. Prove that $\parallel$ is an equivalence.

(c) *Analogously to our treatment of projective planes, prove that all lines in a finite affine plane have the same cardinality n, and that such an affine plane has $n^2 + n$ lines and n^2 points, with $n + 1$ lines passing through each point.

(d) *Show that from any affine plane of order n, one can construct a projective plane of order n.

11. Show that the Fano plane cannot be embedded into the Euclidean plane. That is, there exist no 7 points and 7 lines in the Euclidean plane such that each pair of the points lies on one of the lines and each pair of the lines intersects at one of the points. Use Exercise 5.3.8.

8.2 Existence of finite projective planes

Projective planes of orders $2, 3, 4$, and 5 exist. But there is no projective plane of order 6! (Proving this is not easy; see e.g. Van Lint and Wilson [6].) Projective planes of orders 7, 8, 9 again exist, but none of order 10. Is there any regularity in this? Well, a projective plane of order n exists whenever a field with n elements exists. Here a field is meant in the algebraic sense; that is, it is a set with operations of addition, subtraction, multiplication, and division that satisfy certain axioms—check the Appendix if you are not sure about the definition. As algebra teaches us, an n-element field exists if and only if n is a power of a prime number. This means, in particular, that projective planes of arbitrarily large orders exist.

For an n divisible by at least two distinct primes, no n-element field exists, but nevertheless it is not known whether a projective plane of order n may exist in some such cases or not. There are some partial negative results; for instance, if the number n divided by 4 gives remainder 1 or 2 and it cannot be written as a sum of two squares of integers then no projective plane of order n exists (this is not an easy theorem). This rules out the existence of a projective plane of order 14 and of many other orders, but it doesn't by far cover all possible orders. For instance, it says nothing about orders $n = 6, 10$, or 12.

The existence of projective planes of orders 6 and 10 has also been excluded. These results have an interesting history, For order 6, a proof was already attempted by Euler, but only Tarry gave a convincing argument around 1900. For order 10, a proof was done recently using enormous computer calculations. For the next higher order, 12, the existence of a projective plane remains an open problem. It is clear that this problem can be solved by checking a finite number of configurations, but the number of configurations seems to be too gigantic for contemporary computing technology.

An algebraic construction of a projective plane. For readers with an interest in the subject and with a little background in introductory algebra, we explain how to construct a projective plane from a field. We are particularly interested in finite projective planes, but the construction works in exactly the same way for the real projective plane (and, in general, for any field).

The construction starts with some field K. For the real projective plane (i.e. a suitable extension of the usual Euclidean plane by points at infinity), we take the field $\mathbf{R}$ of all real numbers for K. If we choose an n-element field for K, the construction results in a finite projective plane of order n. Our running example will be the 3-element field K, i.e. the set $\{0, 1, 2\}$ with arithmetic operations modulo 3.

First we consider the set $T = K^3 \setminus \{(0, 0, 0)\}$; that is, the set of all ordered triples (x, y, t) where $x, y, t \in K$ and x, y, t are not all simultaneously equal to 0. On this T, we define an equivalence relation $\approx$ as follows: $(x_1, y_1, t_1) \approx (x_2, y_2, t_2)$ if and only if a nonzero $\lambda \in K$ exists such that $x_2 = \lambda x_1$, $y_2 = \lambda y_1$, and $t_2 = \lambda t_1$ (it is not at all difficult to check that this is indeed an equivalence). Points of the constructed projective plane are the classes of this equivalence. The projective plane thus produced[3] is usually denoted by PK^2 in the literature, where one may write the specific field being used instead of K. For instance, the real projective plane is usually denoted by $P\mathbf{R}^2$.

In order to gain a better intuition about the projective plane, we select one representative triple from each class of the equivalence $\approx$. For these representatives, we choose the triples *whose last nonzero component equals 1*. Hence, the representatives are triples of the types $(x, y, 1)$, $(x, 1, 0)$ (for some $x, y \in K$), and the triple $(1, 0, 0)$. It is easy to convince oneself that any other triple is equivalent to a triple of the above-

---

[3]For a finite field, we haven't yet proved that we obtain a finite projective plane in the sense of Definition 8.1.1, and we haven't even defined what the lines are, so strictly speaking, we shouldn't yet call the object being constructed a projective plane. But, being (hopefully) among friends, there is no reason for such a total strictness, is there?

mentioned form, and also that no two of the triples chosen as representatives are equivalent under $\approx$.

It would be cumbersome to keep speaking about equivalence classes all the time. Thus, we will say "a point (x, y, t)" in the sequel, meaning the whole equivalence class containing (x, y, t).

If K is an n-element field, we can now count how many points we obtain. The number of points of the form $(x, y, 1)$ is n^2, there are n points of the form $(x, 1, 0)$, and, moreover, we have the one point $(1, 0, 0)$—altogether $n^2 + n + 1$ as it should be. For $n = 3$, all the points with their labels are drawn in the following diagram:

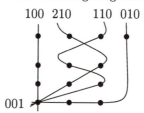

The points with the last coordinate equal to 1 are arranged into a natural 3×3 grid pattern.

Now it is time to define the lines. For each triple $(a, b, c) \in K^3 \setminus \{(0, 0, 0)\}$, we define a line $L(a, b, c)$ as the set of all points (x, y, t) of our projective plane satisfying the equation

$$ax + by + ct = 0. \tag{8.1}$$

Obviously, two equivalent triples (x, y, t) and $(\lambda x, \lambda y, \lambda t)$ either both satisfy this equation or none does, and hence we have indeed defined a certain set of points of the projective plane. Also, one can see that for all nonzero $\lambda \in K$, the triples $(\lambda a, \lambda b, \lambda c)$ define the same line as the triple (a, b, c). Hence, on the triples that define lines we have exactly the same equivalence relation as on the triples that define points. We can select the same representative triples as we did for the points, i.e. triples whose last nonzero component is 1. In the following picture, we have drawn all the lines passing through the point $(0, 0, 1)$ labeled by their representative triples. We have omitted the labels of most of the points (they are as in the preceding diagram):

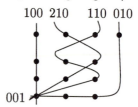

In order to show that for an n-element field, we have really constructed a finite projective plane of order n, we must check conditions

(P0)–(P2). We begin with the condition (P1) (any two distinct lines intersect at a single point). So let (a_1, b_1, c_1) and (a_2, b_2, c_2) be two triples that are not equivalent, i.e. one is not a multiple of the other.

We could now directly calculate the point of intersection of the two considered lines (this is a solution of a small system of linear equations) and verify its uniqueness. We give another proof relying on some basic results of linear algebra. A reader who tries to do the proof by a direct calculation may gain a certain new appreciation of linear algebra.

Let us regard the triples (a_1, b_1, c_1) and (a_2, b_2, c_2) as 3-dimensional vectors over the field K. Both are nonzero vectors, and the fact that one is not a multiple of the other means that they are linearly independent. Hence, the matrix

$$\begin{pmatrix} a_1 & b_1 & c_1 \\ a_2 & b_2 & c_2 \end{pmatrix}$$

has rank 2 (linear independence and rank are over the field K). Let us now view the columns of this matrix as 2-dimensional vectors. We know that 3 vectors in a 2-dimensional space must necessarily be linearly dependent, which means that there exist 3 numbers $x, y, t \in K$, not all of them simultaneously 0, and such that

$$x(a_1, a_2) + y(b_1, b_2) + t(c_1, c_2) = (0, 0). \tag{8.2}$$

If we rewrite this for each coordinate separately, we obtain that the point (x, y, t) lies on both the considered lines.

On the other hand, since the rank of the considered matrix is 2, two linearly independent columns have to exist. Suppose, for instance, that they are (a_1, a_2) and (b_1, b_2). This means that for any vector (u, v), the equation $x(a_1, a_2) + y(b_1, b_2) = (u, v)$ has a unique solution. In other words, if we prescribe the value of t in Eq. (8.2), the values of x and y are already determined uniquely, and so all solutions to this equation are multiples of a single vector. This means that the two considered lines intersect at a single point.

This argument can be expressed in a somewhat more learned and more concise way. The linear mapping sending a vector $(x, y, t) \in K^3$ to the vector $x(a_1, a_2) + y(b_1, b_2) + t(c_1, c_2) \in K^2$ has rank 2, hence it is onto and its kernel is 1-dimensional.

We have proved (P1). The condition (P2) could be proved in a similar way, or we can say right away that the roles of the triples (a, b, c) and (x, y, z) in Eq. (8.1) are completely symmetric, and so we have (again) a duality between lines and points. Finally, verification of the condition (P0) is left to the reader. ∎

Remarks. The construction presented above might give the impression that some points of the projective plane are somewhat special, different from the others. This is not so; no points have any special significance,

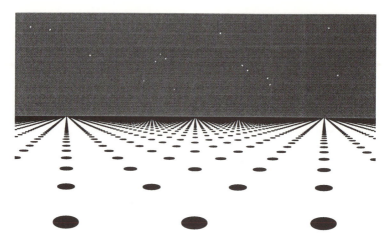

Fig. 8.4   The real projective plane in moonlight.

and the "infinity" can in some sense be imagined wherever convenient—
the projective plane looks "locally everywhere the same".

For the real projective plane, a point $(x, y, 1)$ in the above construction is usually identified with the point (x, y) of the Euclidean plane.
Points whose third coordinate is 0 are the points at infinity. In other
words, a triple (x, y, t) with $t \neq 0$ corresponds to the "Euclidean" point
$(\frac{x}{t}, \frac{y}{t})$. Each point at infinity, $(x, 1, 0)$ or $(1, 0, 0)$, corresponds to one
direction of lines in the Euclidean plane. Points at infinity can be imag-
ined as points "on the horizon" in a drawing of the Euclidean plane in
perspective; see Fig. 8.4, where circles correspond to points with integer
coordinates.

Exercises

1. In the literature, the points of the real projective plane PR^2 are often identified with the lines in the 3-dimensional Euclidean space $\mathbf{R}^3$
   passing through the origin of coordinates. With the knowledge of the
   algebraic construction of PR^2 described in this section, can you explain why? *Would you know how to define lines of the projective
   plane in this geometric interpretation? And how would you embed the
   Euclidean plane into the projective plane in this situation?

2. **,^{CS} Prove the nonexistence of a projective plane of order 6. Write a
   computer program for checking and excluding all configurations coming into consideration. One has to proceed cleverly, since searching all
   systems of 43 7-tuples on 43 points would take way too long. (Clever
   proofs are known that do not need any case analysis but these are not
   easy to discover.)

1	2	3
2	3	1
3	1	2

1	2	3
3	1	2
2	3	1

Fig. 8.5 Two orthogonal Latin squares of order 3.

8.3 Orthogonal Latin squares

A *Latin square of order n* is a square table with n rows and n columns. Each entry is a number from the set $\{1, 2, \ldots, n\}$, and each number in this set occurs exactly once in each row and also in each column. Two 3×3 Latin squares are depicted in Fig. 8.5.

Now we say what it means when two Latin squares of the same order are *orthogonal*. Imagine that one of the squares is printed on a transparency and that we lay it over the other square in such a way that the corresponding entries lie above one another. For example, the squares from Fig. 8.5 overlaid produce

1 1	2 2	3 3
2 3	3 1	1 2
3 2	1 3	2 1

In this way, we get n^2 ordered pairs, each pair being formed by an entry of the square on the transparency and the corresponding entry of the underlying square. The considered squares are orthogonal if no ordered pair appears twice. Since the number of all possible ordered pairs of numbers from 1 to n is n^2 too, each pair has to appear exactly once.

8.3.1 Theorem. *Let M be a set of Latin squares of order n, such that each two of them are orthogonal. Then $|M| \leq n - 1$.*

Proof. We begin with the following observation. Let A and B be orthogonal Latin squares of order n, and let π be some permutation of the numbers $1, 2, \ldots, n$. Let us make a new Latin square A', whose entry at position (i, j) is the number $\pi(a_{ij})$, where a_{ij} is the entry at position (i, j) of the square A. By the definition of orthogonality, it is not hard to see that A' and B are also orthogonal Latin squares. This observation can be summarized by the phrase "the orthogonality of Latin squares doesn't change by renaming the symbols in one of them".[4]

To prove the theorem, imagine we have Latin squares $A_1, A_2, \ldots,$ A_t, each two of them being orthogonal. For each A_i, permute its symbols (i.e. the numbers $1, 2, \ldots, n$) in such a way that the first row of the resulting Latin square A'_i is $(1, 2, \ldots, n)$. By the above observation, the Latin squares $A'_1, \ldots, A'_t$ are still pairwise orthogonal. Let us look at which numbers can occupy the position $(2, 1)$ of the square A'_i. First of all, this entry must not be 1, because the first column already has a 1 in the first row. Further, no two squares A'_i and A'_j may have the same numbers at position $(2, 1)$: if they did, by overlaying A'_i and A'_j we would get a pair of identical numbers, say (k, k), at the position $(2, 1)$, but this pair has already appeared at the kth entry of the first row! Hence each of the numbers $2, 3, \ldots, n$ can only stand at the position $(2, 1)$ in one of the A'_i, and therefore $t \leq n - 1$. $\qquad\qquad\qquad\qquad\qquad\qquad\qquad\qquad\qquad\qquad\qquad\square$

A reader who has been wondering why we, suddenly, started talking about Latin squares in a chapter on projective planes may perhaps be satisfied by the next theorem:

8.3.2 Theorem. *For any $n \geq 2$, a projective plane of order n exists if and only if there exists a collection of $n - 1$ mutually orthogonal Latin squares of order n.*

> **Proof of Theorem 8.3.2.** We will not do it in detail. We only describe how to construct a projective plane from orthogonal Latin squares and vice versa. Given $n - 1$ orthogonal Latin squares $S_1, \ldots, S_{n-1}$ of order n, we will produce a projective plane of order n.
>
> First we define the point set X of the constructed plane. It has $n + 1$ points "at infinity" denoted by r, c, and $s_1, s_2, \ldots, s_{n-1}$, and n^2 points (i, j), $i, j = 1, 2, \ldots, n$. Next, we introduce the lines in several steps. One line $B = \{r, c, s_1, \ldots, s_{n-1}\}$ consists of the points "at infinity". Then we have the n lines $R_1, R_2, \ldots, R_n$, where

---

[4]In fact, one is not really obliged to fill Latin squares only with numbers $1, 2, \ldots, n$. Equally well, one can use n distinct letters, n different kinds of Cognac glasses, and so on. This is also perhaps the appropriate place to address the somewhat puzzling question: why are Latin squares called Latin? It seems that in some traditional problems, the symbols written in the considered square tables used to be Latin letters (while some other squares were filled with Greek letters and called—what else?—Greek squares).

$$R_i = \{r, (i, 1), (i, 2), \ldots, (i, n)\},$$

and the n lines

$$C_j = \{c, (1, j), (2, j), \ldots, (n, j)\}.$$

These are all drawn in the following picture (for $n = 3$):

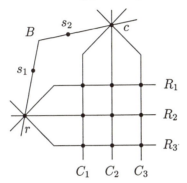

The points and lines of the projective plane we have drawn and labeled so far must look exactly the same in any projective plane of order n (we also haven't yet used any information from the given $n - 1$ orthogonal Latin squares). The squares will now specify the lines of the projective plane passing through the points $s_1, s_2, \ldots, s_{n-1}$ (besides the line B). As the chosen notation suggests, the Latin square S_k determines the lines passing through the point s_k. If $(S_k)_{ij}$ denotes the entry of S_k in the ith row and jth column, we define the lines

$$L_{km} = \{s_k\} \cup \{(i, j) \colon (S_k)_{ij} = m\}$$

for $m = 1, 2, \ldots, n$ and $k = 1, 2, \ldots, n - 1$. For instance, if S_1 were the Latin square in Fig. 8.5 on the left, then the line L_{11} corresponding to the number 1 in the square would be $L_{11} = \{s_1, (1, 1), (2, 3), (3, 2)\}$.

This finishes the description of the finite projective plane corresponding to the collection of $n - 1$ orthogonal Latin squares. It remains to verify the axioms of the projective plane. It is easy to calculate that the total number of both lines and points is $n^2 + n + 1$. By Exercise 8.1.8, it is now sufficient to check that any two lines intersect in at most one point. For this, one has to use both the facts that each S_k is a Latin square and that any two of the Latin squares are orthogonal. We leave this as an exercise.

To prove the equivalence in Theorem 8.3.2, we should also show how to construct $n - 1$ orthogonal Latin squares from a projective plane of order n. This construction follows the same scheme as the converse one. In the projective plane, we choose two distinct points r and c arbitrarily, and we fix the notation further as in the above construction. Then the

kth Latin square S_k is filled out according to the lines passing through the point s_k. So much for the proof of Theorem 8.3.2. $\qquad\qquad\square$

Remark. The proof actually becomes more natural when it is done with finite affine planes (see Exercise 8.1.10).

Exercises

1. Go through the construction in the proof of Theorem 8.3.2 for $n = 2$ (when only one Latin square exists, up to a permutation of the numbers). Check that we obtain the Fano plane in this way.

2. Verify that any two lines constructed in the proof of Theorem 8.3.2 intersect in at most one point (do not forget the lines R_i and C_j!). Distinguish where one uses the definition of a Latin square and where the orthogonality.

3. Show that the construction sketched at the end of the proof of Theorem 8.3.2 indeed produces $n - 1$ orthogonal Latin squares.

4. Define a *liberated square* of order n as an $n \times n$ table with entries belonging to the set $\{1, 2, \ldots, n\}$. Orthogonality of liberated squares is defined in the same way as for Latin squares. For a given number t, consider the following two conditions:

   (i) There exist t mutually orthogonal Latin squares of order n.

   (ii) There exist $t + 2$ mutually orthogonal liberated squares of order n.

   (a) Prove that (i) implies (ii).

   (b) *Prove that (ii) implies (i).

5. Let T be a finite field with n elements. Denote its elements by t_0, $t_1, \ldots, t_{n-1}$, where $t_0 = 0$ and $t_1 = 1$. For $k = 1, 2, \ldots, n - 1$, define an $n \times n$ matrix $S^{(k)}$, where the entry of the matrix $S^{(k)}$ at position (i, j) equals $t_i t_k + t_j$ (the multiplication and addition in this formula are in the field T). Prove that $S^{(1)}$, $S^{(2)}$, $\ldots$, $S^{(n-1)}$ is a collection of mutually orthogonal Latin squares of order n. (Using Theorem 8.3.2, this gives an alternative construction of a projective plane of order n.)

6. For natural numbers $m \leq n$, we define a *Latin $m \times n$ rectangle* as a rectangular table with m rows and n columns with entries chosen from the set $\{1, 2, \ldots, n\}$ and such that no row or column contains the same number twice. Count the number of all possible Latin $2 \times n$ rectangles.

8.4 Combinatorial applications

In combinatorial mathematics, finite projective planes often serve as examples of set systems with various remarkable properties. One can say that if we have some hypothesis about finite set systems and look for a counterexample to it, or if we want an example of a set system with some prescribed properties, then it may be a good idea to try a finite projective plane among the first candidates.

Trying to document the usefulness of projective planes in mathematics and its applications resembles explaining the usefulness of noodles, say, in the kitchen. Certainly one can do lots of great cooking without them. Wonderful foods can also be prepared from noodles, but it's a question of a good recipe (and cook) and of adding various subtle ingredients, spices, etc. Noodles by themselves probably seem exciting to specialists only. Here we haven't accumulated enough mathematical ingredients and spices to prepare a nice but complicated recipe–example, and so we will offer two simple combinatorial dishes only.

Just at the time this section was being written, newspapers reported on the first Swiss bank to introduce on-line banking over the Internet (in the US, it had been around for more than a year). Of course, safety of the information being interchanged over the public network is of utmost importance in such a case. The supposedly indecipherable codes by which the banks expect to achieve safety[5] are based on the so-called elliptic curves over finite fields. The theory of elliptic curves has been developed for a long time in number theory and algebraic geometry, traditionally considered absolutely pure mathematics without any applicability whatsoever. And these elliptic curves are inhabitants (or subsets if you prefer) of finite projective planes. So it goes.

The book by Koblitz [20] can serve both as an introduction to number theory and a gate to the world of mathematical cryptography. But let us return to our combinatorial applications.

Coloring set systems by two colors. Let X be a finite set, and let $\mathcal{M}$ be a system of subsets of X. We say that the set system $\mathcal{M}$ is *2-colorable* if it is possible to color each of the points of X by one of two given colors, say red or white, in such a way that each set $M \in \mathcal{M}$ contains points of both colors (2-colorablility is often called *property B* in the literature.)

For example, if $X = \{1, 2, 3\}$ and $\mathcal{M} = \{\{1, 2\}, \{1, 3\}, \{2, 3\}\}$ then $\mathcal{M}$ is not 2-colorable. More generally, if all the sets of $\mathcal{M}$ have exactly 2 elements, then we can regard $(X, \mathcal{M})$ as a graph, and here being 2-colorable means exactly the same thing as being bipartite. What if

---

[5]Software products using some of these codes are also subject to US export restrictions, just like high-tech weapons.

the sets of $\mathcal{M}$ have more than 2 points each? For instance, what can be said if all the sets of $\mathcal{M}$ have exactly 3 points? It turns out that the situation with 2-colorability becomes much more complicated than for graphs. For instance, the question of whether a given $\mathcal{M}$ is 2-colorable or not becomes algorithmically difficult.

Let us consider the following natural question: what is the smallest number of sets of a set system $\mathcal{M}$ consisting of 3-point sets that is not 2-colorable? It turns out that the answer is 7, and the Fano plane provides one-half of this answer: it has 7 sets consisting of 3 points each, and it is not 2-colorable (we leave a proof as Exercise 1). And, in fact, it is the only set system with 7 triples that is not 2-colorable! We will deal with the second half of the answer, i.e. methods for showing that all systems with 6 or fewer triples are 2-colorable, in Section 9.1 (Theorem 9.1.5).

One could talk much longer about 2-colorability, as it is a quite important concept in combinatorics, but here we only wanted to point out a little problem in this area where the projective planes, surprisingly, come into play.

More on graphs without $K_{2,2}$ with many edges. Theorem 6.3.1 tells us that if G is a graph on m vertices containing no $K_{2,2}$ as a subgraph, then G has at most $\frac{1}{2}\left(m^{3/2} + m\right)$ edges. Using finite projective planes, we show that this bound is nearly the best possible in general:

8.4.1 Theorem. *For infinitely many values of m, there exists a $K_{2,2}$-free graph on m vertices with at least $0.35\, m^{3/2}$ edges.*

Proof. Take a projective plane of order n, and consider its incidence graph (as in the part concerning duality in Section 8.1). The number of vertices of this graph is $m = 2(n^2 + n + 1)$. Each of the $n^2 + n + 1$ lines has $n+1$ points, and this means that the total number of edges is $(n^2 + n + 1)(n + 1) \geq (n^2 + n + 1)^{3/2} = \left(\frac{m}{2}\right)^{3/2} \approx 0.35 m^{3/2}$.

What would it mean if the incidence graph contained a $K_{2,2}$ as a subgraph? Well, in the language of the projective plane, it would say that there exist two points x, x' and two lines L, L' such that $x \in L$, $x' \in L$, $x \in L'$, and $x' \in L'$. This cannot happen in a projective plane. $\square$

Remark. The constant 0.35 in Theorem 8.4.1 can still be improved somewhat. The optimal value is 0.5; see Exercise 2.

Exercises

1. Prove that the Fano plane is not 2-colorable.

2. (Better $K_{2,2}$-free graphs) Let n be a prime power, and let K be an n-element field. Consider the equivalence classes of the equivalence relation $\approx$ on the set of triples $K^3 \setminus \{(0,0,0)\}$ (introduced in Section 8.2). Let these classes be vertices of a graph G, and two vertices (a, b, c) and (x, y, z) are connected by an edge if and only if $ax + by + cz = 0$. Prove that

   (a) the edges are well defined,

   (b) *the graph G contains no $K_{2,2}$ as a subgraph,

   (c) each vertex has degree at least n, and

   (d) if $m = n^2 + n + 1$ denotes the number of vertices, then the number of edges is at least $\frac{1}{2} m^{3/2} - m$.

3. Let G be a bipartite graph with both vertex classes of sizes n and containing no $K_{2,2}$ as a subgraph.

   (a) *By the method of Section 6.3, prove that G has at most

   $$\frac{1}{2} n (1 + \sqrt{4n - 3})$$

   edges.

   (b) *Prove that such a G with precisely this number of edges exists if and only if a projective plane of order q exists with $n = q^2 + q + 1$.

9
Probability and probabilistic proofs

The reader will most likely know various problems about determining the probability of some event (several such problems are also scattered in the other chapters). Textbooks often contain such problems taken "from real life", or at least pretending to be real life problems. They speak about shuffling and drawing cards, tossing coins or even needles, or also about of defunct lightbulbs, defunct phone lines, or decaying radioactive atoms—depending on the author's inclinations and fantasy. In this chapter we want to show a remarkable mathematical application of probability, namely how mathematical statements can be proved using elementary probability theory, although they don't mention any probability or randomness.

Alon and Spencer [10] is an excellent book for studying probabilistic methods in combinatorics in more depth, and Grimmett and Stirzaker [18] can be recommended as a probability theory textbook.

9.1 Proofs by counting

In two introductory examples, probability will not be mentioned; we use a simple counting instead.

9.1.1 Example. Consider a new deck of 52 cards (in such a deck, cards go in a certain fixed order). We will shuffle the cards by so-called *dovetail shuffling*: we divide the deck into two parts of equal size, and we interleave the cards from one part with the cards from the other part, in such a way that the order of cards from each part is unchanged (see Fig. 9.1). We prove that if this procedure is repeated at most 4 times, we cannot get all possible orderings of the cards, and hence 4 rounds of dovetail shuffling certainly won't yield a random order of the cards.

Fig. 9.1 Dovetail shuffling.

Proof. There are 52! possible orderings of the cards. We count how many different orderings can be obtained by the described shuffling procedure. How many ways are there to intermix the two separate parts of the deck (let us refer to them as the left part and the right part)? If we know the ordering in both the left and the right parts, and if we specify which cards in the intermixed deck come from the left part and which from the right part, we can already reconstruct the ordering of the deck after the parts are intermixed. Hence, after the first dividing and interleaving there are $\binom{52}{26}$ possible orderings, and by 4 repetitions of this procedure we can thus obtain at most $\binom{52}{26}^4$ orderings. A pocket calculator, or the estimates in Chapter 2, can tell us that this number is smaller than 52!, and hence there exists some ordering that cannot be obtained by 4 rounds of dovetail shuffling. □

The result just proved doesn't say that we *could* get all possible orderings by 5 rounds. The question of how many random dovetail shuffles are needed to obtain an ordering reasonably close to a random one is considerably more difficult. A very sophisticated paper on the subject is Bayer and Diaconis [31].

Difficult Boolean functions. A Boolean function of n variables is a mapping $f : \{0,1\}^n \to \{0,1\}$, i.e. assigning either 0 or 1 to each possible combination of n 0s and 1s. (Here 1 represents the logical value "true" and 0 the logical value "false".) A Boolean function can be specified by a table of values, but also in many other ways. For example, a computer program that reads n bits of input and computes a YES/NO answer defines a certain Boolean function of n variables. Similarly, an integrated circuit with n input wires and one output wire defines a Boolean function of n variables, assuming that the inputs and the output each have only two possible states.

We want to show that there exist Boolean functions that require a very large description, i.e. a very long program, or a circuit with a huge number of components, etc. Technically, this is probably simplest to do for yet another way of describing Boolean functions, namely by *logical formulas*.

The reader may know logical formulas from predicate calculus, say. A *logical formula in n variables* is a string made up of symbols x_1, $x_2, \ldots, x_n$ for the variables (each of them can be repeated several times), parentheses, and the following symbols for the logical connectives: $\wedge$ (conjunction), $\vee$ (disjunction), $\Rightarrow$ (implication), $\Leftrightarrow$ (equivalence), and $\neg$ (negation). Not every sequence of these symbols is a logical formula, of course; a formula has to satisfy simple syntactical rules, such as proper parenthesizing etc. The details of these rules are not important for us here. One possible formula in 3 variables is, for example, $(x_1 \wedge x_2) \vee (x_3 \wedge \neg x_1)$. Each logical formula in n variables defines a Boolean function of n variables: for given values of the variables $x_1, x_2, \ldots, x_n$, substitute these values into the formula and evaluate the truth value of the formula by the rules for the various connectives. For example, we have $0 \wedge 0 = 0 \wedge 1 = 1 \wedge 0 = 0$ and $1 \wedge 1 = 1$, and so on. It is not too difficult to show that any Boolean function can be defined by a formula. The question now is: how long does such a formula have to be?

We show

9.1.2 Proposition. *There exists a Boolean function of n variables that cannot be defined by any formula with fewer than $2^n / \log_2(n + 8)$ symbols. For example, for 23 variables we may already need a formula with more than a million symbols.*

Proof. The number of all Boolean functions of n variables is 2^{2^n}, while the number of formulas in n variables written by at most m symbols is no more than $(n + 8)^m$, because each of the m positions in the formula can be filled in by one of the $n + 7$ possible symbols, or maybe by a space. In this way, we have also counted lots of nonsense strings of symbols, but a rough upper bound is fully sufficient. If $2^{2^n} > (n + 8)^m$, there exists a Boolean function that cannot be expressed by a formula with at most m symbols. By taking logarithms in the inequality, we get $m \geq 2^n / \log_2(n + 8)$. $\square$

Similarly, one may consider Boolean functions defined by computer programs in some fixed programming language, or by integrated circuits consisting of some given set of components, and so on. In each such case, a counting similar to the above proof shows the existence of functions of n variables that cannot be defined by a program or circuit of size smaller than roughly 2^n. (We did the proof for formulas since their definition and counting seems simplest.)

The two examples just given have a common scheme. We have a set of objects, and we want to show that there exists some "good" object among them (with some required property; in Example 9.1.1, good objects were card orderings unattainable by 4 dovetail shuffles). We count how many objects there are in total, and we upper-bound the number of bad objects. If we can show that the number of bad objects is smaller than the number of all objects, this means that some good object has to exist. Typically, we even show that most of the objects must be good.

A remarkable feature of this method is that we do not construct any particular good object, and we don't even learn anything about what it might look like—we only know that it exists. In Example 9.1.1, we have shown that some ordering of cards cannot be obtained, but we have not exhibited any specific such ordering. In the proof of Proposition 9.1.2 we have not found out how to get a "difficult" Boolean function (with no short formula). This situation is quite usual for proofs of this type. It might look fairly paradoxical, since we are usually in the situation of searching for a piece of hay in a haystack with at most a few needles in it, i.e. we can prove that a large majority of the objects are good. But avoiding the needles proves enormously hard in many situations. For many interesting combinatorial objects, there are relatively easy existence proofs but none or only very difficult explicit constructions are known.

The argument about good and bad objects can be reformulated in the language of probability. Imagine that we choose an object from the considered set at random. If we show that with a nonzero probability, we choose a good object, this means that at least one good object must exist. In more complicated problems, the language of probability becomes simpler than counting the objects, and one can apply various more advanced results of probability theory whose formulation in terms of object counting would be immensely cumbersome. We will use the language of probability in the subsequent example. Any reader who should feel uneasy about some of the notions from probability theory can first read the next section.

Two-coloring revisited. Let X be a finite set and let $\mathcal{M}$ be a system of subsets of X. From Section 8.4, we recall the definition of 2-colorability: we say that $\mathcal{M}$ is 2-colorable if each of the points of X can be colored either red or white in such a way that no set of $\mathcal{M}$ has all points red or all points white. Here we will discuss the following problem (a particular case has been considered in Section 8.4).

9.1.3 Problem. Suppose that each set of $\mathcal{M}$ has exactly k elements. What is the smallest number, $m(k)$, of sets in a system $\mathcal{M}$ that is not 2-colorable?

It is easy to find out that $m(2) = 3$, since we need 3 edges to make a graph that is not bipartite. But the question for $k = 3$ is already much more difficult. In Section 8.4, we met a system of 7 triples that is not 2-colorable, namely the Fano plane, and so $m(3) \leq 7$. In fact, $m(3) = 7$; to prove this, we have to show that all systems with 6 or fewer triples can be 2-colored. We begin with a general statement which gives a weaker bound for $k = 3$. Then, with some more effort, we will improve the result for the particular case $k = 3$.

9.1.4 Theorem. *We have $m(k) \geq 2^{k-1}$, i.e. any system consisting of fewer than 2^{k-1} sets of size k admits a 2-coloring.*

Proof. Let $\mathcal{M}$ be a system of k-element subsets of some set X, and let $|\mathcal{M}| = m$. We color each point of X red or white by the following random procedure. For each point $x \in X$, we toss a fair coin. If we get heads we color x white and if we get tails we color x red.

Let $M \in \mathcal{M}$ be one of the k-tuples in the considered system. What is the probability that all the points of M get the same color in a random coloring? The probability that all the k points are simultaneously white is obviously 2^{-k}, and also the probability that all the points of M turn out red is 2^{-k}. Altogether, the probability of M ending up monochromatic is $2 \cdot 2^{-k} = 2^{1-k}$. Hence the probability that at least one of the m sets in $\mathcal{M}$ is monochromatic is at most $m \, 2^{1-k}$. If this number is strictly smaller than 1, i.e. if $m < 2^{k-1}$, then our random coloring is a 2-coloring for the system $\mathcal{M}$ with a nonzero probability. Hence, at least one 2-coloring exists. Definitely and certainly, no probability is involved anymore! Theorem 9.1.4 is proved. $\square$

How good is the bound on $m(k)$ in the theorem just proved? It is known that for large k, the function $m(k)$ grows roughly as 2^k (more exactly, we have $m(k) = \Omega(2^k k^{1/3})$ and $m(k) = O(2^k k^2)$; see [10]), and so the theorem gives quite a good idea about the behavior of $m(k)$. On the other hand, for $k = 3$ we only get the estimate $m(k) \geq 4$, which is still quite far from the correct value 7. We improve the bound with two more tricks.

9.1.5 Theorem. $m(3) \geq 7$.

We have to show that any system of 6 triples on a finite set X is 2-colorable. We distinguish two cases depending on the number of points of X: $|X| \leq 6$ and $|X| > 6$. Only the first case will be handled by a probabilistic argument.

Lemma. *Let X be a set with at most 6 elements, and let $\mathcal{M}$ be a system of at most 6 triples on X. Then $\mathcal{M}$ is 2-colorable.*

Proof. If needed, we add more points to X so that it has exactly 6 points. We choose 3 of these 6 points at random and color them white, and the remaining 3 points are colored red. Hence, we have $\binom{6}{3} = 20$ possibilities for choosing such a coloring. If M is any triple from $\mathcal{M}$, there are only 2 among the possible colorings that leave M monochromatic: either M is colored red and the remaining 3 points are white, or M is white and the other points are red. Hence the probability that M is monochromatic is $\frac{1}{10}$. The probability that some of the 6 triples of $\mathcal{M}$ become monochromatic is thus no more than $\frac{6}{10} < 1$, and hence a 2-coloring exists. $\qquad\square$

The same proof shows that also any 9 triples on 6 points can be 2-colored.

For the second step, we need the following definition. Let $(X, \mathcal{M})$ be a system of sets, and let x, y be two elements of X. We say that x and y are *connected* if there exists a set $M \in \mathcal{M}$ containing both x and y. If x and y are points that are not connected, we define a new set system $(X', \mathcal{M}')$ arising by "gluing" x and y together. The points x and y are replaced by a single point z, and we put z into all sets that previously contained either x or y. Written formally, $X' = (X \setminus \{x, y\}) \cup \{z\}$, $\mathcal{M}' = \{M \in \mathcal{M}: M \cap \{x, y\} = \emptyset\} \cup \{(M \setminus \{x, y\}) \cup \{z\}: M \in \mathcal{M}, M \cap \{x, y\} \neq \emptyset\}$.

Let us note that if points x and y are not connected and $\mathcal{M}$ is a system of triples, then $(X', \mathcal{M}')$ is again a system of triples, and the set X' has one point fewer than X. Further we claim that *if $(X', \mathcal{M}')$ is 2-colorable then $(X, \mathcal{M})$ is 2-colorable too*. To see this, consider a 2-coloring of the set X', and color X in the same way, where both x and y receive the color the "glued" point z had. It is easy to see that no monochromatic set can arise in this way. Hence, for finishing the proof of Theorem 9.1.5, it suffices to prove the following:

Lemma. *Let $(X, \mathcal{M})$ be a system of 6 triples with $|X| \geq 7$. Then X contains two points that are not connected in $\mathcal{M}$.*

Proof. One triple $M \in \mathcal{M}$ makes 3 pairs of points connected. Hence 6 triples yield at most $3 \cdot 6 = 18$ connected pairs. But the total number of pairs of points on a 7-element set is $\binom{7}{2} = 21$, and hence some pair is not connected (even at least 3). $\square$

Let us remark that the exact value of $m(4)$ is already unknown (and similarly for all the larger k). It is also easy to see that $m(4)$ can in principle be computed by considering finitely many configurations (systems of 4-tuples). But the number of configurations appears sufficient to resist the power of all kinds of supercomputers, at least if that power is not accompanied by enough human ingenuity.

Exercises

1. (a) Prove that any Boolean function of n variables can be expressed by a logical formula.

   (b) Show that the formula in (a) can be made of length at most $Cn2^n$ for a suitable constant C. *Can you improve the order of magnitude of this bound, say to $O(2^n)$ or even better?

2. (a) Prove that $m(4) \geq 15$, i.e. that any system of 14 fourtuples can be 2-colored. Proceed similarly as in the proof of Theorem 9.1.5, distinguishing two cases according to the total number of points.

   (b) *Give as good an upper bound on $m(4)$ as you can! Can you get below 50? Below 30?

3. We have 27 fair coins and one counterfeit coin, which looks like a fair coin but is a bit heavier. Show that one needs at least 4 weighings to determine the counterfeit coin. We have no calibrated weights, and in one weighing we can only find out which of two groups of some k coins each is heavier, assuming that if both groups consist of fair coins only the result is an equilibrium.

4. *In the following diagram, a train with n cars is standing on the rail track labeled A. The cars are being moved to the track B. Each car may or may not be shifted to some of the side tracks I–III but it should visit each side track at most once and it should go through tracks C and D only once.

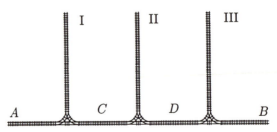

Prove that if n is large enough then there is some order of the cars that cannot be achieved on track B.

9.2   Finite probability spaces

Now it is time to talk about the basic notions of mathematical probability theory. We will restrict ourselves to things we will need for our examples. By no means do we intend to provide a substitute for a proper course in probability theory. Anyone educated in mathematics or in theoretical computer science should know considerably more about probability than what is said here.

Probability is a notion that can appear both in mathematics and outside it: "in real life", "in practice", or whatever else one can call this. Defining the "real" probability is difficult, and it is a philosophical problem. Mathematics avoids its solution, though: it constructs a certain model of the "real" probability, and this model is a purely mathematical object. At its fundamental level, several simple properties of probability derived from knowledge about the real world are built in as axioms, such as the fact that the probability of some event occurring plus the probability of its not occurring add up to 1. But once the axioms are accepted, one works with the mathematical notion of probability as with any other mathematical object, and all its properties and all rules for calculating with probabilities are logically derived from the axioms. This model is very useful and its predictions agree with the behavior of probability in practice, but this doesn't mean that the mathematical probability and the "real" probability are the same notions. In the sequel, we will speak of probability in the mathematical sense, but we will employ examples with "real" probability as a motivation of the notions and axioms.

A basic notion in probability theory is a *probability space*. Here we restrict ourselves to finite probability spaces.

9.2.1 Definition. *By a finite probability space we understand a pair* (Ω, P), *where* Ω *is a finite set and* $P: 2^\Omega \to [0, 1]$ *is a function assigning a number from the interval* $[0, 1]$ *to every subset of* Ω, *such that*

(i) $P(\emptyset) = 0$,

(ii) $P(\Omega) = 1$, *and*

(iii) $P(A \cup B) = P(A) + P(B)$ *for any two disjoint sets* $A, B \subseteq \Omega$.

The set Ω can be thought of as the set of all possible outcomes of some random experiment. Its elements are called the *elementary events*. For instance, if the experiment is rolling one fair die, the elementary events would be "1 was rolled", "2 was rolled",..., "6 was rolled". For brevity, we can denote these elementary events by $\omega_1, \omega_2, \ldots, \omega_6$. Subsets of Ω are called *events*. An example of an event is "an even number was rolled", in other words $\{\omega_2, \omega_4, \omega_6\}$.

Beware: An elementary event is *not* an event, since events are sets of elementary events!

Probability theory has its names for various set-theoretic notions and operations with events. For instance, "$\omega \in A$" can be read as "the event A occurred", "$\omega \in A \cap B$" can be interpreted as "both the events A and B occurred", "$A \cap B = \emptyset$" can be expressed as "the events A and B are incompatible", and so on.

If $A \subseteq \Omega$ is an event, the number $P(A)$ is called the *probability of the event A*. Axioms (i)–(iii) express properties which we naturally expect from probability. From condition (iii), it is easy to see that it suffices to specify the values of the function P on all one-element events (sets), since the probability of any event equals the sum of probabilities of all its elementary events (more precisely, the sum of probabilities of all its one-element subsets, but we allow ourselves the abbreviated formulation).

Concerning axiom (iii) in the definition of a finite probability space, let us remark that for *any two events* $A, B \subseteq \Omega$, not necessarily disjoint ones, we have the inequality $P(A \cup B) \leq P(A) + P(B)$ (Exercise 1).

The simplest finite probability space, and perhaps also the most important one, is that in which all the elementary events have the same probability, i.e. the function P is given by

$$P(A) = \frac{|A|}{|\Omega|}$$

for all events A.

Such a probability space reflects the so-called *classical definition of probability*. In this definition, formulated by Laplace, one assumes that all possible outcomes of some random experiment are equally likely (such an assumption can be based on some symmetry and/or homogeneity in the experiment). If the number of all outcomes (elementary events) of the experiment is n, and, among these, there are m outcomes favorable for some event A, then the probability of the event A is defined as m/n. This is sometimes expressed by the phrase that probability is the number of favorable outcomes divided by the number of possible outcomes. For defining what probability is, the classical definition is not really satisfactory (the catch is in the term "equally likely") and, moreover, it doesn't include infinite probability spaces, but in many specific cases it is at least a useful hint for probability calculation.

On infinite probability spaces. By restricting ourselves to finite probability spaces we have simplified the situation considerably from a mathematical point of view. (A true probability theorist would probably say that we have also excluded everything interesting.) But for modeling many interesting events, it is more natural to work with infinite probability spaces, whose definition is technically more complicated. For example, if we choose 5 points at random in the interval $[0,1]$, what is the probability that some two of them lie at distance at most $\frac{1}{10}$? We can ask thousands of questions of a similar type. We would need to define what a "random point from the interval $[0,1]$" means in the first place. The elementary events should naturally be all the points in $[0,1]$. And the probability of all individual points should be the same, at least if we want the points to be "uniformly distributed", and because there are infinitely many points, the probability of each point–elementary event must be 0. Therefore, it is not possible to specify the probability function P by defining it for the one-element events only (as in the finite case). The probability of an event A has to be a "measure" of A in a suitable sense. This is a more complicated notion, which is closely related to integration and other questions of mathematical analysis. Other examples of infinite probability spaces are used, without calling them so, in Sections 10.5 and 10.6.

Next, we list several important species of finite probability spaces.

9.2.2 Definition (A random sequence of n 0s and 1s). *The elementary events of this probability space are all n-term sequences of 0s and 1s, i.e. elements of the set $\{0,1\}^n$, and all elementary events have the same probability. Since the number of elementary events is 2^n, the probability of any event A equals $|A|/2^n$. We denote this probability space by $\mathcal{C}_n$.*

This probability space models a sequence of n coin tosses, for instance (assuming the coin is symmetric and heads and tails occur with the same probability). If the ith toss yields heads we write 1 in the ith position in the sequence, and for tails we write 0. An example of an event is $A =$ "Heads appear exactly 10×", whose probability is $\binom{n}{10}/2^n$.

9.2.3 Definition (A random permutation). *Elementary events of this probability space are all permutations of the set $\{1,2,\ldots,n\}$, and the probability of an event A equals $|A|/n!$. The set of all permutations of the set $\{1,2,\ldots,n\}$ is traditionally denoted by S_n, and we denote the probability space by $\mathcal{S}_n$.*

This space is a model for arranging some n distinct elements in a random order, e.g. for a well-shuffled card deck.

Problem. What is the probability that in a random ordering of a bridge card deck, the ace of spades precedes the king of hearts?

Reformulated in our probability model, what is the probability of the event $A = \{\pi \in S_{52}: \pi(1) < \pi(2)\}$? We could, of course, honestly count the permutations with $\pi(1) < \pi(2)$, but we can also determine the probability in question by a simple consideration: by symmetry, it is impossible that one of the events "$\pi(1) < \pi(2)$" and "$\pi(1) > \pi(2)$" would be more likely than the other, and so the required probability is $\frac{1}{2}$. More precisely, such a reasoning should perhaps be formulated as follows. We can construct a bijection between the set A and the set $A' = \{\pi \in S_{52}: \pi(1) > \pi(2)\}$. To a permutation $\pi \in A$, we assign the permutation π' with $\pi'(1) = \pi(2)$, $\pi'(2) = \pi(1)$, and $\pi'(i) = \pi(i)$ for $i > 2$. Hence $|A| = |A'|$, and since A and A' are disjoint and together cover the whole probability space, we have $P(A) = \frac{1}{2}$.

To revive the reader's attention, we now include a problem with a surprising solution.

Problem. We play the following game. Our rival has 100 blank cards, and on each of them he writes some quite arbitrary number at will. Then he shuffles the cards (or, a neutral third party should better do it), and the shuffled card deck is laid on a table face down so that the numbers are not visible. We start removing cards from the top one by one and look at their numbers. After turning any of the cards, we can end the game. We win if the last card we turned has the largest number among all the cards (those already turned but also those still lying on the table). If we win we take 40 doublezons, and if we lose we pay 10 doublezons. Can we expect to gain on this game?

At first sight it may seem that this game is not advantageous for us at all. Let us keep the following strategy, though: turn the first 50 cards no matter what, and remember the largest number we saw there; let it be M. Then turn the remaining cards, and finish as soon as we turn a card with a number greater than or equal to M. If we encounter no such card we finish with the last card.

We claim that the probability of winning is greater than $\frac{1}{4}$ for this strategy. Therefore, we can expect to win at least 1 game out of 4 on the average, and so our expected gain in a long series of games will be positive, at least about $\frac{1}{4} \cdot 40 - \frac{3}{4} \cdot 10 = 2.50$ doublezons per game—no fortune but enough for a beer. For simplicity, let us suppose that all the numbers on the cards are distinct (an interested reader can consider the modification for the case of arbitrary numbers). The strategy described surely leads to winning if

- the largest number is among the last 50 cards, and
- the second largest number is among the first 50 cards.

Obviously, for our strategy, the game depends solely on the ordering of the numbers and not on their actual values, and so we can imagine that the cards contain numbers $1, 2, \ldots, 100$, and their random ordering is thus an elementary event from the probability space S_{100}. We are interested in the event $A = \{\pi \in S_{100} \colon \pi(100)' > 50 \text{ and } \pi(99) \leq 50\}$. Here it is useful to think of a permutation as a linear ordering. The position of the number 100 can be chosen in 50 ways, the position of 99 can be chosen in 50 ways independently of the placement of 100, and the remaining numbers can be arranged in 98! ways. Hence

$$P(A) = \frac{50 \cdot 50 \cdot 98!}{100!} = \frac{50 \cdot 50}{99 \cdot 100} \doteq 0.2525 > \frac{1}{4}.$$

Let us finish this example with a few remarks. The event A is not the only situation in which our strategy wins, and hence the probability of A is only a lower bound for the probability of winning. The number 50 used as a threshold in our strategy optimizes the probability of the event A (meaning the event that 100 comes after the threshold and 99 before it). But if we also take other possibilities of winning into account, a somewhat better winning chance is obtained for a different threshold. We will not continue a detailed analysis of the game here; we leave it as a challenge for the reader.

Next, we are going to consider the notion of a "random graph". As we will see later, there are several ways a random graph can be reasonably defined. Here we consider the simplest way. A simple, undirected graph on the vertex set $V = \{1, 2, \ldots, n\}$ is specified by deciding, for each pair $\{i, j\} \in \binom{V}{2}$, whether this pair is an edge or not. Hence there are $2^{\binom{n}{2}}$ graphs (many of them are isomorphic, of course, but this doesn't concern us here). If we select a random graph G on the vertex set V, in such a way that all possible graphs have the same probability, we can view this as $\binom{n}{2}$ symmetric coin tosses. That is, for each pair of vertices we toss a coin to decide whether it becomes an edge or not. This is reflected in the following definition.

9.2.4 Definition (A random graph). *This probability space, denoted by $\mathcal{G}_n$, has all the possible graphs on vertex set $\{1, 2, \ldots, n\}$ as elementary events, and all of them have the same probability, equal to $2^{-\binom{n}{2}}$.*

As examples of events in the probability space $\mathcal{G}_n$ we can study all sorts of natural graph properties, such as $S = $ "the graph G is connected", $B = $ "the graph G is bipartite", and so on. Computing the probability of such events exactly is often very difficult, but we are usually interested in a very rough estimate. For the two mentioned examples of events, it turns that for $n \to \infty$, $P(S)$ rapidly tends to 1 while $P(B)$ approaches 0 very quickly. This is sometimes expressed by saying that "a random graph is almost surely connected, and almost surely it is not bipartite". We prove the first of these two claims, by an approach typical also for many other assertions of this type.

9.2.5 Proposition. *A random graph almost surely isn't bipartite, i.e.* $\lim_{n\to\infty} P(B) = 0.$

Proof. As we know, the vertex set V of a bipartite graph can be partitioned into two parts, U and W, in such a way that all edges go between U and W only. For a given subset $U \subseteq V$, let B_U denote the event that all edges of the random graph G go between vertices of U and vertices of $W = V \setminus U$ only. If $k = |U|$, we have $k(n-k)$ pairs $\{u, w\}$ with $u \in U$ and $w \in V \setminus U$, and so the event (set) B_U consists of $2^{k(n-k)}$ graphs. Therefore $P(B_U) = 2^{k(n-k) - \binom{n}{2}}$. It is not hard to check that the function $k \mapsto k(n-k)$ attains its maximum for $k = \frac{n}{2}$, and the value of this maximum is $n^2/4$; hence $k(n-k) \leq n^2/4$ for all k. So we have, for any U,

$$P(B_U) \leq 2^{n^2/4 - \binom{n}{2}} = 2^{-n(n-2)/4}.$$

Each bipartite graph belongs to some B_U (for a suitable choice of the set U). For different choices of U, the events B_U need not be disjoint, but in any case, the probability of a union of events is always at most the sum of their probabilities, and so

$$P(B) \leq \sum_{U \subseteq V} P(B_U) \leq 2^n \cdot 2^{-n(n-2)/4} = 2^{-n(n-6)/4} \to 0.$$

$\square$

In this problem, we were interested in a certain qualitative property of a "big" random graph. This somewhat resembles the situation in various areas of physics (such as thermodynamics or solid-state physics) where macroscopic properties of a large collection of microscopic particles are studied. It is assumed that the individual particles behave randomly in a suitable sense, and the macroscopic properties are a result of their random interactions. Also mathematical methods in the study of ferromagnetism and other properties of the solid-state matter are similar to methods for random graphs. Analogous approaches are also applied in social sciences, for instance in modeling epidemics etc.

In Definition 9.2.4, we suppose that all graphs have the same probability, or in other words, that the edge probability is $\frac{1}{2}$. In interesting problems and applications, the edge probability is usually chosen as a parameter p generally distinct from $\frac{1}{2}$. This means that a random graph is constructed by $\binom{n}{2}$ coin tosses, one toss for each pair of vertices (heads means an edge and tails no edge), but the coin has the probability of heads equal to p and the probability of tails equal to $1 - p$. The properties of the random graph are often investigated in dependence on this p. For instance, if p grows from 0 to 1, for which value does the random graph typically start to be connected? This is again conceptually not too far from physics questions like: at what temperature does this crystal start to melt?

Independent events. We have to cover one more key notion. Two events A, B in a probability space (Ω, P) are called *independent* if we have

$$P(A \cap B) = P(A)P(B).$$

Independence means that if Ω is divided into two parts, A and its complement, the event B "cuts" both these parts in the same ratio. In other words, if an elementary event ω were chosen at random not in the whole Ω but among the elementary events from A, the probability of $\omega \in B$ would be exactly equal to $P(B)$ (assuming $P(A) \neq 0$).

Independence does *not* mean that $A \cap B = \emptyset$ as some might perhaps think.

Most often, we encounter independent events in the following situation. The elements of Ω, i.e. the elementary events, can be viewed as ordered pairs, so Ω models the possible results of a "compound" experiment consisting of two experiments. Let us assume that the course and result of the first of these two experiments cannot possibly influence the result of the second experiment and vice versa (the experiments are separated by a thick enough wall or something). If $A \subseteq \Omega$ is an event depending on the outcome of the first experiment only (i.e if we know this outcome we can already decide whether A occurred or not), and similarly if B only depends on the outcome of the second experiment, then the events A and B are independent.

The space $\mathcal{C}_n$ (a random n-term sequence of 0s and 1s) is a typical source of such situations. Here we have a compound experiment consisting of n subsequent coin tosses, and we assume that these do not influence one another in any way; for instance, the coin is not deformed or stolen during the first experiment. So, for example, if an event A only depends on the first 5 tosses ("heads appeared at least $3\times$ in the first 5 tosses") and an event B only depends on the 6th and subsequent tosses ("heads appeared an odd number of times in tosses number 6 through 10"), such events are independent.

Similarly, in the probability space $\mathcal{G}_n$ (a random graph), the edges are independent, and so for example the events "the graph G has at least one triangle on the vertices $1, 2, \ldots, 10$" and "the graph G contains an odd-length cycle with vertices among $11, 12, \ldots, 20$" are independent.

A subtler situation can be demonstrated in the probability space $\mathcal{S}_n$ (a random permutation). The events $A = \{\pi(1) = 1\}$ and $B = \{\pi(2) = 1\}$ are clearly *not* independent because $P(A) > 0$ and $P(B) > 0$, but $A \cap B = \emptyset$ and thus $P(A \cap B) = 0$. If we define another event $C = \{\pi(2) = 2\}$, it is equally easy to see that B and C aren't independent either. But maybe it is not so obvious anymore that even A and C are not independent: we have $P(A) = P(C) = \frac{1}{n}$ but $P(A \cap C) = \frac{1}{n(n-1)} \neq P(A)P(C)$. Intuitively, if we know that A occurred, i.e. that $\pi(1) = 1$, we have also excluded one of the n possibilities for the value of $\pi(2)$, and hence $\pi(2)$ has a slightly bigger chance of being equal to 2. On the other hand, the events A and $D = \{\pi(2) < \pi(3)\}$ are independent, as one can check by computing the relevant probabilities. One has to be careful about such subtleties. (Perhaps the most frequent source of error in probabilistic proofs is that some events are assumed to be independent although in reality they are not.)

The notion of independence can also be extended to several events $A_1, A_2, \ldots, A_n$.

9.2.6 Definition. *Events $A_1, A_2, \ldots, A_n \subseteq \Omega$ are called independent if we have, for each set of indices $I \subseteq \{1, 2, \ldots, n\}$,*

$$P\left(\bigcap_{i \in I} A_i\right) = \prod_{i \in I} P(A_i).$$

In particular, this definition requires that each two of these events be independent, but we have to warn that the independence of each pair doesn't in general imply the independence of all events!

In the probability spaces $\mathcal{C}_n$ (random 0/1 sequence) and $\mathcal{G}_n$ (random graph) we have typical situations with many independent events. Let us define an event A_i in the probability space $\mathcal{C}_n$, consisting of all sequences with 1 in the ith position. Then the events $A_1, A_2, \ldots, A_n$ are independent (we omit a proof—it is not difficult).

In various probabilistic calculations and proofs, the probability spaces are usually not described explicitly—one only works with them without saying. Nevertheless, it is important to clarify these basic notions.

We conclude this section with a nice probability proof (historically, it is one of the first proofs by this method).

Let us consider a tournament of n players, say a tennis tournament, in which each player plays against everyone else and each match has a winner. If there are big differences among the players' strength, we may expect that the best player beats everyone, the second best beats all but the first one, etc., so that the tournament determines the order of players convincingly. A tournament with a more leveled strength of players can give a more complicated outcome. Of course, it can happen that every player is beaten by someone. And, mathematicians wouldn't be mathematicians if they didn't ask the following generalization:

Problem. Is it possible that in some tournaments, every two players are both beaten by some other player? And, more generally, for which numbers k can there be a tournament in which for every k players there exists another player who has beaten them all?

For $k = 2$, one can construct such a tournament "by hand". But for larger values of k, the construction of such tournaments seemed difficult, and a solution has been sought in vain for quite some time. But the method using probability shows the existence of such tournaments (with many players) quite easily. For simplicity, we show the solution for $k = 3$ only, so we want a tournament outcome in which every 3 players x, y, z are all beaten by some other player, w.

Let us consider a random tournament, where we imagine that the result of each match is determined by lot, say by tossing a fair coin. Let us look at some fixed triple $\{x, y, z\}$ of players. The probability that some other player, w, wins against all of them is $2^{-3} = \frac{1}{8}$. Hence the probability that w loses against at least one of x, y, z is $\frac{7}{8}$. What is the probability that each of the $n - 3$ players who can appear in the role of w loses against at least one among x, y, z? For distinct players w, the results of their matches with x, y, z are mutually independent, and so the desired probability is $(\frac{7}{8})^{n-3}$. The triple $\{x, y, z\}$ can be selected in $\binom{n}{3}$ ways, and hence the probability that for at least one of the triples $\{x, y, z\}$, no player beats x, y, and z simultaneously, is at most $\binom{n}{3}(\frac{7}{8})^{n-3}$. Using a pocket calculator we can check that for $n \geq 91$, this probability is smaller than 1.

Therefore, there exists at least one result of a tournament of 91 players in which any 3 players are simultaneously beaten by some other player. This is the desired property. □

Exercises

1. Prove that for any two events A, B in a (finite) probability space, we have $P(A \cup B) \leq P(A) + P(B)$. Generalize this to the case of n events.

2. (Probabilistic formulation of inclusion–exclusion)

   (a) Formulate the inclusion–exclusion principle (Theorem 2.7.2) in the language of probability theory. Let $A_1, A_2, \ldots, A_n$ be events in some finite probability space. Assuming that all elementary events in this probability space have the same probability, express the probability $P(A_1 \cup \cdots \cup A_n)$ using the probabilities of various intersections of the A_i.

   (b) Show that the formula as in (a) holds for events in an arbitrary finite probability space. (The finiteness is not really needed here.)

3. Prove that a random graph in the sense of Definition 9.2.4 almost surely contains a triangle (this provides another proof for Proposition 9.2.5).

4. *Show that a random graph is almost surely connected.

5. Find an example of 3 events in some probability space, such that each 2 of them are independent but all 3 are not independent.

6. Show that if A, B are independent events then also their complements, $\Omega \setminus A$ and $\Omega \setminus B$, are independent.

7. Let (Ω, P) be a finite probability space in which all elementary events have the same probability. Show that if $|\Omega|$ is a prime number then no two nontrivial events (distinct from $\emptyset$ and Ω) can be independent.

8. (a) Show that the events $A_1, A_2, \ldots, A_n$ in the probability space $\mathcal{C}_n$ defined in the text following Definition 9.2.6 are really independent.

   (b) *Let (Ω, P) be a finite probability space. Suppose that n independent events $A_1, A_2, \ldots, A_n \subseteq \Omega$ exist such that $0 < P(A_i) < 1$ for each i. Show that then $|\Omega| \geq 2^n$.

9. For simplicity, assume that the probabilities of the birth of a boy and of a girl are the same (which is not quite so in reality). For a certain family, we know that they have exactly two children, and that at least one of them is a boy. What is the probability that they have two boys?

10. (a) ^{CS} Write a program to generate a random graph with a given edge probability p and to find its connected components. For a given number n of vertices, determine experimentally at which value of p the

random graph starts to be connected, and at which value of p it starts to have a "giant component" (a component with at least $\frac{n}{2}$ vertices, say).

(b) **Can you find theoretical explanations for the findings in (a)? You may want to consult the book [10].

9.3 Random variables and their expectation

9.3.1 Definition. *Let (Ω, P) be a finite probability space. By a random variable* on Ω, we mean any mapping $f: \Omega \to \mathbf{R}$.

A random variable f thus assigns some real number $f(\omega)$ to each elementary event $\omega \in \Omega$. Let us give several examples of random variables.

9.3.2 Example (Number of 1s). If $\mathcal{C}_n$ is the probability space of all n-term sequences of 0s and 1s, we can define a random variable f_1 as follows: for a sequence s, $f_1(s)$ is the number of 1s in s.

9.3.3 Example (Number of surviving rabbits). Each of n hunters selects a rabbit at random from a group of n rabbits, aims a gun at it, and then all the hunters shoot at once. (We feel sorry for the rabbits but this is what really happens sometimes.) A random variable f_2 is the number of rabbits that survive (assuming that no hunter misses). Formally, the probability space here is the set of all mappings $\alpha: \{1, 2, \ldots, n\} \to \{1, 2, \ldots, n\}$, each of them having the probability n^{-n}, and $f_2(\alpha) = |\{1, 2, \ldots, n\} \setminus \alpha(\{1, 2, \ldots, n\})|$.

9.3.4 Example (Number of left maxima). On the probability space $\mathcal{S}_n$ of all permutations of the set $\{1, 2, \ldots, n\}$, we define a random variable f_3: $f_3(\pi)$ is the number of *left maxima* of a permutation π, i.e. the number of the i such that $\pi(i) > \pi(j)$ for all $j < i$.

Imagine a long-jump contest, and assume for simplicity that each competitor has a very stable performance, i.e. always jumps the same distance, and these distances are different for different competitors (these, admittedly unrealistic, assumptions can be relaxed significantly). In the first series of jumps, n competitors jump in a random order. Then f_3 means the number of times the current longest jump changes during the series.

9.3.5 Example (Sorting algorithm complexity). This random variable is somewhat more complicated. Let A be some sorting algorithm, meaning that the input of A is an n-tuple $(x_1, x_2, \ldots, x_n)$ of numbers, and the output is the same numbers in a sorted order. Suppose that the number of steps made by algorithm A only depends on the ordering of the input numbers (so that we can imagine that the input is some permutation π of the set $\{1, 2, \ldots, n\}$). This condition is satisfied by many algorithms that only use pairwise comparisons of the input numbers for sorting; some of them are frequently used in practice. We define a random variable f_4 on the probability space $\mathcal{S}_n$: we let $f_4(\pi)$ be the number of steps made by algorithm A for the input sequence $(\pi(1), \pi(2), \ldots, \pi(n))$.

9.3.6 Definition. *Let (Ω, P) be a finite probability space, and let f be a random variable on it. The expectation of f is a real number denoted by $\mathbf{E}[f]$ and defined by the formula*

$$\mathbf{E}[f] = \sum_{\omega \in \Omega} P(\{\omega\}) f(\omega).$$

In particular, if all the elementary events $\omega \in \Omega$ have the same probability (as is the case in almost all of our examples), then the expectation of f is simply the arithmetic average of the values of f over all elements of Ω:

$$\mathbf{E}[f] = \frac{1}{|\Omega|} \sum_{\omega \in \Omega} f(\omega).$$

The expectation can be thought of as follows: if we repeat a random choice of an elementary event ω from Ω many times, then the average of f over these random choices will approach $\mathbf{E}[f]$.

Example 9.3.2 (Number of 1s) continued. For an illustration, we compute the expectation of the random variable f_1, the number of 1s in an n-term random sequence of 0s and 1s, according to the definition. The random variable f_1 attains a value 0 for a single sequence (all 0s), value 1 for n sequences, ..., value k for $\binom{n}{k}$ sequences from C_n. Hence

$$\mathbf{E}[f_1] = \frac{1}{2^n} \sum_{s \in \{0,1\}^n} f_1(s)$$

$$= \frac{1}{2^n} \sum_{k=0}^{n} \binom{n}{k} k.$$

As we will calculate in Example 10.1.1, the final sum equals $n2^{n-1}$, and so $\mathbf{E}[f_1] = \frac{n}{2}$. Since we expect that for n coin tosses, heads should occur about $\frac{n}{2}$ times, the result agrees with intuition.

The value of $\mathbf{E}[f_1]$ can be determined in a simpler way, by the following trick. For each sequence $s \in C_n$ we consider the sequence $\bar{s}$ arising from s by exchanging all 0s for 1s and all 1s for 0s. We have $f_1(s) + f_1(\bar{s}) = n$, and so

$$\mathbf{E}[f_1] = \frac{1}{2^n} \sum_{s \in \{0,1\}^n} f_1(s) = \frac{1}{2^n \cdot 2} \sum_{s \in \{0,1\}^n} (f_1(s) + f_1(\bar{s}))$$

$$= 2^{-n-1} 2^n n = \frac{n}{2}.$$

We now describe a method that often allows us to compute the expectation in a surprisingly simple manner (we saw that the calculation according to the definition can be quite laborious even in very simple cases). We need a definition and a simple theorem.

9.3.7 Definition. *Let $A \subseteq \Omega$ be an event in a probability space (Ω, P). By the* indicator *of the event A we understand the random variable $I_A: \Omega \to \{0, 1\}$ defined in the following way:*

$$I_A(\omega) = \begin{cases} 1 & \text{for } \omega \in A \\ 0 & \text{for } \omega \notin A. \end{cases}$$

(So the indicator is just another name for the characteristic function of A.)

9.3.8 Observation. *For any event A, we have $\mathbf{E}[I_A] = P(A)$.*

Proof. By the definition of expectation we get

$$\mathbf{E}\left[I_A\right] = \sum_{\omega \in \Omega} I_A(\omega) P(\{\omega\}) = \sum_{\omega \in A} P(\{\omega\}) = P(A).$$

□

The following result almost doesn't deserve to be called a theorem since its proof from the definition is immediate (and we leave it to the reader). But we will find this statement extremely useful in the sequel.

9.3.9 Theorem (Linearity of expectation). *Let f, g be arbitrary random variables on a finite probability space (Ω, P), and let α be a real number. Then we have $\mathbf{E}\left[\alpha f\right] = \alpha \mathbf{E}\left[f\right]$ and $\mathbf{E}\left[f + g\right] = \mathbf{E}\left[f\right] + \mathbf{E}\left[g\right]$.*

□

Let us emphasize that f and g can be totally arbitrary, and need not be independent in any sense or anything like that. (On the other hand, this nice behavior of expectation *only* applies to adding random variables and multiplying them by a constant. For instance, it is not true in general that $\mathbf{E}\left[fg\right] = \mathbf{E}\left[f\right]\mathbf{E}\left[g\right]$!) Let us continue with a few examples of how 9.3.7–9.3.9 can be utilized.

Example 9.3.2 (Number of 1s) continued again. We calculate $\mathbf{E}\left[f_1\right]$, the average number of 1s, in perhaps the most elegant way. Let the event A_i be "the ith coin toss gives heads", so A_i is the set of all n-term sequences with a 1 in the ith position. Obviously, $P(A_i) = \frac{1}{2}$ for all i. We note that for each sequence $s \in \{0, 1\}^n$ we have $f_1(s) = I_{A_1}(s) + I_{A_2}(s) + \cdots + I_{A_n}(s)$ (this is just a rather complicated way to write down a trivial statement). By linearity of expectation and then using Observation 9.3.8 we obtain

$$\mathbf{E}\left[f_1\right] = \mathbf{E}\left[I_{A_1}\right] + \mathbf{E}\left[I_{A_2}\right] + \cdots + \mathbf{E}\left[I_{A_n}\right]$$
$$= P(A_1) + P(A_2) + \cdots + P(A_n) = \frac{n}{2}.$$

□

Example 9.3.3 (Number of surviving rabbits) continued. We will compute $\mathbf{E}\left[f_2\right]$, the expected number of surviving rabbits. This time, let A_i be the event "the ith rabbit survives"; formally, A_i is the set of all mappings α that map no element to i. The probability that

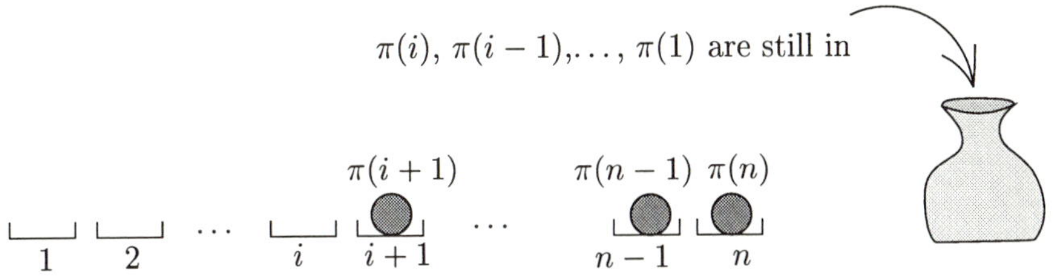

$$\pi(i),\ \pi(i-1),\dots,\ \pi(1) \text{ are still in}$$

Fig. 9.2 A procedure for selecting a random permutation.

the jth hunter shoots the ith rabbit is $\frac{1}{n}$, and since the hunters select rabbits independently, we have $P(A_i) = (1 - 1/n)^n$. The remaining calculation is as in the preceding example:

$$\mathbf{E}\left[f_2\right] = \sum_{i=1}^{n} \mathbf{E}\left[I_{A_i}\right] = \sum_{i=1}^{n} P(A_i) = \left(1 - \frac{1}{n}\right)^n n \approx \frac{n}{e}$$

(since $(1-1/n)^n$ converges to e^{-1} for $n \to \infty$; see Exercise 2.5.2). About 37% of the rabbits survive on the average. □

Example 9.3.4 (Number of left maxima) continued. Now we will calculate the expected number of left maxima of a random permutation, $\mathbf{E}\left[f_3\right]$. Let us define A_i as the event "i is a left maximum of π", meaning that $A_i = \{\pi \in S_n \colon \pi(i) > \pi(j) \text{ for } j = 1, 2, \dots, i-1\}$. We claim that $P(A_i) = \frac{1}{i}$. Perhaps the most intuitive way of deriving this is to imagine that the random permutation π is produced by the following method. We start with a bag containing the numbers $1, 2, \dots, n$. We draw a number from the bag at random and declare it to be $\pi(n)$. Then we draw another random number from the bag which becomes $\pi(n-1)$ etc., as in Fig. 9.2. The value of $\pi(i)$ is selected at the moment the bag contains exactly i numbers. The probability that we choose the largest one of these i numbers for $\pi(i)$ (which is exactly the event A_i) thus equals $\frac{1}{i}$. The rest is again the same as in previous examples:

$$\mathbf{E}\left[f_3\right] = \sum_{i=1}^{n} \mathbf{E}\left[I_{A_i}\right] = \sum_{i=1}^{n} P(A_i) = 1 + \frac{1}{2} + \frac{1}{3} + \cdots + \frac{1}{n}.$$

The value of the sum of reciprocals on the right-hand side is roughly $\ln n$; see Section 2.4. □

Exercises

1. Show with examples that if f and g are arbitrary random variables then none of the following equalities must necessarily hold: $\mathbf{E}\,[fg] = \mathbf{E}\,[f]\,\mathbf{E}\,[g]$, $\mathbf{E}\,[f^2] = \mathbf{E}\,[f]^2$, $\mathbf{E}\,[1/f] = 1/\mathbf{E}\,[f]$.

2. Prove that $\mathbf{E}\,[f^2] \geq \mathbf{E}\,[f]^2$ holds for any random variable f.

3. Let $f(\pi)$ be the number of fixed points of a permutation π (see Section 2.8). Compute $\mathbf{E}\,[f]$ for a random permutation π in the space $\mathcal{S}_n$.

4. Let π be a random permutation of the set $\{1, 2, \ldots, n\}$.

   (a) *Determine the expected length of the cycle of π containing the number 1 (see Section 2.2 for the definition of a cycle).

   (b) *Determine the expected number of cycles of π.

5. A bus route connects downtown Old Holstein with the local university campus. Mr. X., a student at the university, takes the bus from downtown every weekday after he wakes up, which happens at a random time of the day (24 hours). According to his records, he has to wait for the bus for 30 minutes on the average. At the same time, the bus company claims that the average interval between two buses during the day (over the period of 24 hours) is 15 minutes. Can you construct a schedule such that both Mr. X. and the bus company are right?

6. *We toss a fair coin n times. What is the expected number of "runs"? Runs are consecutive tosses with the same result. For instance, the toss sequence HHHTTHTH has 5 runs.

7. (Markov inequality) Let X be a random variable on some probability space attaining nonnegative values only. Let $\mu = \mathbf{E}\,[X]$ be its expectation, and let $t \geq 1$ be a real number. Prove that the probability that X attains a value $\geq t\mu$ is at most $\frac{1}{t}$; in sympols,

$$P\left(\{\omega \in \Omega \colon X(\omega) \geq t\mu\}\right) \leq \frac{1}{t}.$$

(This is a simple but quite important inequality. It is often used if we want to show that the probability of some quantity getting too big is small.)

8. (a) What is the expected number of surviving rabbits in Example 9.3.3 if there are m rabbits and n hunters?

   (b) *Using Exercise 7 and suitable estimates, show that if we have $n > m(\ln m + 5)$ then with probability at least 0.99, no rabbit survives. (In other words, most of the mappings from an n-element set into an m-element set are onto.)

   (c) *Solve part (b) differently: use the derivation of the formula for the number of mappings onto via inclusion–exclusion (Exercise 2.8.7) and the Bonferroni inequality (2.20) with $q = 1$.

9.4   Several applications

In this section, we have collected examples of using the probabilistic method and the linearity of expectation in particular. They are not routine examples but rather small mathematical gems.

Existence of large bipartite subgraphs.   Given a graph $G = (V, E)$, we would like to partition the vertex set into two parts in such a way that as many edges as possible go between these parts. Moreover, we often need that the parts have an approximately equal size. The following theorem shows that we can always make at least half of the edges go between the parts, and, moreover, that the parts can be chosen with an equal size (if the number of vertices is even).

9.4.1 Theorem. *Let G be a graph with an even number, $2n$, of vertices and with $m > 0$ edges. Then the set $V = V(G)$ can be divided into two disjoint n-element subsets A and B in such a way that more than $\frac{m}{2}$ edges go between A and B.*

Proof.   Choose A as a random n-element subset of V, all the $\binom{2n}{n}$ n-element subsets having the same probability, and let us put $B = V \setminus A$. Let X denote the number of edges of G going "across", i.e. edges $\{a, b\}$ with $a \in A$ and $b \in B$. We calculate the expectation $\mathbf{E}[X]$ of the random variable X. For each edge $e = \{u, v\} \in E(G)$, we define the event C_e that occurs whenever the edge e goes between A and B; formally, $C_e = \{A \in \binom{V}{n}: |A \cap e| = 1\}$. Then we have $X = \sum_{e \in E(G)} I_{C_e}$, and hence $\mathbf{E}[X] = \sum_{e \in E(G)} P(C_e)$. So we need to determine the probability $P(C_e)$.

Altogether, there are $\binom{2n}{n}$ possible choices of A. If we require that $u \in A$ and $v \notin A$, the remaining $n - 1$ elements of A can be selected in $\binom{2n-2}{n-1}$ ways. Similar reasoning works for the symmetric situation $u \notin A$, $v \in A$. Thus

$$P(C_e) = \frac{2\binom{2n-2}{n-1}}{\binom{2n}{n}} = \frac{n}{2n-1} > \frac{1}{2}.$$

From this we get $\mathbf{E}[X] = \sum_{e \in E(G)} P(C_e) > \frac{m}{2}$. The expectation of X is the arithmetic average of the values of X over all choices of the set A. An average cannot be greater than the maximum of all these values, and therefore a choice of A exists with more than half of the edges going across. $\square$

Independent sets.   This is related to the subject of Section 6.3 (extremal graph theory). Here we ask: how many edges can a graph on n vertices have if it doesn't contain the complete graph on k vertices as a subgraph? In particular, for $k = 3$, how many edges can a graph on n vertices with no triangles have? This question is answered by Turán's theorem, one of the celebrated results of extremal graph theory. This theorem can be formulated in various ways. The strongest version describes exactly how a graph with the maximum possible number of edges looks. Here we demonstrate a very cute probabilistic proof. We only give the bound on the maximum number of edges. With some more work, one can also derive the structure of the graph with the maximum possible number of edges, but we omit that part.

Turán's theorem is most often applied in a "reverse" form: *if a graph on n vertices has more than a certain number of edges then it has to contain a K_k.* If we consider the complement of the graph G, i.e. edges in the new graph are exactly at the positions where G has no edges, Turán's theorem says *if a graph on n vertices has fewer than a certain number of edges then it contains an independent set of size at least k* (an independent set is a set of vertices such that no two of them are connected by an edge). This is perhaps the most useful version for applications, and it is also the one we state and prove here.

9.4.2 Theorem (Turán's theorem). *For any graph G on n vertices, we have*

$$\alpha(G) \geq \frac{n^2}{2|E(G)| + n},$$

where $\alpha(G)$ denotes the size of the largest independent set of vertices in the graph G.

The probabilistic method is used in the proof of the next lemma:

Lemma. *For any graph G, we have*

$$\alpha(G) \geq \sum_{v \in V(G)} \frac{1}{\deg_G(v) + 1}$$

(where $\deg_G(v)$ denotes the degree of a vertex v in the graph G).

Proof.   Suppose that the vertices of G are numbered $1, 2, \ldots, n$, and let us pick a random permutation π of the vertices. We define a set $M = M(\pi) \subseteq V(G)$ consisting of all vertices v such that all neighbors u of v satisfy $\pi(u) > \pi(v)$; that is, the vertex v precedes all neighbors in the ordering given by the permutation π. Note that the set $M(\pi)$ is an independent set in G, and so $|M(\pi)| \leq \alpha(G)$ for any permutation π. Hence also $\mathbf{E}\left[|M|\right] \leq \alpha(G)$. We now calculate the expected size of M in a different way.

For a vertex v, let A_v be the event "$v \in M(\pi)$". If N_v denotes the set of all neighbors of the vertex v, then all the orderings of the set $N_v \cup \{v\}$ by the permutation π have the same probability, and so the probability of v being the smallest element of this set equals $1/(|N_v|+1) = 1/(\deg_G(v)+1)$. Therefore $P(A_v) = 1/(\deg_G(v)+1)$, and we can calculate as we already did several times before:

$$\alpha(G) \geq \mathbf{E}\left[|M|\right] = \sum_{v \in V(G)} \mathbf{E}\left[I_{A_v}\right]$$

$$= \sum_{v \in V(G)} P(A_v) = \sum_{v \in V(G)} \frac{1}{\deg_G(v)+1}.$$

$\square$

Proof of Theorem 9.4.2. This is now a mere calculation with inequalities. The number of edges $e = |E(G)|$ is half of the sum of the vertex degrees. So we have the following situation: for nonnegative real numbers $d_1, d_2, \ldots, d_n$, we know that $\sum_{i=1}^{n} d_i = 2e$, and we ask what the smallest possible value is of the sum

$$\sum_{i=1}^{n} \frac{1}{d_i + 1}.$$

It can be shown that this sum is minimized for $d_1 = d_2 = \cdots = d_n = 2e/n$ (we leave this as an exercise), and in such a case, its value is $n^2/(2e+n)$ as in the theorem. $\square$

Number of intersections of level $\leq k$. This is a geometric problem arising in the analysis of certain geometric algorithms. Let us consider a set L consisting of n lines in the plane, such that no three lines of L meet at a common point and no two of them are parallel. Let o be a point lying on none of the lines of L. We will consider the pairwise intersections of the lines of L. Each pair of lines has one intersection, so there are $\binom{n}{2}$ intersections altogether. Let us say that an intersection v has *level k* if the segment ov intersects, in addition to the two lines defining the intersection v, exactly k more lines of L (Fig. 9.3 depicts all intersections of level 1). What is the maximum possible number of intersections of level at most k, for a given

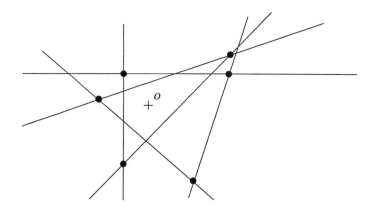

Fig. 9.3 Intersections of level 1 for a set of lines.

n and k? The following theorem gives an upper bound which is the best possible up to the value of the constant of proportionality:

9.4.3 Theorem. *For any set of n lines, there exist at most $3(k+1)n$ intersections of level at most k.*

Proof. First we look at the particular case $k = 0$. We find that the intersections of level 0 are exactly the vertices of the convex polygon containing the point o. Since each line contributes at most one side of this polygon, the number of its sides is at most n, and hence also the number of intersections of level 0 is at most n. We will use this in the proof for an arbitrary k, which we do next.

Let p denote a certain suitable number in the interval $(0, 1)$, whose value will be determined at the end of the proof. Let us imagine the following random experiment. We choose a subset $R \subseteq L$ at random, by choosing each line $\ell \in L$ with probability p, these choices being independent for distinct lines ℓ.

Here we should perhaps say a little more about the underlying probability space. It is a generalization of the space C_n from Example 9.2.2. That example models tosses of a symmetric coin, where both heads and tails have the same probability, namely $\frac{1}{2}$. In the case considered here, we would use an asymmetric coin of some kind, where heads have probability p and tails probability $1 - p$. For each line $\ell \in L$, we toss this asymmetric coin once, and we include ℓ into our sample R if the toss yields heads. Formally, the probability space is the set of all subsets of L, and the probability of an r-element set $R \subseteq L$ is $p^r(1 - p)^{n-r}$, where $r = |R|$ (in order to get exactly the set R by the coin tosses, certain r tosses must give heads and the remaining $n - r$ tosses tails). This is also an example of a probability space where the elementary events need not have all the same probability.

Let us return to our geometric situation. We have thus chosen a random set $R \subseteq L$ of lines. Let us imagine that only the lines of R are drawn in the plane. We define a random variable $f = f(R)$ as the number of intersections that have level 0 with respect to the lines of R; that is, intersections of lines of R such that no line of R obscures their view of the point o. We are going to estimate the expectation $\mathbf{E}[f]$ in two ways. On one hand, by the remark at the beginning of the proof, we have $f(R) \leq |R|$ for any specific set R, and hence $\mathbf{E}[f] \leq \mathbf{E}[|R|]$. It is not difficult to calculate (we leave this to Exercise 8) that $\mathbf{E}[|R|] = pn$.

Now we will count $\mathbf{E}[f]$ in a different way. For each intersection v of the lines of L, we define an event A_v that occurs if and only if v is one of the intersections of level 0 with respect to the lines of R, i.e. it contributes 1 to the value of $f(R)$. The event A occurs if and only if the following two conditions are satisfied:

- Both the lines determining the intersection v lie in R.
- None of the lines intersecting the segment ov at an interior point (and thus obscuring the view from the point v to the point o) falls into R.

From this, we deduce that $P(A_v) = p^2(1-p)^{\ell(v)}$, where $\ell(v)$ denotes the level of the intersection v.

Let M denote the set of all intersections of the lines of L, and let $M_k \subseteq M$ be the set of intersections at level at most k. We have

$$\mathbf{E}[f] = \sum_{v \in M} \mathbf{E}[I_{A_v}] = \sum_{v \in M} P(A_v) \geq \sum_{v \in M_k} P(A_v)$$

$$= \sum_{v \in M_k} p^2(1-p)^{\ell(v)} \geq \sum_{v \in M_k} p^2(1-p)^k = |M_k|p^2(1-p)^k.$$

Altogether we have derived $np \geq \mathbf{E}[f] \geq |M_k|p^2(1-p)^k$, in other words

$$|M_k| \leq \frac{n}{p(1-p)^k}.$$

Let us now choose the number p in such a way that the value of the right-hand side is as small as possible. A suitable choice is, for instance, $p = 1/(k+1)$. By Exercise 2.5.2, we have $(1 - \frac{1}{k+1})^k \geq$

$e^{-1} > \frac{1}{3}$ for any $k \geq 1$. This leads to $|M_k| \leq 3(k+1)n$ as the theorem claims.   □

Let us remark that the problem of estimating the maximum possible number of intersections of level *exactly* k is much more difficult and still unsolved. The branch of mathematics studying problems of a similar nature, i.e. combinatorial questions about geometric configurations, is called *combinatorial geometry*. A highly recommendable book for studying this subject is Pach and Agarwal [25]. A more specialized book considering problems closely related to estimating the number of intersections of level k is Sharir and Agarwal [42].

Average number of comparisons in QUICKSORT. Algorithm QUICKSORT, having received a sequence $(x_1, x_2, \ldots, x_n)$ of numbers as input, proceeds as follows. The numbers $x_2, x_3, \ldots, x_n$ are compared to x_1 and divided into two groups: those smaller than x_1 and those at least as large as x_1. In both groups, the order of numbers remains the same as in the input sequence. Each group is then sorted by a recursive invocation of the same method. The recursion terminates with trivially small groups (at most one-element ones, say). For example, the input sequence $(4, 3, 6, 1, 5, 2, 7)$ would be sorted as follows:

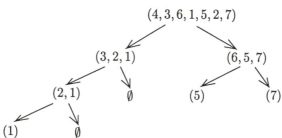

This algorithm may need about n^2 steps in the worst case (the worst thing we can do to the algorithm is to confront it with an already sorted sequence). In practice, however, QUICKSORT is very popular, and it is considered one of the fastest sorting algorithms. It behaves very well "on average". This is partially expressed in the following theorem.

9.4.4 Theorem. *Let $x_1 < x_2 < \cdots < x_n$ be a sequence of real numbers in an increasing order. Let π be a permutation of the set $\{1, 2, \ldots, n\}$, and let $T(\pi)$ be the number of comparisons (of pairs of elements) made by QUICKSORT for the input sequence*

$(x_{\pi(1)}, x_{\pi(2)}, \ldots, x_{\pi(n)})$. *Then the expectation of* $T(\pi)$ *for a random permutation* π *is at most* $2n \ln n$.

Let us remark that *any* algorithm that sorts every n-tuple of distinct real numbers and uses only pairwise comparisons of these numbers has to make at least $\log_2 n!$ comparisons in the worst case. This is because the algorithm must select one of the $n!$ possible permutations according to the outcomes of the comparisons, and k comparisons only have 2^k different outcomes. As we know from Section 2.5, we have $n! \geq \left(\frac{n}{e}\right)^n$, and so $\log_2 n! = (\log_2 e) \ln n! \geq (\log_2 e)(n-1) \ln n \approx 1.443(n-1) \ln n$. Therefore, the average behavior of QUICKSORT guaranteed by Theorem 9.4.4 is quite good.

Reasons for the $O(n \log n)$ average behavior of the QUICKSORT algorithm are not difficult to see. When the input elements are randomly ordered, we expect that the first element usually divides the remaining elements into two groups of roughly comparable size—it is unlikely that one group is much smaller than the other. If this happens in most cases, the recursion in the algorithm will have roughly $\log n$ levels, and at each level of recursion we need $O(n)$ comparisons in total.

This may sound quite convincing, but certainly it is not a proof. Next, we give a rigorous proof, based on a rather different idea.

Proof of Theorem 9.4.4. Let $T_i = T_i(\pi)$ be the number of elements compared to the element $x_{\pi(i)}$ at the moment $x_{\pi(i)}$ is the dividing element. For instance, we always have $T_1 = n - 1$, since $x_{\pi(1)}$ is the first element in the input sequence, and all the other elements are compared to it. If $\pi(2) < \pi(1)$, T_2 is $\pi(1) - 2$, and for $\pi(2) > \pi(1)$ we have $T_2 = n - \pi(1) - 1$. In general, T_i can be interpreted according to the following diagram:

$$x_{\pi(i)}$$

The small circles in this diagram depict the elements $x_1, x_2, \ldots, x_n$ in the sorted order. The full circles are the elements with indices $\pi(1), \pi(2), \ldots, \pi(i-1)$, i.e. the first $i-1$ elements in the input sequence. The element $x_{\pi(i)}$ is marked by a double circle, and the remaining elements have empty circles. It is not difficult to see that T_i is exactly the number of empty circles "seen" by the element $x_{\pi(i)}$, if the full circles are considered opaque.

We will investigate the expectation of T_i. The key idea is to think of the algorithm running backwards in time, as if we were watching a movie going backwards. Imagine that we consider the indices i in the backward order $n, n-1, \ldots, 1$, and we look at the corresponding picture with full and empty circles. Initially, all the circles are full. The random permutation π is not yet defined, and we create it by random choices as we go along. First we choose one full circle at random and make it empty. The index of this circle becomes $\pi(n)$. Then we randomly select another full circle and make it empty, which determines $\pi(n - 1)$, and so on; this is the same way of generating a random permutation as in Fig. 9.2. At the moment when i full circles remain, we choose one of them at random, which corresponds to $x_{\pi(i)}$. The quantity T_i is the number of empty circles seen by $x_{\pi(i)}$ at this moment.

To estimate the expected value of T_i, we use double-counting. Each of the $n - i$ *empty* circles sees at most two full circles. Hence the total number of pairs (full circle,empty circle) whose members see each other is at most $2(n - i)$. On the average, one of the i full circles sees at most $\frac{2(n-i)}{i}$ empty circles, and hence $\mathbf{E}\,[T_i] \leq \frac{2(n-i)}{i}$. Since $T(\pi) = \sum_{i=1}^{n} T_i(\pi)$, we get

$$\mathbf{E}\,[T] = \sum_{i=1}^{n} \mathbf{E}\,[T_i] \leq \sum_{i=1}^{n} \frac{2(n - i)}{i} = 2n \sum_{i=1}^{n} \frac{1}{i} - 2n \leq 2n \ln n.$$

$\square$

Remark.   It can be shown that the probability of the running time of QUICKSORT deviating from the expectation by a significant amount is quite small (exponentially small), i.e. bad permutations are quite rare. We will not consider this here.

Exercises

1. From Theorem 9.4.2, prove that any graph on n vertices with no triangles has at most $n^2/4$ edges.

2. *Show that if $d_1, \ldots, d_n$ are nonnegative real numbers adding up to 1, then the expression $\sum_{i=1}^{n} 1/(d_i + 1)$ is minimal for $d_1 = d_2 = \cdots = d_n = 1/n$.

3. Consider the following algorithm for finding an independent set in a graph G. Keep deleting a vertex of the maximum degree from the current graph, together with edges containing it, until the current graph

has no edges. Show that this algorithm always finds an independent set of size at least $F(G) = \sum_{v \in V(G)} 1/(1 + \deg_G(v))$. Hint: Check that if H arises from G by deleting a maximum degree vertex then $F(H) \geq F(G)$.

4. *Prove the following version of Turán's theorem. Let $r \geq 3$ be an integer, and let n be divisible by $r - 1$. Then any graph on n vertices containing no K_r as a subgraph has at most $\frac{n^2}{2}\left(1 - \frac{1}{r-1}\right)$ edges. Proceed by induction on the number $\frac{n}{r-1}$.

5. Let G be a graph. Consider the following algorithm. Divide $V(G)$ into two parts A and B arbitrarily, and repeat the following step while applicable: if there is a vertex with more neighbors in the part containing it than in the other part, select some such vertex and move it to the other part (than the one currently containing it). Show that this algorithm always finishes (and give as good a bound on the running time as possible), and that at least half of the edges of G go between the resulting parts A, B.

6. (a) Modify the above algorithm QUICKSORT as follows. Let the current input sequence be $(x_1, x_2, \ldots, x_n)$. Choose a random number i among $1, 2, \ldots, n$ (all values have probability $\frac{1}{n}$), and use x_i as the dividing element. Show that for any fixed n-tuple of input numbers $(x_1, x_2, \ldots, x_n)$, assuming that no two of them are equal, the expected number of comparisons is the same and is bounded by $2n \ln n$ as in Theorem 9.4.4.

(b) *Consider another modification of the algorithm: given an input sequence $(x_1, x_2, \ldots, x_n)$, choose two independent random numbers i_1, i_2 from $\{1, 2, \ldots, n\}$, and use $\max(x_{i_1}, x_{i_2})$ as the dividing element. Suppose that $(x_1, x_2, \ldots, x_n)$ are all distinct. Show that the expected number of comparisons for a given input sequence is the same as in (a). We do not count the comparisons needed to determine $\max(x_{i_1}, x_{i_2})$!

7. *Consider n empty circles in a row. An observer sits in one of the $n - 1$ gaps between them. A random subset of r circles is colored black. Show that the expected number of empty circles the observer sees is at most $2(n - r)/(r + 1)$. (The observer can see through empty circles but not through black ones.) Hint: Instead of coloring r circles color $r + 1$ of them black first and then make a random black circle empty again.

8. Let the subset $R \subseteq L$ be chosen at random as in the proof of Theorem 9.4.3. Prove that $\mathbf{E}\left[\|R\|\right] = pn$.

9. *Consider n lines as in Theorem 9.4.3, assuming, moreover, that none of them is vertical. We say that an intersection v is a *peak* if one of the lines defining it has a positive slope and the other one a negative slope. Prove that at most $6(k + 1)^2$ peaks of level at most k exist.

10

Generating functions

In this chapter, we are going to present a useful calculation technique. The basic idea, quite a surprising one, is to consider an infinite sequence of real numbers and associate a certain continuous function with it, the so-called generating function of the sequence. Problems about sequences can then be solved by calculations with functions.

In this introductory text, we will only get to simpler examples, whose solution can mostly be found also without generating functions. In some problems, various tricks can reach the goal even quicker than the method of generating functions (but it may not be easy to discover such tricks). This should not discourage the reader from learning about generating functions, since these are a powerful tool also for more advanced problems where other methods fail or become too complicated.

In some solutions, alternative methods are mentioned, for others they are indicated in the exercises. Sometimes the reader will perhaps find even simpler solutions.

10.1 Combinatorial applications of polynomials

How do we multiply the polynomials $p(x) = x + x^2 + x^3 + x^4$ and $q(x) = x + x^3 + x^4$? Here is a simple rule: multiply each term of $p(x)$ by each term of $q(x)$ and add all these products together. Adding the products is simple, since all the products have coefficient 1. In this way, we calculate that $p(x)q(x) = x^8 + 2x^7 + 2x^6 + 3x^5 + 2x^4 + x^3 + x^2$.

Let us now ask a different question. We pick some power of x, say x^5, and we want to know its coefficient in $p(x)q(x)$, without calculating the whole product. In our case, x^5 appears by multiplying the term x in $p(x)$ by x^4 in $q(x)$, and also by multiplying x^2 in $p(x)$ by x^3 in $q(x)$, and finally by multiplying x^4 in $p(x)$ by x in $q(x)$. Each of these possibilities adds 1 to the resulting coefficient, and hence the coefficient of x^5 in the product $p(x)q(x)$ is 3.

This is as if we had 4 silver coins with values 1, 2, 3, and 4 doublezons (these are the exponents of x in the polynomial $p(x)$) and

3 golden coins with values 1, 3, and 4 doublezons (corresponding to the exponents of x in $q(x)$), and asked how may ways there are to pay 5 doublezons by one silver and one golden coin. Mathematically speaking, the coefficient of x^5 is the number of ordered pairs (i, j), where $i + j = 5$, $i \in \{1, 2, 3, 4\}$, and $j \in \{1, 3, 4\}$.

Let us express the consideration just made for the two particular polynomials somewhat more generally. Let I and J be finite sets of natural numbers. Let us form the polynomials $p(x) = \sum_{i \in I} x^i$ and $q(x) = \sum_{j \in J} x^j$ (note that the coefficients in such polynomials are 0s and 1s). Then, for any natural number r, the number of solutions (i, j) of the equation

$$i + j = r$$

with $i \in I$ and $j \in J$ equals the coefficient of x^r in the product $p(x)q(x)$.

A further, more interesting generalization of this observation deals with a product of 3 or more polynomials. Let us illustrate it on a particular example first.

Problem. How many ways are there to pay the amount of 21 doublezons if we have 6 one-doublezon coins, 5 two-doublezon coins, and 4 five-doublezon coins?

Solution. The required number equals the number of solutions of the equation

$$i_1 + i_2 + i_3 = 21$$

with

$$i_1 \in \{0, 1, 2, 3, 4, 5, 6\}, \ i_2 \in \{0, 2, 4, 6, 8, 10\}, \ i_3 \in \{0, 5, 10, 15, 20\}.$$

Here i_1 is the amount paid by coins of value 1 doublezon, i_2 the amount paid by 2-doublezon coins, and i_3 the amount paid by 5-doublezon coins.

This time we claim that the number of solutions of this equation equals the coefficient of x^{21} in the product

$$\left(1 + x + x^2 + x^3 + \cdots + x^6\right)\left(1 + x^2 + x^4 + x^6 + x^8 + x^{10}\right)$$

$$\times \left(1 + x^5 + x^{10} + x^{15} + x^{20}\right)$$

(after multiplying out the parentheses and combining the terms with the same power of x, of course). Indeed, a term with x^{21} is obtained by taking some term x^{i_1} from the first parentheses, some term x^{i_2} from the second, and x^{i_3} from the third, in such a way that $i_1 + i_2 + i_3 = 21$. Each such possible selection of i_1, i_2, and i_3 contributes 1 to the considered coefficient of x^{21} in the product.

How does this help us in solving the problem? From a purely practical point of view, this allows us to get the answer easily by computer, provided that we have a program for polynomial multiplication at our disposal. In this way, the authors have also found the result: 9. Since we only deal with relatively few coins, the solution can also be obtained by listing all possibilities, but it could easily happen that we forget some. However, the method explained above is most significant as a prelude to handling more complicated situations.

A combinatorial meaning of the binomial theorem. The binomial theorem asserts that

$$(1+x)^n = \binom{n}{0} + \binom{n}{1}x + \binom{n}{2}x^2 + \cdots + \binom{n}{n}x^n. \qquad (10.1)$$

On the left-hand side, we have a product of n polynomials, each of them being $1+x$. Analogously to the above considerations with coins, the coefficient of x^r after multiplying out the parentheses equals the number of solutions of the equation

$$i_1 + i_2 + \cdots + i_n = r,$$

with $i_1, i_2, \ldots, i_n \in \{0, 1\}$. But each such solution to this equation means selecting r variables among $i_1, i_2, \ldots, i_n$ that equal 1—the $n - r$ remaining ones must be 0. The number of such selections is the same as the number of r-element subsets of an n-element set, i.e. $\binom{n}{r}$. This means that the coefficient of x^r in the product $(1+x)^n$ is $\binom{n}{r}$. We have just proved the binomial theorem combinatorially!

If we play with the polynomial $(1+x)^n$ and similar ones skillfully, we can derive various identities and formulas with binomial coefficients. We have already seen simple examples in Section 2.3, namely the formulas $\sum_{k=0}^{n} \binom{n}{k} = 2^n$ and $\sum_{k=0}^{n} (-1)^k \binom{n}{k} = 0$ obtained by substituting $x = 1$ and $x = -1$ into (10.1), respectively.

For the next example, the reader should be familiar with the notion of a derivative (of a polynomial).

10.1.1 Example. For all $n \geq 1$, we have

$$\sum_{k=0}^{n} k \binom{n}{k} = n2^{n-1}.$$

Proof. This equality can be proved by differentiating both sides of the formula (10.1) as a function of the variable x. On both sides, we must obtain the same polynomial. By differentiating the left-hand side we get $n(1+x)^{n-1}$, and differentiating the right-hand side yields $\sum_{k=0}^{n} k\binom{n}{k}x^{k-1}$. By setting $x = 1$ we get the desired identity. $\square$

An example of a different type is based on the equality of coefficients in two different expressions for the same polynomial.

Another proof of Proposition 2.3.4. We want to prove

$$\sum_{i=0}^{n}\binom{n}{i}^2 = \binom{2n}{n}.$$

Consider the identity

$$(1+x)^n(1+x)^n = (1+x)^{2n}.$$

The coefficient of x^n on the right-hand side is $\binom{2n}{n}$ by the binomial theorem. On the left-hand side, we can expand both the powers $(1+x)^n$ according to the binomial theorem, and then we multiply the two resulting polynomials. The coefficient of x^n in their product can be expressed as $\binom{n}{0}\binom{n}{n} + \binom{n}{1}\binom{n}{n-1} + \binom{n}{2}\binom{n}{n-2} + \cdots + \binom{n}{n}\binom{n}{0}$, and this must be the same number as the coefficient of x^n on the right-hand side. This leads to

$$\sum_{i=0}^{n}\binom{n}{i}\binom{n}{n-i} = \binom{2n}{n}.$$

So we have proved Proposition 2.3.4 in a different way. $\square$

A number of other sums and formulas can be handled similarly. But if we try some more complicated calculations, we soon encounter, as an unpleasant restriction, the fact that we work with polynomials having finitely many terms only. It turns out that the "right" tool for such calculations is an object analogous to a polynomial but with possibly infinitely many powers of x, the so-called power series. This is the subject of the next section.

Exercises

1. Let $a(x) = a_0 + a_1 x + a_2 x^2 + \cdots + a_n x^n$ and $b(x) = b_0 + b_1 x + b_2 x^2 + \cdots + b_m x^m$ be two polynomials. Write down a formula for the coefficient of x^k in the product $a(x)b(x)$, where $0 \le k \le n + m$.

2. A coffee shop sells three kinds of cakes—Danish cheese cakes, German chocolate cakes, and brownies. How many ways are there to buy 12 cakes in such a way that at least 2 cakes of each kind are included, but no more than 3 German chocolate cakes? Express the required number as a coefficient of a suitable power of x in a suitable product of polynomials.

3. How many ways are there to distribute 10 identical balls among 2 boys and 2 girls, if each boy should get at least 1 ball and each girl should get at least 2 balls? Express the answer as a coefficient of a suitable power of x in a suitable product of polynomials.

4. Prove the multinomial theorem 2.3.5 in a similar way as the binomial theorem was proved in the text.

5. Calculate the sum in Example 10.1.1 by a suitable manipulation of the expression $k\binom{n}{k}$ and by using the binomial theorem.

6. Compute the sum $\sum_{i=0}^{n}(-1)^i\binom{n}{i}\binom{n}{n-i}$.

10.2   Calculation with power series

Properties of power series. A *power series* is an infinite series of the form $a_0 + a_1 x + a_2 x^2 + \cdots$, where $a_0, a_1, a_2, \ldots$ are real numbers and x is a variable attaining real values.[1] This power series will usually be denoted by $a(x)$.

A simple example of a power series is

$$1 + x + x^2 + x^3 + \cdots \tag{10.2}$$

(all the a_i are 1). If x is a real number in the interval $(-1, 1)$, then this series converges and its sum equals $\frac{1}{1-x}$ (this is the well-known formula for the sum of an infinite geometric series; if you're not familiar with it you should certainly consider doing Exercise 1). In this sense, the series (10.2) determines the function $\frac{1}{1-x}$. Conversely, this function contains all the information about the series (10.2). Indeed, if we differentiate the function k times and then substitute $x = 0$ into the result, we obtain exactly $k!$ times the coefficient of x^k. In other words, the series (10.2) is the Taylor series of the function $\frac{1}{1-x}$ at $x = 0$. Hence the function $\frac{1}{1-x}$ can be understood as an incarnation of the infinite sequence $(1, 1, 1, \ldots)$ and vice versa. Such

---

[1] It is extremely useful to consider also complex values of x and to apply methods from the theory of functions of a complex variable. But we are not going to get this far in our introductory treatment.

a transmutation of infinite sequences into functions and back is a
key step in the technique of generating functions.

In order to explain what generating functions are, we have to use
some notions of elementary calculus (convergence of infinite series, derivative, Taylor series), as we have already done in the example just
given. If you're not familiar with enough calculus to understand this
introduction, you still need not give up reading, because almost nothing from calculus is usually needed to solve problems using generating
functions. For example, you can accept as a matter of faith that the
infinite series $1 + x + x^2 + \cdots$ and the function $\frac{1}{1-x}$ mean the same
thing, and use this and similar facts listed below for calculations.

The following proposition says that if the terms of a sequence
$(a_0, a_1, a_2, \ldots)$ do not grow too fast, then the corresponding power
series $a(x) = a_0 + a_1 x + a_2 x^2 + \cdots$ indeed defines a function of the
real variable x, at least in some small neighborhood of 0. Further,
from the knowledge of the values of this function we can reconstruct
the sequence $(a_0, a_1, a_2, \ldots)$ uniquely.

10.2.1 Proposition. *Let $(a_0, a_1, a_2, \ldots)$ be a sequence of real numbers, and let us suppose that for some real number K, we have
$|a_n| \leq K^n$ for all $n \geq 1$. Then for any number $x \in (-\frac{1}{K}, \frac{1}{K})$, the
series $a(x) = \sum_{i=0}^{\infty} a_i x^i$ converges (even absolutely), and hence the
value of its sum defines a function of the real variable x on this interval. This function will also be denoted by $a(x)$. The values of the
function $a(x)$ on an arbitrarily small neighborhood of 0 determine all
the terms of the sequence $(a_0, a_1, a_2, \ldots)$ uniquely. That is, the function $a(x)$ has derivatives of all orders at 0, and for all $n = 0, 1, 2, \ldots$
we have*

$$a_n = \frac{a^{(n)}(0)}{n!}$$

*($a^{(n)}(0)$ stands for the nth derivative of the function $a(x)$ at the
point 0).*

A proof follows from basic results of mathematical analysis, and we
omit it here (as we do for proofs of a few other results in this section).
Most natural proofs are obtained from the theory of functions of a
complex variable, which is usually covered in more advanced courses
only. But we need very little for our purposes, and this can be proved,
somewhat more laboriously, also from basic theorems on limits and
derivatives of functions of a real variable. In the subsequent examples,
we will not explicitly check that the power series in question converge
in some neighborhood of 0 (i.e. the assumptions of Proposition 10.2.1).
Usually it is easy. Moreover, in many cases, this can be avoided in

the following way: once we find a correct solution of a problem using generating functions, in a possibly very suspicious manner, we can verify the solution by some other method, say by induction. And, finally, we should remark that there is also a theory of the so-called *formal power series*, which allows one to work even with power series that never converge (except at 0) in a meaningful way. So convergence is almost never a real issue in applications of generating functions.

Now, finally, we can say what a generating function is:

10.2.2 Definition. *Let* $(a_0, a_1, a_2, \ldots)$ *be a sequence of real numbers. By the* generating function *of this sequence[2] we understand the power series* $a(x) = a_0 + a_1 x + a_2 x^2 + \cdots$.

If the sequence $(a_0, a_1, a_2, \ldots)$ has only finitely many nonzero terms, then its generating function is a polynomial. Thus, in the preceding section, we have used generating functions of finite sequences without calling them so.

Manufacturing generating functions. In applications of generating functions, we often encounter questions like "What is the generating function of the sequence $(1, \frac{1}{2}, \frac{1}{3}, \frac{1}{4}, \ldots)$?" Of course, the generating function is $1 + \frac{1}{2}x + \frac{1}{3}x^2 + \frac{1}{4}x^3 + \cdots$ by definition, but is there a nice closed formula? In other words, doesn't this power series define some function well known to us from calculus, say? (This one does, namely $-\frac{\ln(1-x)}{x}$; see below.) The answer can often be found by an "assembling" process, like in a home workshop. We have a supply of parts, in our case sequences like $(1, 1, 1, \ldots)$, for which we know the generating function right away. We also have a repertoire of simple operations on sequences and their corresponding operations on the generating functions. For example, knowing the generating function $a(x)$ of some sequence $(a_0, a_1, a_2, \ldots)$, the generating function of the sequence $(0, a_0, a_1, a_2, \ldots)$ equals $x\, a(x)$. With some skill, we can usually "assemble" the sequence we want.

Let us begin with a supply of the "parts". Various examples of Taylor (or Maclaurin) series mentioned in calculus courses belong to

---

[2] A more detailed name often used in the literature is the *ordinary generating function*. This suggests that other types of generating functions are also used in mathematics. We briefly mention the so-called *exponential generating functions*, which are particularly significant for combinatorial applications. The exponential generating function of a sequence $(a_0, a_1, a_2, \ldots)$ is the power series $\sum_{i=0}^{\infty} (a_i/i!)x^i$. For example, the sequence $(1, 1, 1, \ldots)$ has the exponential generating function e^x. In the sequel, except for a few exercises, we restrict ourselves to ordinary generating functions.

such a supply. For instance, we have

$$\frac{x}{1} + \frac{x^2}{2} + \frac{x^3}{3} + \cdots = -\ln(1-x) \tag{10.3}$$

(valid for all $x \in (-1, 1)$) and

$$1 + \frac{x}{1!} + \frac{x^2}{2!} + \frac{x^3}{3!} + \cdots = e^x$$

(holds for all real x). Lots of other examples can be found in calculus textbooks. Here is another easy calculus result, which is used particularly often:

10.2.3 Proposition (Generalized binomial theorem). *For an arbitrary real number r and for any nonnegative integer k, we define the binomial coefficient $\binom{r}{k}$ by the formula*

$$\binom{r}{k} = \frac{r(r-1)(r-2)\ldots(r-k+1)}{k!}$$

(in particular, we set $\binom{r}{0} = 1$). Then the function $(1+x)^r$ is the generating function of the sequence $(\binom{r}{0}, \binom{r}{1}, \binom{r}{2}, \binom{r}{3}, \ldots)$. The power series $\binom{r}{0} + \binom{r}{1}x + \binom{r}{2}x^2 + \cdots$ always converges for all $|x| < 1$.

A proof again belongs to the realm of calculus, and it is easily done via the Taylor series.

For combinatorial applications, it is important to note that for r being a negative integer, the binomial coefficient $\binom{r}{k}$ can be expressed using the "usual" binomial coefficient (involving nonnegative integers only): $\binom{r}{k} = (-1)^k \binom{-r+k-1}{k} = (-1)^k \binom{-r+k-1}{-r-1}$. Hence for negative integer powers of $1 - x$ we obtain

$$\frac{1}{(1-x)^n} = \binom{n-1}{n-1} + \binom{n}{n-1}x + \binom{n+1}{n-1}x^2 + \cdots + \binom{n+k-1}{n-1}x^k + \cdots.$$

Note that the equality $\frac{1}{1-x} = 1 + x + x^2 + \cdots$ is a particular case for $n = 1$.

Operations with sequences and with their generating functions. To facilitate the above-mentioned process of "assembling" generating functions for given sequences, we list some important operations. In the sequel, let $(a_0, a_1, a_2, \ldots)$ and $(b_0, b_1, b_2, \ldots)$ be sequences and let $a(x)$ and $b(x)$ be their respective generating functions.

A. If we add the sequences term by term, then the corresponding operation with generating functions is simply their *addition*. That is, the sequence $(a_0 + b_0, a_1 + b_1, a_2 + b_2, \ldots)$ has generating function $a(x) + b(x)$.

B. Another simple operation is *multiplication by a fixed real number* α. The sequence $(\alpha a_0, \alpha a_1, \alpha a_2, \ldots)$ has generating function $\alpha \cdot a(x)$.

C. If n is a natural number, then the generating function $x^n a(x)$ corresponds to the sequence

$$\underbrace{(0, 0, \ldots, 0}_{n\times}, a_0, a_1, a_2, \ldots).$$

This is very useful for *shifting the sequence to the right* by a required number of positions.

D. What do we do if we want to *shift the sequence to the left*, i.e. to gain the generating function for the sequence $(a_3, a_4, a_5, \ldots)$, say? Obviously, we have to divide by x^3, but we must not forget to subtract the 3 initial terms first. The correct generating function of the above sequence is

$$\frac{a(x) - a_0 - a_1 x - a_2 x^2}{x^3}.$$

E. *Substituting αx for x.* Let α be a fixed real number, and let us consider the function $c(x) = a(\alpha x)$. Then $c(x)$ is the generating function of the sequence $(a_0, \alpha a_1, \alpha^2 a_2, \ldots)$. For example, we know that $\frac{1}{1-x}$ is the generating function of the sequence of all 1s, and so by the rule just given, $\frac{1}{1-2x}$ is the generating function of the sequence of powers of 2: $(1, 2, 4, 8, \ldots)$. This operation is also used in the following trick for replacing all terms of the considered sequence with an odd index by 0: as the reader can easily check, the function $\frac{1}{2}(a(x) + a(-x))$ corresponds to the sequence $(a_0, 0, a_2, 0, a_4, 0, \ldots)$.

F. Another possibility is a *substitution of x^n for x*. This gives rise to the generating function of the sequence whose term number nk equals the kth term of the original sequence, and all of whose other terms are 0s. For instance, the function $a(x^3)$ produces the sequence $(a_0, 0, 0, a_1, 0, 0, a_2, 0, 0, \ldots)$. A more complicated operation generalizing both E and F is the substitution of one power series for x into another power series. We will meet only a few particular examples in the exercises.

Let us see some of the operations listed so far in action.

Problem. What is the generating function of the sequence

$$(1, 1, 2, 2, 4, 4, 8, 8, \ldots),$$

i.e. $a_n = 2^{\lfloor n/2 \rfloor}$?

Solution. As was mentioned in E, the sequence $(1, 2, 4, 8, \ldots)$ has generating function $1/(1 - 2x)$. By F we get the generating function $1/(1 - 2x^2)$ for the sequence $(1, 0, 2, 0, 4, 0, \ldots)$, and by C, the sequence $(0, 1, 0, 2, 0, \ldots)$ has generating function $x/(1 - 2x^2)$. By addition we finally get the generating function for the given sequence; that is, the answer is $(1 + x)/(1 - 2x^2)$. □

G. Popular operations from calculus, *differentiation* and *integration* of generating functions, mean the following in the language of sequences. The derivative of the function $a(x)$, i.e. $a'(x)$, corresponds to the sequence

$$(a_1, 2a_2, 3a_3, \ldots).$$

In other words, the term with index k is $(k + 1)a_{k+1}$ (a power series is differentiated term by term in the same way as a polynomial). The generating function $\int_0^x a(t)dt$ gives the sequence $(0, a_0, \frac{1}{2}a_1, \frac{1}{3}a_2, \frac{1}{4}a_3, \ldots)$; that is, for all $k \geq 1$, the term with index k equals $\frac{1}{k}a_{k-1}$. For instance, we can derive the power series (10.3) for $\ln(1 - x)$ by integrating the function $\frac{1}{1-x}$.

Here is an example where differentiation helps:

10.2.4 Problem. What is the generating function for the sequence $(1^2, 2^2, 3^2, \ldots)$ of squares, i.e. for the sequence $(a_0, a_1, a_2, \ldots)$ with $a_k = (k + 1)^2$?

Solution. We begin with the sequence of all 1s with the generating function $\frac{1}{1-x}$. The first derivative of this function, $1/(1 - x)^2$, gives the sequence $(1, 2, 3, 4, \ldots)$ by G. The second derivative is $2/(1 - x)^3$, and its sequence is $(2 \cdot 1, 3 \cdot 2, 4 \cdot 3, \ldots)$, again according to G; the term with index k is $(k + 2)(k + 1) = (k + 1)^2 + k + 1$. But we want $a_k = (k + 1)^2$, and so we subtract the generating function of the sequence $(1, 2, 3, \ldots)$. We thus get

$$a(x) = \frac{2}{(1 - x)^3} - \frac{1}{(1 - x)^2}.$$

□

H. The last operation in our list is perhaps the most interesting one: *multiplication of generating functions*. The product $a(x)b(x)$ is the generating function of the sequence $(c_0, c_1, c_2, \ldots)$, where the numbers c_k are given by the equations

$$
\begin{aligned}
c_0 &= a_0 b_0 \\
c_1 &= a_0 b_1 + a_1 b_0 \\
c_2 &= a_0 b_2 + a_1 b_1 + a_2 b_0 \\
&\vdots
\end{aligned}
$$

and in general we can write

$$
c_k = \sum_{i,j \geq 0:\, i+j=k} a_i b_j. \tag{10.4}
$$

This is easy to remember—the terms in the product $a(x)b(x)$ up to the kth one are the same as in the product of the polynomials $(a_0 + a_1 x + \cdots + a_k x^k)$ and $(b_0 + b_1 x + \cdots + b_k x^k)$.

Multiplication of generating functions has a combinatorial interpretation which we now explain in a somewhat childish example. A natural example comes in Section 10.4. Suppose that we have a supply of indistinguishable wooden cubes, and we know that we can build a tower in a_i different ways from i cubes, $i = 0, 1, 2, \ldots$, and also that we can build a pyramid in b_j different ways from j cubes, $j = 0, 1, 2, \ldots$.

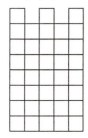 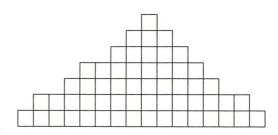

If we now have k cubes altogether, then $c_k = a_0 b_k + a_1 b_{k-1} + \cdots + a_k b_0$ is the number of ways to build one tower and one pyramid at the same time (assuming that a tower and a pyramid never have any cubes in common). In short, the generating function for the number of ordered pairs (tower, pyramid) is obtained as the product of the generating functions for the number of towers and for the number of pyramids.

Remark. The operations listed above are useful not only for finding the generating function for a given sequence, but often also for the reverse task, namely finding the sequence corresponding to a given generating function. In principle, the sequence can be found by computing the Taylor series (according to Proposition 10.2.1), i.e. by a repeated differentiation, but in practice this is seldom a good method.

We conclude this section with an example of an application of generating functions. More examples are supplied in the exercises, and several more sophisticated problems follow in subsequent sections.

Problem. A box contains 30 red, 40 blue, and 50 white balls; balls of the same color are indistinguishable. How many ways are there of selecting a collection of 70 balls from the box?

Solution. Armed with the results of Section 10.1, we find that the number we seek equals the coefficient of x^{70} in the product

$$(1+x+x^2+\cdots+x^{30})(1+x+x^2+\cdots+x^{40})(1+x+x^2+\cdots+x^{50}).$$

We need not multiply this out. Instead, we can rewrite

$$1+x+x^2+\cdots+x^{30} = \frac{1-x^{31}}{1-x}. \tag{10.5}$$

To see this equality, we can recall a formula for the sum of the first n terms of a geometric series. In case we cannot remember it, we can help ourselves as follows. We begin with the generating function of the sequence $(1,1,1,\ldots)$, which is $\frac{1}{1-x}$, and we subtract from it the generating function of the sequence

$$(\underbrace{0,0,\ldots,0}_{31\times},1,1,\ldots),$$

which is $x^{31}/(1-x)$ by item C above. The result is $(1-x^{31})/(1-x)$, which is the generating function for the sequence

$$(\underbrace{1,1,\ldots,1}_{31\times},0,0,\ldots).$$

This shows that (10.5) holds.

Hence we can rewrite the whole product as

$$\frac{1-x^{31}}{1-x}\cdot\frac{1-x^{41}}{1-x}\cdot\frac{1-x^{51}}{1-x}=\frac{1}{(1-x)^3}(1-x^{31})(1-x^{41})(1-x^{51}).$$

The factor $(1-x)^{-3}$ can be expanded according to the generalized binomial theorem 10.2.3. In the product of the remaining factors $(1-x^{31})(1-x^{41})(1-x^{51})$, it suffices to find the coefficients of powers up to x^{70}, which is quite easy. We get

$$\left(\binom{2}{2}+\binom{3}{2}x+\binom{4}{2}x^2+\cdots\right)(1-x^{31}-x^{41}-x^{51}+\cdots),$$

where the dots $\cdots$ in the second parentheses stand for powers higher than x^{70}. The coefficient of x^{70} in this product is $\binom{70+2}{2}-\binom{70+2-31}{2}-\binom{70+2-41}{2}-\binom{70+2-51}{2}=1061$.

Exercises

1. (a) Show that $(1-x)(1+x+x^2+\cdots+x^n)$ and $1-x^{n+1}$ are the same polynomials, and deduce the formula $1+x+x^2+\cdots+x^n=(1-x^{n+1})/(1-x)$. For which values of x is it correct?

   (b) Using the formula in (a), show that the infinite series $1+x+x^2+\cdots$ converges to $\frac{1}{1-x}$ for all $x\in(-1,1)$, and diverges for all x outside this interval. (This requires basic notions from calculus.)

2. (a) Determine the coefficient of x^{15} in $(x^2+x^3+x^4+\cdots)^4$.

   (b) Determine the coefficient of x^{50} in $(x^7+x^8+x^9+x^{10}+\cdots)^6$.

   (c) Determine the coefficient of x^5 in $(1-2x)^{-2}$.

   (d) Determine the coefficient of x^4 in $\sqrt[3]{1+x}$.

   (e) Determine the coefficient of x^3 in $(2+x)^{3/2}/(1-x)$.

   (f) Determine the coefficient of x^4 in $(2+3x)^5\sqrt{1-x}$.

   (g) Determine the coefficient of x^3 in $(1-x+2x^2)^9$.

3. Find generating functions for the following sequences (express them in a closed form, without infinite series!):

   (a) $0,0,0,0,-6,6,-6,6,-6,\ldots$

   (b) $1,0,1,0,1,0,\ldots$

   (c) $1,2,1,4,1,8,\ldots$

   (d) $1,1,0,1,1,0,1,1,0,\ldots$

4. Find the probability that we get exactly 12 points when rolling 3 dice.

5. Let a_n be the number of ordered triples (i, j, k) of integer numbers such that $i \geq 0$, $j \geq 1$, $k \geq 1$, and $i + 3j + 3k = n$. Find the generating function of the sequence $(a_0, a_1, a_2, \ldots)$ and calculate a formula for a_n.

6. Let a_n be the number of ordered r-tuples $(i_1, \ldots, i_r)$ of nonnegative integers with $i_1 + i_2 + \cdots + i_r = n$; here r is some fixed natural number.

   (a) Find the generating function of the sequence $(a_0, a_1, a_2, \ldots)$.

   (b) Find a formula for a_n. (This has been solved by a different method in Section 2.3.)

7. Solve Problem 10.2.4 without using differentiation—apply the generalized binomial theorem instead.

8. Let a_n be the number of ways of paying the sum of n doublezons using coins of values 1, 2, and 5 doublezons.

   (a) Write down the generating function for the sequence $(a_0, a_1, \ldots)$.

   (b) *Using (a), find a formula for a_n (reading the next section can help).

9. (a) Check that if $a(x)$ is the generating function of a sequence $(a_0, a_1, a_2, \ldots)$ then $\frac{1}{1-x} a(x)$ is the generating function of the sequence of partial sums $(a_0, a_0 + a_1, a_0 + a_1 + a_2, \ldots)$.

   (b) Using (a) and the solution to Problem 10.2.4, calculate the sum $\sum_{k=1}^{n} k^2$.

   (c) By a similar method, calculate the sum $\sum_{k=1}^{n} k^3$.

   (d) For natural numbers n and m, compute the sum $\sum_{k=0}^{m} (-1)^k \binom{n}{k}$.

   (e) Now it might seem that by this method we can calculate almost any sum we can think of, but it isn't so simple. What happens if we try computing the sum $\sum_{k=1}^{n} \frac{1}{k}$ in this manner?

10. *Let n, r be integers with $n \geq r \geq 1$. Pick a random r-element subset of the set $\{1, 2, \ldots, n\}$ and call it R (all the $\binom{n}{r}$ possible subsets have the same probability of being picked for R). Show that the expected value of the smallest number in R is $\frac{n+1}{r+1}$.

11. (Calculus required) *Prove the formula for the product of power series. That is, show that if $a(x)$ and $b(x)$ are power series satisfying the assumptions of Proposition 10.2.1, then on some neighborhood of 0, the power series $c(x)$ with coefficients given by the formula (10.4) converges to the value $a(x)b(x)$.

12. (Calculus required) Let $a(x) = a_0 + a_1 x + a_2 x^2 + \cdots$ be a power series with nonnegative coefficients, i.e. suppose that $a_i \geq 0$ for all i. We define its *radius of convergence* ρ by setting

$$\rho = \sup\{x \geq 0 \colon a(x) \text{ converges}\}.$$

   (a) *Prove that $a(x)$ converges for each real number $x \in [0, \rho)$, and that the function $a(x)$ is continuous in the interval $[0, \rho)$.

(b) Find an example of a sequence $(a_0, a_1, a_2, \ldots)$ with $\rho = 1$ such that the series $a(\rho)$ diverges.

(c) Find an example of a sequence $(a_0, a_1, a_2, \ldots)$ with $\rho = 1$ such that the series $a(\rho)$ converges.

13. (A warning example; calculus required) Define a function f by

$$f(x) = \begin{cases} e^{-1/x^2} & \text{for } x \neq 0 \\ 0 & \text{for } x = 0. \end{cases}$$

(a) *Show that all derivatives of f at 0 exist and equal 0.

(b) Prove that f is not given by a power series at any neighborhood of 0.

14. (Exponential generating functions) In a footnote, we have defined the *exponential generating function* of a sequence $(a_0, a_1, a_2, \ldots)$ as the power series $A(x) = \sum_{i=0}^{\infty}(a_i/i!)x^i$. Here we are going to sketch some combinatorial applications.

For a group of n people, one can consider various types of arrangements. For instance, one type of arrangement of n people would be to select one of the people as a leader, and the others as a crowd. For each group of n people, there are n arrangements of this type. Other examples of types of arrangements are: arranging the n people in a row from left to right ($n!$ possibilities), arranging the n people in a ring for a folk dance (there are $(n-1)!$ possibilities), letting the people be just an unorganized crowd (1 possibility), arranging the people into a jury (12 possibilities for $n = 12$, with electing one chairman, and 0 possibilities otherwise).

(a) For each type of arrangement mentioned above, write down the exponential generating function of the corresponding sequence (i.e. for the sequence $(a_0, a_1, a_2, \ldots)$, where a_i is the number of arrangements of the considered type of i people).

(b) Let a_n be the number of possible arrangements of some type A for a group of n people, and let b_n be the number of possible arrangements of some type B for a group of n people. Let an arrangement of type C mean the following: divide the given group of n people into two groups, the First Group and the Second Group, and arrange the First Group by an arrangement of type A and the Second Group by an arrangement of type B. Let c_n be the number of arrangements of type C for a group of n people. Check that if $A(x)$, $B(x)$, and $C(x)$ are the corresponding exponential generating functions, then $C(x) = A(x)B(x)$. Discuss specific examples with the arrangement types mentioned in the introduction to this exercise.

(c) *Let $A(x)$ be the exponential generating function for arrangements of type A as in (b), and, moreover, suppose that $a_0 = 0$ (no empty group is allowed). Let an arrangement of type D mean dividing the given n people into k groups, the First Group, Second Group, ..., kth Group ($k = 0, 1, 2, \ldots$), and arranging each group by an arrangement of type A. Express the exponential generating function $D(x)$ using $A(x)$.

(d) *Let $A(x)$ be as in (c), and let an arrangement of type E mean dividing the given n people into some number of groups and organizing each group by an arrangement of type A (but this time it only matters who is with whom, the groups are not numbered). Express the exponential generating function $E(x)$ using $A(x)$.

(e) How many ways are there of arranging n people into pairs (it only matters who is with whom, the pairs are not numbered)? Solve using (d).

15. *Using part (d) of Exercise 14, find the exponential generating function for the *Bell numbers*, i.e. the number of equivalences on n given points. Calculate the first few terms of the Taylor series to check the result numerically (see also Exercise 2.8.8).

16. Twelve students should be assigned work on 5 different projects. Each student should work on exactly one project, and for every project at least 2 and at most 4 students should be assigned. In how many ways can this be done? Use the idea of part (b) in Exercise 14.

17. (Hatcheck lady strikes again)

(a) *Using parts (a) and (d) of Exercise 14, write down the exponential generating function for arranging n people into one or several rings for a dance (a ring can have any number of persons including 1). Can you derive the number of such arrangements in a different way?

(b) Consider arrangements as in (a), but with each ring having at least 2 people. Write down the exponential generating function.

(c) *Using the result of (b), find the number of arrangements of n people into rings with at least 2 people each. If your calculations are correct, the result should equal the number of permutations without a fixed point (see Section 2.8). Explain why. (This actually gives an alternative solution to the hatcheck lady problem, without inclusion–exclusion!)

10.3 Fibonacci numbers and the golden section

We will investigate the sequence $(F_0, F_1, F_2, \ldots)$ given by the following rules:

$$F_0 = 0, \quad F_1 = 1, \quad F_{n+2} = F_{n+1} + F_n \quad \text{for } n = 0, 1, 2, \ldots.$$

This sequence has been studied by Leonardo of Pisa, called Fibonacci, in the 13th century, and it is known as the Fibonacci numbers. Its first few terms are

$$0, 1, 1, 2, 3, 5, 8, 13, 21, 34, 55, 89, 144, \ldots.$$

Fibonacci illustrated this sequence with a not too realistic example concerning the reproduction of rabbits in his treatise, but the sequence does have a certain relevance in biology (for instance, the number of petals of various flowers is very often a Fibonacci number; see Stewart [43] for a possible explanation). In mathematics and computer science it surfaces in various connections and quite frequently.

We show how to find a formula for the nth Fibonacci number using generating functions. So let $F(x)$ denote the generating function of this sequence. The main idea is to express the generating function of a sequence whose kth term equals $F_k - F_{k-1} - F_{k-2}$ for all $k \geq 2$. By the relation defining the Fibonacci numbers, this sequence must have all terms from the second one on equal to 0. On the other hand, such a generating function can be constructed from $F(x)$ using the operations discussed in Section 10.2. In this way, we get an equation from which we can determine $F(x)$.

Concretely, let us take the function $F(x) - xF(x) - x^2 F(x)$ corresponding to the sequence

$$(F_0, F_1 - F_0, F_2 - F_1 - F_0, F_3 - F_2 - F_1, F_4 - F_3 - F_2, \ldots)$$

$$= (0, 1, 0, 0, 0, \ldots).$$

In the language of generating functions, this means $(1 - x - x^2)F(x) = x$, and hence

$$F(x) = \frac{x}{1 - x - x^2}. \tag{10.6}$$

If we now tried to compute the Taylor series of this function by differentiation, we wouldn't succeed (we would only recover the definition of the Fibonacci numbers). We must bring in one more trick, which is well known in calculations of integrals by the name *decomposition into partial fractions*. In our case, this method guarantees that the rational function on the right-hand side of (10.6) can be rewritten in the form

$$\frac{x}{1 - x - x^2} = \frac{A}{x - x_1} + \frac{B}{x - x_2},$$

where x_1, x_2 are the roots of the quadratic polynomial $1 - x - x^2$ and A and B are some suitable constants. For our purposes, a slightly different form will be convenient, namely

$$\frac{x}{1 - x - x^2} = \frac{a}{1 - \lambda_1 x} + \frac{b}{1 - \lambda_2 x}, \tag{10.7}$$

where $\lambda_1 = \frac{1}{x_1}$, $\lambda_2 = \frac{1}{x_2}$, $a = -\frac{A}{x_1}$, and $b = -\frac{B}{x_2}$. From (10.7), it is easy to write down a formula for F_n. As the reader is invited to check, we get $F_n = a\lambda_1^n + b\lambda_2^n$.

We omit the calculation of the roots x_1, x_2 of the quadratic equation here, as well as the calculation of the constants a and b, and we only present the result:

$$F_n = \frac{1}{\sqrt{5}} \left[\left(\frac{1 + \sqrt{5}}{2} \right)^n - \left(\frac{1 - \sqrt{5}}{2} \right)^n \right].$$

It is remarkable that this expression full of irrational numbers should give an integer for each natural number n—but it does.

With approximate numerical values of the constants, the formula looks as follows:

$$F_n = (0.4472135\ldots) \left[(1.6180339\ldots)^n - (-0.6180339\ldots)^n \right].$$

We see that for large n, the numbers F_n behave roughly as $\frac{1}{\sqrt{5}}\lambda_1^n$, and that the ratio F_n/F_{n+1} converges to the limit $\frac{1}{\lambda_1} = 0.6180339\ldots$. This ratio was highly valued in ancient Greece; it is called the *golden section*. A rectangle with side ratio equal to the golden section was considered the most beautiful and proportional one among all rectangles. (Using the picture below, you can check whether your aesthetic criteria for rectangles agree with those of the ancient Greeks.) If we cut off a square from such a golden-section rectangle, the sides of the remaining rectangle again determine the golden section:

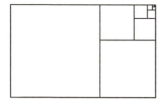

Similar to the Fibonacci numbers, the golden section is encountered surprisingly often in mathematics.

Another derivation: a staircase. Consider a staircase with n stairs. How many ways are there to ascend the staircase, if we can climb by 1 stair or by 2 stairs in each step?

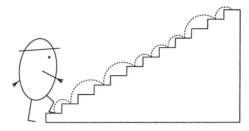

In other words, how many ways are there to write the number n as a sum of 1s and 2s, or how many solutions are there to the equation

$$s_1 + s_2 + \cdots + s_k = n,$$

with $s_i \in \{1, 2\}$, $i = 1, 2, \ldots, k$, $k = 0, 1, 2, \ldots$? If we denote the number of such solutions by S_n, we get $S_1 = 1$, $S_2 = 2$, and it is not difficult to see that for $n \geq 1$, we have $S_{n+2} = S_{n+1} + S_n$ (think it over!). This implies that S_n is exactly the Fibonacci number F_{n-1}.

We derive the generating function of the sequence $(S_0, S_1, S_2, \ldots)$ in a different way. According to the recipe from Section 10.1, we get that for a given k, the number of solutions to the equation $s_1 + s_2 + \cdots + s_k = n$ with $s_i \in \{1, 2\}$ equals the coefficient of x^n in the product $(x + x^2)^k$. We can choose k arbitrarily, however (it is not prescribed by how many steps we should ascend the staircase), and therefore S_n is the coefficient of x^n in the sum $\sum_{k=0}^{\infty}(x + x^2)^k$. Thus, this sum is the generating function of the sequence $(S_0, S_1, S_2, \ldots)$. The generating function can be further rewritten: the sum is a geometric series with quotient $x + x^2$ and so it equals $1/(1 - x - x^2)$. Hence the generating function for the Fibonacci numbers is $x/(1 - x - x^2)$, as we have already derived by a different approach.

Recipes for some recurrences. By the above method, one can find a general form of a sequence $(y_0, y_1, y_2, \ldots)$ satisfying the equation

$$y_{n+k} = a_{k-1}y_{n+k-1} + a_{k-2}y_{n+k-2} + \cdots + a_1 y_{n+1} + a_0 y_n \qquad (10.8)$$

for all $n = 0, 1, 2, \ldots$, where k is a natural number and $a_0, a_1, \ldots, a_{k-1}$ are real (or complex) numbers. For instance, for the Fibonacci numbers we would set $k = 2$, $a_0 = a_1 = 1$. Let us denote the set of all sequences $(y_0, y_1, y_2, \ldots)$ satisfying (10.8) by the symbol $\mathcal{Y}$ (so $\mathcal{Y}$ depends on k and on $a_0, a_1, \ldots, a_{k-1}$). This set contains many sequences in general, because the first k terms of the sequence $(y_0, y_1, y_2, \ldots)$ can be chosen at will (while all the remaining terms are determined by the relation (10.8)). In the sequel, we will describe what the sequences of $\mathcal{Y}$ look like, but first we take a detour into terminology.

A learned name for Eq. (10.8) is a *homogeneous linear recurrence of kth degree with constant coefficients*. Let us try to explain the parts of this complicated name.

- A *recurrence* or *recurrent relation* is used as a general notion denoting a relation (usually a formula) expressing the nth term of a sequence via several previous terms of the sequence.[3]
- *Homogeneous* appears in the name since whenever $(y_0, y_1, y_2, \ldots) \in \mathcal{Y}$ then also $(\alpha y_0, \alpha y_1, \alpha y_2, \ldots) \in \mathcal{Y}$ for any real number α. (In mathematics, "homogeneity" usually means "invariance to scaling".) On the other hand, an example of an inhomogeneous recurrence is $y_{n+1} = y_n + 1$.
- The word *linear* here means that the values of y_j always appear in the first power in the recurrence and are not multiplied together. A nonlinear recurrence is, for example, $y_{n+2} = y_{n+1} y_n$.
- Finally, the phrase *with constant coefficients* expresses that a_0, $a_1, \ldots, a_{k-1}$ are fixed numbers independent of n. One could also consider a recurrence like $y_{n+1} = (n - 1)y_n$, where the coefficient on the right-hand side is a function of n.

So much for the long name. Now we are going to formulate a general result about solutions of the recurrent relation of the considered type. We define the *characteristic polynomial* of the recurrence (10.8) as the polynomial

$$p(x) = x^k - a_{k-1}x^{k-1} - a_{k-2}x^{k-2} - \cdots - a_1 x - a_0.$$

For example, the characteristic polynomial of the recurrence relation for the Fibonacci numbers is $x^2 - x - 1$. Let us recall that any polynomial of degree k with coefficient 1 at x^k can be written in the form

$$(x - \lambda_1)(x - \lambda_2) \ldots (x - \lambda_k),$$

where $\lambda_1, \ldots, \lambda_k$ are (generally complex) numbers called the *roots* of the given polynomial.

10.3.1 Proposition. *Let $p(x)$ be the characteristic polynomial of the homogeneous linear recurrence (10.8).*

(i) *(Simple roots) Suppose that $p(x)$ has k pairwise distinct roots $\lambda_1, \ldots, \lambda_k$. Then for any sequence $y = (y_0, y_1, \ldots) \in \mathcal{Y}$ satisfying (10.8), complex constants $C_1, C_2, \ldots, C_k$ exist such that for all n, we have*

$$y_n = C_1 \lambda_1^n + C_2 \lambda_2^n + \cdots + C_k \lambda_k^n.$$

(ii) *(The general case) Let $\lambda_1, \ldots, \lambda_q$ be pairwise different complex numbers, and let $k_1, \ldots, k_q$ be natural numbers with $k_1 + k_2 + \cdots + k_q = k$ such that*

---

[3]Another name sometimes used with this meaning is *difference equation*. This is related to the so-called difference of a function, which is a notion somewhat similar to derivative. The theory of difference equations is often analogous to the theory of differential equations.

$$p(x) = (x - \lambda_1)^{k_1} (x - \lambda_2)^{k_2} \ldots (x - \lambda_q)^{k_q}.$$

Then for any sequence $y = (y_0, y_1, \ldots) \in \mathcal{Y}$ satisfying (10.8), complex constants C_{ij} exist $(i = 1, 2, \ldots, q,\ j = 0, 1, \ldots, k_i - 1)$ such that for all n, we have

$$y_n = \sum_{i=1}^{q} \sum_{j=0}^{k_i - 1} C_{ij} \binom{n}{j} \lambda_i^n.$$

How do we solve a recurrence of the form (10.8) using this proposition? Let us give two brief examples. For the recurrence relation $y_{n+2} = 5y_{n+1} - 6y_n$, the characteristic polynomial is $p(x) = x^2 - 5x + 6 = (x - 2)(x - 3)$. Its roots are $\lambda_1 = 2$ and $\lambda_2 = 3$, and Proposition 10.3.1 tells us that we should look for a solution of the form $C_1 2^n + C_2 3^n$. Being given some initial conditions, say $y_0 = 2$ and $y_1 = 5$, we must determine the constants C_1, C_2 in such a way that the formula gives these required values for $n = 0, 1$. In our case, we would set $C_1 = C_2 = 1$.

And one more example, a quite artificial one, with multiple roots: the equation $y_{n+5} = 8y_{n+4} + 25y_{n+3} - 38y_{n+2} + 28y_{n+1} - 8y_n$ has characteristic polynomial[4] $p(x) = (x - 1)^2 (x - 2)^3$, and so Proposition 10.3.1 says that the solution should be sought in the form $y_n = C_{10} + C_{11}n + C_{20}2^n + C_{21}n2^n + C_{22}\binom{n}{2}2^n$. The values of the constants should again be calculated according to the values of the first 5 terms of the sequence $(y_0, y_1, y_2, \ldots)$.

The procedure just demonstrated for solving the recurrence (10.8) can be found in all kinds of textbooks, and the generating functions are seldom mentioned in this connection. Indeed, Proposition 10.3.1 can be proved in a quite elegant manner using linear algebra (as indicated in Exercise 16), and once we know from somewhere that the solution should be of the indicated form, generating functions become unnecessary. However, the method explained above for Fibonacci numbers really finds the correct form of the solution. Moreover, it is a nice example of an application of generating functions, and a similar approach can sometimes be applied successfully for recurrences of other types, where no general solution method is known (or where it is difficult to find such a method in the literature, which is about the same in practice).

---

[4]Of course, the authors have selected the coefficients so that the characteristic polynomial comes out very nicely. The instructions given above for solving homogeneous linear recurrence relations with constant coefficients (as well as the method with generating functions) leave aside the question of finding the roots of the characteristic polynomial. In examples in various textbooks, the recurrences mostly have degree 1 or 2, or the coefficients are chosen in such a way that the roots are small integers.

Exercises

1. *Determine the number of n-term sequences of 0s and 1s containing no two consecutive 0s.

2. Prove that any natural number $n \in \mathbf{N}$ can be written as a sum of mutually distinct Fibonacci numbers.

3. Express the nth term of the sequences given by the following recurrence relations (generalize the method used for the Fibonacci numbers, or use Proposition 10.3.1).

   (a) $a_0 = 2,\ a_1 = 3,\ a_{n+2} = 3a_n - 2a_{n+1}\ (n = 0, 1, 2, \ldots)$,

   (b) $a_0 = 0,\ a_1 = 1,\ a_{n+2} = 4a_{n+1} - 4a_n\ (n = 0, 1, 2, \ldots)$,

   (c) $a_0 = 1, a_{n+1} = 2a_n + 3\ (n = 0, 1, 2, \ldots)$.

4. In a sequence $(a_0, a_1, a_2, \ldots)$, each term except for the first two is the arithmetic mean of the preceding two terms, i.e. $a_{n+2} = (a_{n+1} + a_n)/2$. Determine the limit $\lim_{n \to \infty} a_n$ (as a function of a_0, a_1).

5. Solve the recurrence relation $a_{n+2} = \sqrt{a_{n+1}a_n}$ with initial conditions $a_0 = 2,\ a_1 = 8$, and find $\lim_{n \to \infty} a_n$.

6. (a) Solve the recurrence $a_n = a_{n-1} + a_{n-2} + \cdots + a_1 + a_0$ with the initial condition $a_0 = 1$.

   (b) *Solve the recurrence $a_n = a_{n-1} + a_{n-3} + a_{n-4} + a_{n-5} + \cdots + a_1 + a_0$ $(n \geq 3)$ with $a_0 = a_1 = a_2 = 1$.

7. *Express the sum

$$S_n = \binom{2n}{0} + 2\binom{2n-1}{1} + 2^2\binom{2n-2}{2} + \cdots + 2^n\binom{n}{n}$$

as a coefficient of x^{2n} in a suitable power series, and find a simple formula for S_n.

8. Calculate $\sum_{k=0}^{\lfloor n/2 \rfloor} \binom{n-k}{k}(-4)^{-k}$.

9. *Show that the number $\frac{1}{2}[(1 + \sqrt{2})^n + (1 - \sqrt{2})^n]$ is an integer for all $n \geq 1$.

10. *Show that the number $(6 + \sqrt{37})^{999}$ has at least 999 zeros following the decimal point.

11. *Show that for any $n \geq 1$, the number $(\sqrt{2} - 1)^n$ can be written as the difference of the square roots of two consecutive integers.

12. *Find the number of n-term sequences consisting of letters a, b, c, d such that a is never adjacent to b.

13. ("Mini-Tetris") How many ways are there to fill completely without overlap an $n \times 2$ rectangle with pieces of the following types? The sides of the pieces are 1 and 2; the pieces can be rotated by a multiple of the right angle. In (b) and (c), it suffices to calculate the generating function or reduce the problem to solving a recurrence of the form (10.8), and determine the order of growth (an asymptotic estimate for large n).

(a) ☐ ☐,

(b) * ☐ └,

(c) ** ☐ └.

14. ^{CS} This exercise is related to the theory of formal languages. We give the necessary definitions but to appreciate the context, the reader should consult a textbook on automata and formal languages.

Let Σ be some fixed finite alphabet (such as $\Sigma = \{a, b, c\}$). A *word* over Σ is a finite sequence of letters (such as *babbaacccba*). The empty word having no letters is denoted by ε. A *language* over Σ is a set of words over Σ. If u and v are words then uv denotes the *concatenation* of u and v, i.e. we write first u and then v. The *generating function* of a language L is the generating function of the sequence $n_0, n_1, n_2, \ldots$, where n_i is the number of words of length i in L.

Let us say that a language L is *very regular* if it can be obtained by finitely many applications of the following rules:

1. The languages $\emptyset$ and $\{\varepsilon\}$ are very regular, and the language $\{\ell\}$ is very regular for every letter $\ell \in \Sigma$.
2. If L_1, L_2 are very regular languages, then also the language $L_1.L_2$ is very regular, where $L_1.L_2 = \{uv: u \in L_1, v \in L_2\}$.
3. If L is a very regular language then also L^* is very regular, where $L^* = \{\varepsilon\} \cup L \cup L.L \cup L.L.L \cup \ldots$.
4. If L_1, L_2 are very regular languages with $L_1 \cap L_2 = \emptyset$, then also $L_1 \cup L_2$ is very regular.

(a) Show that the following languages over $\Sigma = \{a, b\}$ are all very regular: the language consisting of all words of odd length, the language consisting of all words beginning with *aab* and having an even number of letters a, and *the language consisting of all words having no two consecutive letters a.

(b) *Show that the generating function of any very regular language is a rational function (a polynomial divided by a polynomial), and describe how to calculate it.

(c) **Show that if L_1, L_2 are very regular languages (not necessarily disjoint ones) then also $L_1 \cup L_2$ is very regular. (Therefore, regular languages have rational generating functions.)

(d) (A programming project) Look up information on regular languages and their relation to finite automata. Write a program that accepts a specification of a regular language (by a regular expression or by a nondeterministic finite automaton) and computes a formula for the number of words of length n in the language.

15. *Prove Proposition 10.3.1 by generalizing the method shown for the Fibonacci numbers (use a general theorem on decomposing a rational function into partial fractions from integral calculus).

16. Prove Proposition 10.3.1 directly, using linear algebra, according to the following hints.

    (a) Check that the set $\mathcal{Y}$ is a vector space with respect to componentwise addition of sequences and componentwise multiplication by a complex number.

    (b) Show that the dimension of $\mathcal{Y}$ equals k.

    (c) *Show that in the situation of part (i) of Proposition 10.3.1, the sequences $(\lambda_i^0, \lambda_i^1, \lambda_i^2, \ldots)$ belong to $\mathcal{Y}$ (where $i = 1, 2, \ldots, k$), and they are all linearly independent in $\mathcal{Y}$ (hence they form a basis of the vector space $\mathcal{Y}$ by (b)). This proves part (i).

    (d) Verify that in the situation of part (ii) of Proposition 10.3.1, each of the sequences $(\binom{n}{j}\lambda_i^n)_{n=0}^\infty$ belongs to the solution set $\mathcal{Y}$.

    (e) **Prove that the sequences considered in part (d) are linearly independent in $\mathcal{Y}$, i.e. they form a basis.

10.4 Binary trees

We are going to consider the so-called binary trees, which are often used in data structures. Fig. 10.1 depicts several different binary trees with 5 vertices. For our purposes, a binary tree can concisely be defined as follows: a binary tree either is empty (it has no vertex), or consists of one distinguished vertex called the *root*, plus an ordered pair of binary trees called the *left subtree* and *right subtree*.[5]

Let b_n denote the number of binary trees with n vertices. Our goal is to find a formula for b_n. By listing all small binary trees, we find that $b_0 = 1$, $b_1 = 1$, $b_2 = 2$, and $b_3 = 5$. This can serve for checking the result obtained below.

As usual, we let $b(x) = b_0 + b_1 x + b_2 x^2 + \cdots$ be the corresponding generating function. For $n \geq 1$, the number of binary trees with n

---

[5]This is a definition by induction. First we say what a binary tree with 0 vertices is and then we define a binary tree with n vertices using the already defined binary trees with fewer vertices. By admitting the empty binary tree, we avoid dealing with various special cases, such as when the root has only a left subtree or only a right subtree.

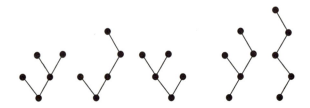

Fig. 10.1   Several different binary trees on 5 vertices.

vertices equals the number of ordered pairs of the form (B, B'), where B and B' are binary trees having together $n - 1$ vertices. That is, if B has k vertices then B' has $n - k - 1$ vertices, $k = 0, 1, \ldots, n - 1$. Therefore, the number of such ordered pairs is

$$b_n = b_0 b_{n-1} + b_1 b_{n-2} + \cdots + b_{n-1} b_0. \tag{10.9}$$

Comparing this with the definition of multiplication of power series, we see that b_n is exactly the coefficient of x^{n-1} in the product $b(x) \cdot b(x) = b(x)^2$. (This is the natural example for the combinatorial meaning of multiplication of generating functions promised in Section 10.2.) Hence b_n is the coefficient of x^n in the function $xb(x)^2$. And so $xb(x)^2$ is the generating function of the same sequence as $b(x)$, *except* that $b(x)$ has constant term $b_0 = 1$ while the power series $xb(x)^2$ has constant term 0; this is because the formula (10.9) is only correct for $n \geq 1$. We can write the following equality of generating functions:

$$b(x) = 1 + xb(x)^2.$$

Suppose that x is such a real number that the power series $b(x)$ converges. Then $b(x)$ is also a real number, and it must satisfy the quadratic equation $b(x) = 1 + xb(x)^2$. By a well-known formula for the roots of a quadratic equation, $b(x)$ must be one of the numbers

$$\frac{1 + \sqrt{1 - 4x}}{2x} \quad \text{or} \quad \frac{1 - \sqrt{1 - 4x}}{2x}.$$

This seems as if there are two possible solutions. But we know that the sequence $(b_0, b_1, b_2, \ldots)$, and thus also its generating function, are determined uniquely. Since $b(x)$ is continuous as a function of x (whenever it converges), we must take either the first solution (the one with "+") for all x, or the second solution (the one with "−")

for all x. If we look at the first solution, we find that for x tending to 0 it goes to ∞, while the generating function $b(x)$ must approach $b_0 = 1$. So whenever $b(x)$ converges, it must converge to the second solution $b(x) = (1 - \sqrt{1 - 4x})/2x$.

It remains to calculate the coefficients of this generating function. To this end, we make use of the generalized binomial theorem 10.2.3. This theorem gives us the expansion

$$\sqrt{1 - 4x} = \sum_{k=0}^{\infty} (-4)^k \binom{1/2}{k} x^k.$$

The coefficient of x^0 is 1, and hence the power series $1 - \sqrt{1 - 4x}$ has zero constant term. We can thus divide it by $2x$ (by shifting to the left by one position and dividing all coefficients by 2). From this we get that for all $n \geq 1$,

$$b_n = -\frac{1}{2}(-4)^{n+1}\binom{1/2}{n+1}. \tag{10.10}$$

By further manipulations, which we leave as an exercise, one can obtain a nicer form:

$$b_n = \frac{1}{n+1}\binom{2n}{n}.$$

The numbers b_n thus defined are known by the name *Catalan numbers* in the literature (and they are named after Eugène Charles Catalan rather than after Catalonia).

Besides giving the number of binary trees, they have many other combinatorial meanings. Some of them are indicated in the exercises below.

More about methods for counting various types of graphs, trees, etc., can be found, for example, in the book by Harary and Palmer [19]. Publicly available computer software which can solve many such tasks automatically is described by Flajolet, Salavy, and Zimmermann [34], together with the underlying theory.

Exercises

1. Convert the expression on the right-hand side of (10.10) to the form given below that equation.

2. *As the reader might have noticed, we have left out a discussion of convergence of the series $b(x)$. Prove that $b(x)$ converges for some number $x \neq 0$ (but do not use the above-derived formula for b_n since this was obtained under the assumption that $b(x)$ converges).

3. Consider an $n \times n$ chessboard:

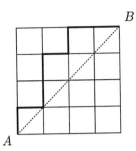

Consider the shortest paths from the corner A to the corner B following the edges of the squares (each of them consists of $2n$ edges).

(a) How many such paths are there?

(b) *Show that the number of paths that never go below the diagonal (the line AB) is exactly b_n, i.e. the Catalan number. One such path is drawn in the figure.

(c) **Give an elementary proof of the formula $b_n = \frac{1}{n+1}\binom{2n}{n}$, without generating functions.

4. Consider a product of 4 numbers, $abcd$. It can be "parenthesized" in 5 ways: $((ab)c)d$, $(a(bc))d$, $(ab)(cd)$, $a((bc)d)$, and $a(b(cd))$. Prove that the number of such parenthesizings of a product of n numbers is the Catalan number b_{n-1}.

5. There are $2n$ people standing in a queue for tickets costing 5 double-zons each. Everyone wants to buy one ticket; n of the people have a 10-doublezon banknote and n of them a 5-doublezon coin. Initially, the cashier has no money.

(a) How many orderings of the people in the queue are there for which the cashier can always give a 5-doublezon coin back to each 10-doublezon note owner?

(b) What is the probability that the cashier will be able to return cash to everyone for a random ordering of the queue?

6. *Consider a regular n-gon. Divide it into triangles by drawing $n-3$ nonintersecting diagonals (a diagonal is a segment connecting two vertices). Show that the number of such triangulations is b_{n-2}.

7. There are $2n$ points marked on a circle. We want to divide them into pairs and connect the points in each pair by a segment (chord) in such a way that these segments do not intersect. Show that the number of ways to do this is b_n.

8. In this exercise and several subsequent ones, we are going to use the terms *rooted tree* and *planted tree* introduced in Section 4.2 (both are

trees with root; for planted trees the left-to-right order of sons of each vertex is important, for rooted trees it isn't).

(a) Let c_n denote the number of (nonisomorphic) planted trees with n vertices, where each vertex either is an end-vertex or has exactly 2 sons (examples of such nonisomorphic trees are the first and the third trees in Fig. 10.1). Show that the corresponding generating function $c(x)$ satisfies the equation $c(x) = x + xc(x)^2$, and calculate a formula for c_n.

(b) *Can you derive the value of c_n from the knowledge of b_n (the number of binary trees)?

(c) Find d_n, the number of nonisomorphic planted trees with n vertices, where each vertex has 0, 1, or 2 sons. (Note that this is something *different* from binary trees!)

9. Let t_n denote the number of nonisomorphic planted trees with n vertices.

(a) Show that the corresponding generating function $t(x)$ satisfies $t(x) = \frac{x}{1-t(x)}$ and calculate a formula for t_n.

(b) *Can you derive the value of t_n from the knowledge of b_n?

10. We say that a planted tree is *old* if it has no *young vertex*, where a young vertex is a leaf adjacent to the root. Let s_n be the number of old planted trees with n vertices. Express the generating function $s(x)$ using $t(x)$ from the previous exercise.

11. *Now let us consider rooted trees where each nonleaf vertex has exactly 2 sons (for rooted trees, the left-to-right order of sons is immaterial, so the first and third trees in Fig. 10.1 are now isomorphic). Let $\bar{b}(x)$ be the corresponding generating function. Derive the equation

$$\bar{b}(x) = 1 + \frac{x}{2}\left(\bar{b}(x)^2 + \bar{b}(x^2)\right).$$

12. Consider *alkane radicals*; these are acyclic hydrocarbons where each carbon atom has 4 single bonds, except for one that has 3 single bonds and 1 "free" bond. Such molecules can be imagined as rooted trees. The vertices are the carbon atoms and each vertex has 0, 1, 2, or 3 sons. The order of sons of a vertex is not important. Let r_n be the number of such trees with n vertices (i.e. of alkane radicals with n carbon atoms), and let $r(x)$ be the generating function.

(a) **Derive the equation $r(x) = 1 + \frac{x}{6}\left(r(x)^3 + 3r(x^2)r(x) + 2r(x^3)\right)$.

(b) Using this equation, calculate a table of values of r_n for small values of n. Compare to values given in some chemistry handbook.

(c) ^{CS} Write a program to do the calculation in (b). A more ambitious project is a program that calculates coefficients of generating functions given by equations of various forms.

10.5   On rolling the dice

Problem. We roll a die until a 6 appears for the first time. How many times do we need to roll, on the average?[6]

The probability of getting a 6 in the first round is $p = \frac{1}{6}$. The probability of a 6 not appearing in the first round and showing up in the second round is $(1 - p)p$, and in general, the probability of a 6 appearing in the ith round for the first time is $q_i = (1 - p)^{i-1}p$. The average number of rounds (i.e. the expectation) is then

$$S = \sum_{i=0}^{\infty} iq_i = \sum_{i=1}^{\infty} i(1 - p)^{i-1}p.$$

To sum this series, we introduce the generating function $q(x) = q_1x + q_2x^2 + \cdots$. By differentiating this series term by term, we get $q'(x) = 1 \cdot q_1 + 2 \cdot q_2x + 3 \cdot q_3x^2 + \cdots$, and hence the desired S equals the value of $q'(1)$.

With some effort, we calculate that our generating function is given by the expression

$$q(x) = \frac{p}{1 - p} \cdot \frac{1}{1 - (1 - p)x} - \frac{p}{1 - p}.$$

A small exercise in differentiation gives $q'(x) = p/(1 - (1 - p)x)^2$, and hence $S = q'(1) = \frac{1}{p}$. In our case, for $p = \frac{1}{6}$, the average number of rounds is thus 6.

Here is a way of reasoning giving a much shorter solution. In any case, we roll the die at least once. With probability $1 - p$, we do not roll 6 in the first round, and then still S more rounds await us on the average (the die has no memory, and so the situation after the first unsuccessful attempt is the same as if nothing happened). Therefore

$$S = 1 + (1 - p)S,$$

and $S = 1/p$ results immediately.

The method with generating functions shown in this section has many applications in probability theory. If X is some random variable attaining value i with probability q_i, $i = 0, 1, 2, \ldots$, and if $q(x) = \sum_{i=0}^{\infty} q_ix^i$ is the generating function, then the expectation of X is $q'(1)$.

---

[6]After some consideration, we decided on this formulation of the problem, as opposed to another one which offers itself, involving the so-called Russian roulette.

Exercises

1. Using generating functions as we did above, calculate
   (a) the average number of 6s obtained by rolling a die n times,
   (b) *the average value of the expression $(X - 6)^2$, where X is the number of times a die must be rolled until the first 6 appears (this is a certain measure of the "typical deviation" from the average number of rounds; in probability theory, it is called the *variance*).
   (c) Let X be some random variable attaining values $0, 1, 2, \ldots$, where i is attained with probability q_i, and let $q(x)$ be the generating function of the q_i. Express the the variance of X, i.e. the quantity $\operatorname{Var}[X] = \mathbf{E}\left[(X - \mathbf{E}[X])^2\right]$, using the derivatives of $q(x)$ at suitable points.

10.6 Random walk

Imagine the real axis drawn in the plane with integers marked by circles. A frog leaps among these circles according to the following rules of random walk:

- Initially (before the first move) the frog sits at number 1.

- In each move, the frog leaps either by two circles to the right (from i to $i + 2$) or by one circle to the left (from i to $i - 1$). It decides on one of these possibilities at random, and both possibilities have the same probability (as if it tossed a fair coin and decided according to the result heads/tails).

Problem. What is the probability that the frog ever reaches the number 0?

First, we have to clarify what we mean by such a probability. It is easy to define the probability that the frog reaches 0 within the first 7 leaps, say (let us denote this probability by P_7). The first 7 leaps give 2^7 different possible trajectories, since in each leap, the frog makes a decision between two possibilities, and these decisions can be combined arbitrarily. By the rules in our problem, all these trajectories have the same probability. The above-defined probability P_7 equals the number of trajectories that pass through 0 (there are 75 of them) divided by the total number of trajectories, i.e. by 2^7.

The probability P required in the problem can be defined as the limit $P = \lim_{i \to \infty} P_i$, where the definition of P_i has been explained above for $i = 7$. This limit certainly exists because, clearly, $P_1 \leq P_2 \leq \ldots$.

Let a_i denote the number of trajectories of the first i leaps such that the frog reaches 0 by the ith leap and never before. We thus have

$$P = \sum_{i=1}^{\infty} \frac{a_i}{2^i}.$$

If we introduce the generating function $a(x) = a_1 x + a_2 x^2 + a_3 x^3 + \cdots$, we get $P = a(\frac{1}{2})$.

For solving the problem, it will be useful to look also at trajectories starting at numbers other than 1 (but continuing according to the same rule). For instance, what is the number b_i of trajectories starting at the number 2 that first reach 0 by the ith leap? In order that such a trajectory reaches 0, it has to reach 1 first. Let j be the number of the leap by which 1 is first reached. If j is determined, there are a_j possibilities to avoid 1 in leaps $1, 2, \ldots, j - 1$ and reach it by the jth leap. Then, $i - j$ leaps remain to move from 1 to 0, and the number of ways for these $i - j$ leaps is a_{i-j}. So for a given j, there are $a_j a_{i-j}$ possibilities, and altogether we get[7]

$$b_i = \sum_{j=1}^{i-1} a_j a_{i-j}.$$

In the language of generating functions this means $b(x) = a(x)^2$.

Analogously, let c_i be the number of trajectories starting at the number 3 and first reaching 0 by the ith leap. As before, we can see that $c(x) = a(x)b(x) = a(x)^3$.

Let us investigate trajectories starting at 1 from a different point of view. By the first move , either the frog reaches 0 directly (which gives $a_1 = 1$), or it leaps to the number 3. In the latter case, it has c_{i-1} possibilities of reaching 0 for the first time after the next $i - 1$ leaps. Hence for $i > 1$, we have $a_i = c_{i-1}$. Converted to a relation between the generating functions, this reads

$$a(x) = x + xc(x) = x + xa(x)^3. \tag{10.11}$$

In particular, for $x = \frac{1}{2}$, the following equation results (where we write $P = a(\frac{1}{2})$):

---

[7]Note that we use the fact that leaps to the left are always by 1, and so the frog cannot reach 0 from 2 without leaping to 1 first.

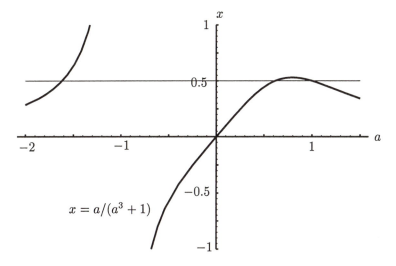

Fig. 10.2 The function $x = a/(a^3 + 1)$.

$$P = \frac{1}{2} + \frac{1}{2}P^3.$$

This has three solutions: 1, $\frac{1}{2}(\sqrt{5}-1)$, and $-\frac{1}{2}(\sqrt{5}+1)$. The negative root can be excluded right away. The right value is given by the second root, $P = \frac{1}{2}(\sqrt{5} - 1) = 0.618033988\ldots$ (the golden section again!).

Why can't 1 be the answer? Here is one possible argument. Since the series $a(x)$ converges for $x = \frac{1}{2}$ and since all the a_i are nonnegative, $a(x)$ is a nondecreasing continuous function of x on the interval $(0, \frac{1}{2}]$. At the same time, $a(x)$ has to be the root of the equation (10.11), or in other words, we must have $x = a(x)/(a(x)^3 + 1)$ for all such x. If we plot the function $a \mapsto a/(a^3 + 1)$ as in Fig. 10.2, we see that $a(\frac{1}{2})$ cannot be given by the rightmost intersection of the $x = \frac{1}{2}$ line with the plotted curve since then we would have $a(x) > 1$ for all x slightly smaller than $\frac{1}{2}$. This can be made rigorous by using some elementary calculus (computing the tangent to the curve at the point $(1,1)$) but we hope the picture is convincing enough.

From Eq. (10.11) we could, in principle, calculate the function $a(x)$, and then try to express the numbers a_i using the Taylor series, say (which is quite laborious). A charming feature of the presented solution is that we don't need to do anything like that.

Exercises

1. Consider a random walk where we start at the number 0 and in each step we move from i to $i+1$ or to $i-1$ with equal probability.

   (a) *Prove that we eventually return to 0 with probability 1.

   (b) Prove that each number k is visited at least once with probability 1.

2. Consider a random walk as in the preceding exercise.

   (a) *Let S_n denote the expected number of steps needed to reach n (from the preceding exercise, we know that n will be reached sooner or later with probability 1). What is wrong with the following argument? We claim that $S_n = cn$ for some constant c. This is true for $n = 0$, so let $n > 0$. On the average, we need S_1 steps to reach 1, and then S_{n-1} more steps to reach n starting from 1. Hence $S_n = S_1 + S_{n-1} = c + c(n-1) = cn$, where $c = S_1$.

   (b) **What is the expected number of steps needed to get at least n steps away from 0 (i.e. to reach n or $-n$)?

10.7   Integer partitions

How many ways are there to write a natural number n as a sum of several natural numbers? The answer is not too difficult if we count *ordered partitions* of n; that is, if we regard the expressions $3 = 2 + 1$ and $3 = 1 + 2$ as two different ways of expressing 3 as a sum (Exercise 1). The problem becomes much harder and more interesting if we consider identical expressions differing only by the order of addends (in this case we will simply speak of *partitions of n* throughout this section). For instance, for $n = 5$, all the possible partitions are $5 = 1 + 1 + 1 + 1 + 1$, $5 = 1 + 1 + 1 + 2$, $5 = 1 + 2 + 2$, $5 = 1 + 1 + 3$, $5 = 2 + 3$, $5 = 1 + 4$, and $5 = 5$. Let p_n stand for the number of partitions of n in this sense.

To make the way of writing the partitions unique, we can for instance insist that the addends be written in a nondecreasing order, as we wrote them in the listing of the partitions of 5. So another formulation of the question is: how many ways are there to build a "nondecreasing wall" using n bricks, as in the following picture (corresponding to the partitions $10 = 1 + 1 + 2 + 2 + 4$ and $10 = 1 + 1 + 2 + 6$)?

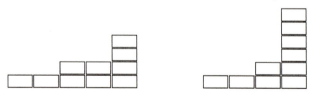

(Such a picture is called the *Ferrers diagram* of a given partition.)

The definition of p_n looks quite simple, but, perhaps surprisingly, there is no "simple" formula for p_n (like a binomial coefficient, say). The problem of estimating p_n has been considered in number theory and it was solved with an almost unbelievable accuracy by Hardy and Ramanujan in 1918. (The history is described in Littlewood's book [21], which can be recommended as great reading about mathematics.) In fact, Hardy and Ramanujan discovered an amazing exact (and complicated) formula for p_n. Currently, proving or even understanding these precise results requires knowledge of quite a deep theory (see e.g. Andrews [30] for a presentation). Here we will only scratch the surface of this problem; reasonably good asymptotic estimates for p_n can be derived in a simple but clever way using generating functions, and this is what we will demonstrate.

We know how to express the number of solutions to an equation of the form

$$i_1 + i_2 + \cdots + i_k = n,$$

with the values of each i_j lying in some prescribed set, as the coefficient of x^n in a suitable expression. Here the order of the i_j does matter, though, so it is not obvious how to relate this to the (unordered) partitions of n. The trick is to let i_j express *the contribution of the addends equal to j* in a partition of n, or, in other words, the number of bricks in columns of height exactly j in our nondecreasing wall. Therefore, the partitions of n are in a bijective correspondence with the solutions of the equation

$$i_1 + i_2 + \cdots + i_n = n$$

with

$$i_1 \in \{0, 1, 2, 3, \ldots\}, \quad i_2 \in \{0, 2, 4, 6, \ldots\}, \quad \ldots, \quad i_j \in \{0, j, 2j, 3j, \ldots\}.$$

For example, the partition $5 = 1 + 2 + 2$ would correspond to the solution $i_1 = 1$, $i_2 = 4$, $i_3 = i_4 = i_5 = 0$.

The next step is standard. From the new formulation we immediately get that p_n is the coefficient of x^n in the product

$$P_n(x) = (1 + x + x^2 + \cdots)(1 + x^2 + x^4 + x^6 + \cdots) \ldots (1 + x^n + x^{2n} + \cdots)$$

$$= \prod_{k=1}^{n} \frac{1}{1 - x^k}.$$

This finite product is *not* the generating function of the sequence $(p_n)_{n=0}^{\infty}$. To make it the actual generating function (having the coefficient p_n at x^n for all n simultaneously), we have to use the infinite product

$$P(x) = \prod_{k=1}^{\infty} \frac{1}{1 - x^k}.$$

But since an infinite product could perhaps look more frightening we stick to the finite product $P_n(x)$ in estimating p_n.

Now we are going to upper-bound p_n using the method already demonstrated in the proof of Theorem 2.6.1 (estimating binomial coefficients). For all numbers $x \in (0, 1)$, we have

$$p_n \leq \frac{1}{x^n} P_n(x) = \frac{1}{x^n} \prod_{k=1}^{n} \frac{1}{1 - x^k}.$$

We want to choose x in such a way that the right-hand side is as small as possible. A suitable way of estimating $P_n(x)$ is another clever part of the proof. First, for dealing with a product, it is often advisable to consider the logarithm; in our case this leads to

$$\ln p_n \leq \ln \left(\frac{1}{x^n} P_n(x) \right) = -n \ln x - \sum_{k=1}^{n} \ln(1 - x^k).$$

We recall the power series (10.3) for logarithm:

$$-\ln(1 - y) = \frac{y}{1} + \frac{y^2}{2} + \frac{y^3}{3} + \frac{y^4}{4} + \cdots$$

for all $y \in (-1, 1)$. Therefore, we can write

$$-\sum_{k=1}^{n} \ln(1 - x^k) = \sum_{k=1}^{n} \sum_{j=1}^{\infty} \frac{x^{kj}}{j} = \sum_{j=1}^{\infty} \frac{1}{j} \sum_{k=1}^{n} x^{jk}$$

$$\leq \sum_{j=1}^{\infty} \frac{1}{j} \sum_{k=1}^{\infty} x^{jk} = \sum_{j=1}^{\infty} \frac{1}{j} \frac{x^j}{1 - x^j}.$$

Now, perhaps unexpectedly, we use the formula for the sum of a geometric progression in a "reverse direction". We have

$$1 - x^j = (1 - x)(1 + x + x^2 + \cdots + x^{j-1}) \geq (1 - x)jx^{j-1}$$

(recall that $0 < x < 1$) and so

$$\sum_{j=1}^{\infty} \frac{1}{j} \frac{x^j}{1 - x^j} \leq \sum_{j=1}^{\infty} \frac{1}{j} \frac{x^j}{(1 - x)jx^{j-1}} = \frac{x}{1 - x} \sum_{j=1}^{\infty} \frac{1}{j^2}.$$

Next, we need the following remarkable

10.7.1 Fact. $\displaystyle\sum_{j=1}^{\infty} \frac{1}{j^2} = \frac{\pi^2}{6}.$

As a mathematical dessert, we reproduce a proof (going back to Euler) of this fact at the end of this section. Continuing with our estimates for $\ln p_n$, we get

$$\ln p_n \leq -n \ln x + \frac{\pi^2}{6} \frac{x}{1-x}.$$

For the subsequent calculation, it is convenient to introduce a new variable $u = x/(1-x)$ (hence u can be any number in $[0, \infty)$, and $x = u/(1+u)$). After this substitution, and using the inequality $\ln(1 + \frac{1}{u}) \leq \frac{1}{u}$ (which is readily derived from Fact 2.5.4), we get

$$\ln p_n < n \ln\left(1 + \frac{1}{u}\right) + \frac{\pi^2}{6} u \leq \frac{n}{u} + \frac{\pi^2}{6} u.$$

By substituting $u = \sqrt{6n}/\pi$ into the last expression, we arrive at $\ln p_n \leq \pi\sqrt{\frac{2}{3}n}$. (Why is $u = \sqrt{6n}/\pi$ the right value to use? A little calculus shows that this is the value where the considered upper bound for $\ln p_n$ attains its minimum as a function of u.) Hence we have proved the following:

10.7.2 Theorem. *For all $n \geq 1$, we have*

$$p_n < e^{\pi\sqrt{\frac{2}{3}n}} = e^{(2.5650...)\sqrt{n}}.$$

A lower bound. How good is the upper bound just derived? The results of Hardy and Ramanujan imply that $p_n \sim \frac{1}{4\sqrt{3}\,n} e^{\pi\sqrt{\frac{2}{3}n}}$, and so the upper bound we have derived is quite good—it even has the correct constant in the exponent. But what if we are in a wilderness (in a broad sense, meaning cut off from mathematical libraries) and want to know whether the bound in Theorem 10.7.2 is roughly correct? Here is a quick way of deriving a weaker lower bound which at least tells us we are not too far off the mark.

As the reader is invited to calculate in Exercise 1, the number of ordered partitions on n into k addends is $\binom{n-1}{k-1}$. Since every (unordered) partition with k addends gives rise to at most $k!$ ordered partitions, we have

$$p_n \geq \frac{\binom{n-1}{k-1}}{k!} \geq \frac{(n-1)(n-2)\ldots(n-k+1)}{(k!)^2}$$

for any $k \in \{1, 2, \ldots, n\}$. How large should k be to give the best lower bound? If we increase k by 1, the denominator is multiplied by the factor

$(k+1)^2$ and the numerator by $n-k$. Hence if $(k+1)^2 < n-k$ then $k+1$ is better than k, and for the best k, $(k+1)^2$ should be approximately equal to $n-k$. To keep the expressions simple, we set $k = \lfloor\sqrt{n}\rfloor$. The intuition is that then $(n-1)(n-2)\ldots(n-k+1)$ behaves roughly like $n^{\sqrt{n}}$. Here is a rigorous calculation supporting this intuition. We have

$$(n-1)(n-2)\ldots(n-k+1) \geq (n-k)^{k-1} = n^{k-1}\left(1-\frac{k}{n}\right)^{k-1},$$

and since $\frac{k}{n} = \lfloor\sqrt{n}\rfloor/n \leq 1/\lfloor\sqrt{n}\rfloor = \frac{1}{k}$ and $(1-\frac{1}{k})^{k-1} > e^{-1}$, we further get

$$n^{k-1}\left(1-\frac{k}{n}\right)^{k-1} \geq n^{k-1}\left(1-\frac{1}{k}\right)^{k-1} \geq \frac{n^k}{en}.$$

By Theorem 2.5.5, we upper-bound $k! \leq ek(k/e)^k$, and so we get

$$p_n \geq \left(\frac{n}{k^2}\right)^k \frac{e^{2k-3}}{nk^2} \geq \frac{e^{2k-3}}{n^2} \geq \frac{1}{e^5 n^2}e^{2\sqrt{n}}.$$

If n is large enough then n^2 is much smaller than $e^{\sqrt{n}}$, and hence $n^{-2}e^{2\sqrt{n}} = \Omega(e^{\sqrt{n}})$. So we have shown that for n large enough, p_n lies between $e^{c_1\sqrt{n}}$ and $e^{c_2\sqrt{n}}$, where $c_2 > c_1 > 0$ are suitable constants. Such information about the order of growth of a function may often be sufficient.

Proof of Fact 10.7.1. We begin with the de Moivre formula (see also Exercise 1.3.4): $(\cos\alpha + i\sin\alpha)^n = \cos(n\alpha) + i\sin(n\alpha)$, where i is the imaginary unit, $i^2 = -1$. Expanding the left-hand side by the binomial theorem and considering only the imaginary part of both sides, we get the identity

$$\binom{n}{1}\sin\alpha\cos^{n-1}\alpha - \binom{n}{3}\sin^3\alpha\cos^{n-3}\alpha + \binom{n}{5}\sin^5\alpha\cos^{n-5}\alpha - \cdots$$

$$= \sin(n\alpha).$$

Using the function $\cot\alpha = \frac{\cos\alpha}{\sin\alpha}$, we can rewrite the left-hand side as

$$\sin^n\alpha\left[\binom{n}{1}\cot^{n-1}\alpha - \binom{n}{3}\cot^{n-3}\alpha + \binom{n}{5}\cot^{n-5}\alpha - \cdots\right].$$

From now on, let $n = 2m+1$ be odd. Then the expression in brackets can be written as $P(\cot^2\alpha)$, where $P(x)$ is the polynomial $\binom{n}{1}x^m - \binom{n}{3}x^{m-1} + \binom{n}{5}x^{m-2} - \cdots$.

We claim that the roots of $P(x)$ are the m numbers $r_1, r_2, \ldots, r_m$, where $r_k = \cot^2\frac{k\pi}{n}$. Indeed, for $\alpha = \frac{k\pi}{n}$, $\sin\alpha$ is nonzero while $\sin(n\alpha) =$

0, and so $P(\cot^2 \alpha)$ must be 0. Because $r_1, \ldots, r_m$ are all different and $P(x)$ has degree m, this exhausts all the roots. Therefore,

$$P(x) = n(x - r_1)(x - r_2) \ldots (x - r_m).$$

By comparing the coefficients of x^{m-1} on both sides of the last equality, we find that $n(r_1 + r_2 + \cdots + r_m) = \binom{n}{3}$. In this way, we have derived the identity

$$\sum_{k=1}^{m} \cot^2 \frac{k\pi}{2m+1} = \frac{m(2m-1)}{3}. \tag{10.12}$$

For $0 < \alpha < \frac{\pi}{2}$, we have $\cot \alpha = 1/\tan \alpha < 1/\alpha$, and so (10.12) provides the inequality

$$\sum_{k=1}^{m} \frac{(2m+1)^2}{\pi^2 k^2} > \frac{m(2m-1)}{3}$$

or

$$\sum_{k=1}^{m} \frac{1}{k^2} > \frac{\pi^2}{6} \frac{2m(2m-1)}{(2m+1)^2}.$$

Letting $m \to \infty$ gives $\sum_{k=1}^{\infty} 1/k^2 \geq \pi^2/6$. An upper bound can be derived similarly. Using the identity $\cot^2 \alpha = \sin^{-2} \alpha - 1$ and the inequality $\sin \alpha < \alpha$ valid for $0 < \alpha < \frac{\pi}{2}$, (10.12) gives

$$\frac{m(2m-1)}{3} - m = \sum_{k=1}^{m} \frac{1}{\sin^2 \frac{k\pi}{2m+1}} > \sum_{k=1}^{m} \frac{(2m+1)^2}{\pi^2 k^2},$$

which for $m \to \infty$ leads to $\sum_{k=1}^{\infty} 1/k^2 \leq \pi^2/6$. $\qquad\square$

Exercises

1. (a) Check that the number of ordered partitions of n with k summands, i.e. the number of solutions to $i_1 + i_2 + \cdots + i_k = n$ with all $i_j \geq 1$, is $\binom{n-1}{k-1}$.

   (b) Calculate the number of all ordered partitions of n. For instance, for $n = 3$ we have the 4 ordered partitions $3 = 1 + 1 + 1$, $3 = 1 + 2$, $3 = 2 + 1$, and $3 = 3$. Give three solutions: a "direct" combinatorial one, one based on (a), and one using generating functions.

2. (a) ^{CS} Write a program to list all the partitions of n, each exactly once. (This is a nice programming exercise.)

   (b) ^{CS} Write a program to compute p_n for a given n. If you don't want to use multiple precision arithmetic, the program should at least be able to calculate p_n for n up to 5000 (where p_n is well below 10^{100}) to 8 significant digits, say. A contest can be organized for the fastest correct program. (A consultation of the literature can give a great advantage in such a competition.)

3. This time we consider the number $\bar{p}_n$ of partitions of n with *all summands distinct*. For instance, for $n = 5$ we only admit the partitions $5 = 1 + 4$ and $5 = 2 + 3$.

(a) Express $\bar{p}_n$ as a coefficient of x^n in a suitable expression.

(b) Following the proof method of Theorem 10.7.2, prove that $\bar{p}_n \leq e^{2\sqrt{n}}$.

(c) *Prove a lower bound of the form $\bar{p}_n \geq e^{c\sqrt{n}}$ for all sufficiently large values of n, with $c > 0$ some constant. What is the largest c you can get?

4. (a) Write down the generating functions for the numbers $\bar{p}_n$ from Exercise 3 (the generating function is an infinite product).

(b) Find a generating function for the numbers of unordered partitions of n into *odd* summands (also an infinite product).

(c) *Verify that the generating functions in (a) and (b) are the same, and hence that the number of partitions into distinct summands equals the number of partitions into odd summands. **Can you find a direct argument (a bijection)?

5. The Royal Mint of Middle Coinland issues a_1 types of coins of value 1 doublezon, a_2 types of coins of value 2 doublezons, etc. Show that the number of ways of paying the sum of n doublezons by such coins equals the coefficient of x^n in

$$\prod_{i=1}^{n} \frac{1}{(1 - x^i)^{a_i}}.$$

6. (a) Express the number of nonisomorphic rooted trees of height at most 2 with m leaves using the numbers p_n. The notion of rooted trees and their isomorphism are introduced in Section 4.2. The root is not counted as a leaf, and the height of a rooted tree is the maximum distance of a leaf from the root.

(b) This time consider rooted trees of height at most 2 with m vertices. Express their number using the numbers p_n.

(c) *Express the number r_n of rooted trees with n leaves, all of them at height 3, as the coefficient of x^n in a suitable expression (possibly involving the numbers p_i; Exercise 5 can serve as an inspiration).

(d) **Show that the numbers r_n in (c) satisfy $r_n = e^{O(n/\log n)}$. Use (c) and the proof of Theorem 10.7.2. (The calculation may be quite challenging.)

(e) *Show that the estimate in (d) is roughly correct, i.e. that $r_n \geq e^{cn/\log n}$ for some constant $c > 0$.

11
Applications of linear algebra

Linear algebra is a part of algebra dealing with systems of linear equations, matrices, determinants, vector spaces, and similar things. We have already seen several proofs using linear algebra in previous chapters, most notably in Section 7.5. Here we will demonstrate few more methods and applications. First, we present two problems, one concerning the existence of so-called block designs and the other about covering a complete graph by complete bipartite graphs. Hardly anyone would suspect that these problems are related to matrices, and yet an elegant solution can be given based on the notion of rank of a matrix. An interested reader can find much more about similar proofs in the very vividly written textbook by Babai and Frankl [12]. In the subsequent two sections of this chapter, we assign several vector spaces to each graph; this leads to a compact and insightful description of seemingly very complicated sets. Finally in Section 11.6, we consider two elegant algorithms where linear algebra blends with a probabilistic method.

Throughout this chapter, we assume some basic knowledge of linear algebra. All the required material is summarized in the Appendix.

11.1   Block designs

Here we consider very regular systems of finite sets, the so-called block designs. Finite projective planes are a special case of this concept but the general notion of block designs no longer has a geometric motivation.

Let V be a finite set and let $\mathcal{B}$ be a system of subsets[1] of the set V. In order to emphasize that the set system $\mathcal{B}$ lives on the set V, we write it as an ordered pair $(V, \mathcal{B})$. (Let us remark that such

---

[1]This notation differs from what we have been mostly using in previous chapters, but it is traditional in the theory of blocks designs.

a pair $(V, \mathcal{B})$ can also be regarded as a generalization of the notion of graph, and then it may also be called a *hypergraph*; the points of V are then called the *vertices* and the sets of $\mathcal{B}$ the *hyperedges*.) If all the sets $B \in \mathcal{B}$ have the same cardinality k, we say that $(V, \mathcal{B})$ is *k-uniform*.

We have already met an important example of a k-uniform set system in Chapter 8 on finite projective planes. We have shown that if V denotes the set of points of a finite projective plane and $\mathcal{B}$ stands for the set of its lines, then the set system $(V, \mathcal{B})$ is $(k + 1)$-uniform for a suitable k, and, moreover, we have $|V| = |\mathcal{B}| = k^2 + k + 1$. This example will hopefully help also in understanding the following definition, which may look quite technical at first sight (but we will illustrate it with several more examples later on).

11.1.1 Definition. *Let v, k, t, and λ be integers. We suppose that $v > k \geq t \geq 1$ and $\lambda \geq 1$. A block design of type t-(v, k, λ) is a set system $(V, \mathcal{B})$ satisfying the following conditions:*

(1) V has v elements.

(2) Each set of $B \in \mathcal{B}$ has k elements. The sets of $\mathcal{B}$ are called the blocks.

(3) Each t-element subset of the set V is contained in exactly λ blocks of $\mathcal{B}$.

Here are a few basic examples illustrating this definition.

Example. Let V be a finite set and k an integer. We put $\mathcal{B} = \binom{V}{k}$ (in words, $\mathcal{B}$ consists of all the k-element subsets of V). The pair $(V, \mathcal{B})$ is called the *trivial block design*.

It is easy to check that $(V, \mathcal{B})$ is a t-(v, k, λ) block design, where $t \in \{1, 2, \ldots, k\}$ can be chosen arbitrarily, $v = |V|$, and $\lambda = \binom{v-t}{k-t}$. (To see this, convince yourself that any t-element subset of V is contained in exactly $\binom{v-t}{k-t}$ blocks $B \in \mathcal{B}$.)

Example. Let V be a v-element set, and let $k \geq 1$ be an integer dividing v. Partition the elements of V into disjoint k-element subsets $B_1, B_2, \ldots, B_{v/k}$, and put $\mathcal{B} = \{B_1, B_2, \ldots, B_{v/k}\}$. Then $(V, \mathcal{B})$ is a block design of type 1-$(v, k, 1)$.

Example. Let V be the set of points of a projective plane of order n, and let $\mathcal{B}$ denote the set of its lines. Such a pair $(V, \mathcal{B})$ is a block design of type 2-$(n^2 + n + 1, n + 1, 1)$. This nontrivial fact was proved in Section 8.1. Conversely, it can be shown that any block design of type 2-$(n^2+n+1, n+1, 1)$, for $n \geq 2$, is a projective plane of order n (Exercise 1).

11.1.2 Example. Let $V = \{0,1,2,3,4,5\}$ and let $\mathcal{B}$ consist of the following triples: $\{0,1,2\}$, $\{0,2,3\}$, $\{0,3,4\}$, $\{0,4,5\}$, $\{0,1,5\}$, $\{1,2,4\}$, $\{2,3,5\}$, $\{1,3,4\}$, $\{2,4,5\}$, $\{1,3,5\}$. Then $(V, \mathcal{B})$ is a 2-$(6,3,2)$ block design. (This can be simply checked by definition with no cleverness involved.)

This block design can also be defined in a more "structured" manner. Consider a cycle with vertices $1, 2, \ldots, 5$ and one extra vertex 0. The system $\mathcal{B}$ then consists of all triples of vertices containing exactly one edge of the cycle; see the diagram below:

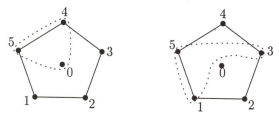

These examples should invoke the (correct) impression that block designs constitute a certain kind of regularity. Usually it is not easy to construct block designs of a given type, and the basic question in this whole area is the question of existence.

Basic problem. *For given numbers v, k, λ, t, decide whether a block design of type t-(v, k, λ) exists or not.*

Here we derive some necessary conditions by algebraic means.

At the end of this brief introduction, let us mention that block designs arose and are still being used in mathematical statistics in the design of experiments. This motivation has also influenced the notation introduced above.

Imagine that we want to assess several different ways of treating a certain plant (for suppressing its insect parasites, say). There are v types of treatment to be compared (v for "variety"). We will compare the treatments by a series of experiments, and in each experiment we can apply k types of treatment; this is given by the technical conditions of the experiment. Each experiment will form a *block* of the tested treatments. We could in principle test all possible k-tuples, or blocks, of treatments, but in the situation of field experiments, this trivial way of testing (hence the name "trivial block design") is far too demanding even for small values of k and v. For this reason, statisticians started using designs of experiments where one doesn't test all possible k-tuples but only some selected blocks. This may of course lead to errors, since the experiments are incomplete: some possible blocks are not considered, and hence some possible mutual influences of the treatments are neglected. In order to compensate for this (enforced) incompleteness of the testing, we require that at least each pair of treatments appear together in the same number of experiments–blocks. The scheme of such a series of experiments is exactly a block design of type 2-(v, k, λ). If we require that each triple of treatments appear in the same number, λ, of experiments, we get a block design of type 3-(v, k, λ), and so on.

Various block designs appear under different special names in the literature: For instance, *balanced incomplete block design* (or BIBD) for designs of type 2-(v, k, λ), *Steiner systems* (for $\lambda = 1$), *tactical configurations* (for $t > 2$), etc.

Integrality conditions. It should be clear that a block design of type t-(v, k, λ) doesn't exist on each set, i.e. for all values of v. For instance, a 1-$(v, k, 1)$ design is a partition into k-element sets, and hence v has to be divisible by k. Another, less trivial example is obtained for the case of projective planes, where v is determined uniquely by the size of the lines. The following theorem describes the most important class of necessary conditions for the existence of a block design of type t-(v, k, λ).

11.1.3 Theorem (Integrality conditions). *Suppose that a block design of type t-(v, k, λ) exists. Then the following fractions must be integers:*

$$\lambda \frac{v(v-1)\ldots(v-t+1)}{k(k-1)\ldots(k-t+1)}, \quad \lambda \frac{(v-1)\ldots(v-t+1)}{(k-1)\ldots(k-t+1)}, \quad \ldots, \quad \lambda \frac{v-t+1}{k-t+1}.$$

Proof. This is an application of double-counting. Let $(V, \mathcal{B})$ be a block design of type t-(v, k, λ). Fix an integer number s with $0 \leq s \leq t$ and an s-element set $S \subseteq V$. Let us count the number N of pairs (T, B) with $S \subseteq T \in \binom{V}{t}$ and $T \subseteq B \in \mathcal{B}$:

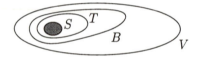

On the one hand, T can be picked in $\binom{v-s}{t-s}$ ways, and each T is a subset of exactly λ blocks $B \in \mathcal{B}$; hence $N = \lambda \binom{v-s}{t-s}$.

On the other hand, let M be the number of blocks containing the set S. Since each block B containing S contains $\binom{k-s}{t-s}$ t-element sets T with $S \subseteq T$, we also have $N = M\binom{k-s}{t-s}$. Hence

$$M = \lambda\frac{\binom{v-s}{t-s}}{\binom{k-s}{t-s}} = \lambda\frac{(v-s)\dots(v-t+1)}{(k-s)\dots(k-t+1)},$$

and so the fraction on the right-hand side has to be an integer as the theorem asserts. □

Remark. By considering the above proof for $s = 0$ and for $s = 1$, we see that $\lambda\frac{v(v-1)\dots(v-t+1)}{k(k-1)\dots(k-t+1)}$ determines the total number of blocks and $\lambda\frac{(v-1)\dots(v-t+1)}{(k-1)\dots(k-t+1)}$ is the number of blocks containing a given element $x \in V$, i.e. the "degree of x". In the statistical interpretation explained above, this number expresses the number of times the treatment x has been applied, and it is usually denoted by r (for "repetitions").

11.1.4 Example (Steiner triple systems). The "first" nontrivial example of a block design of type t-(v, k, λ) is obtained for $t = 2$, $\lambda = 1$, and $k = 3$. This is a system of triples where each pair of points is contained in exactly one triple. In other words, it is a way of covering all edges of a complete graph by edge-disjoint triangles.

In this case, the integrality conditions from Theorem 11.1.3 require that the numbers

$$\frac{v(v-1)}{6} \quad \text{and} \quad \frac{v-1}{2}$$

be integers. From this, it is easy to derive that either $v \equiv 1 \pmod{6}$ or $v \equiv 3 \pmod{6}$. Hence v has to be one of the numbers 3, 7, 9, 13, 15, 19, 21, 25, 27, ... For all such values of v, a block design of type 2-$(v, 3, 1)$ exists. Such designs are called *Steiner triple systems* (see Exercise 4). For $v = 7$, a Steiner triple system is a projective plane of order 2 (the Fano plane). For $v = 9$, we have the following Steiner system:

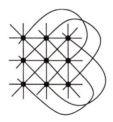

This can be regarded as an *affine plane*, which arises from a projective plane of order 3 by deleting one line and all its points (see Exercise 8.1.10).

> **Balanced incomplete block designs.** For $t = 2$ (i.e. if we require that each pair is in exactly λ k-tuples from $\mathcal{B}$), the integrality conditions look as follows:

$$\begin{array}{rcll} \lambda v(v - 1) & \equiv & 0 & (\mathrm{mod}\ k(k - 1)) \\ \lambda(v - 1) & \equiv & 0 & (\mathrm{mod}\ k - 1). \end{array} \tag{11.1}$$

These conditions are no longer sufficient in general. But the following difficult and important result holds:

11.1.5 Theorem (Wilson's theorem). *For any choice of numbers $k \geq 1$ and $\lambda \geq 1$, there exists a number $v_0(k, \lambda)$ such that for all $v \geq v_0(k, \lambda)$ satisfying the integrality conditions (11.1), a block design of type 2-(v, k, λ) exists.*

In other words, the integrality conditions are also sufficient for $t = 2$ if the ground set is large enough. The theorem says nothing about small values of v: for instance, about the existence of block designs of type 2-$(k^2 + k + 1, k + 1, 1)$, i.e. of finite projective planes.

We have mentioned this theorem for completeness, and we are not going to prove it here (a proof can be found in Beth, Jungnickel, and Lenz [32]).

Exercises

1. (a) Verify that a projective plane of order q is a block design of type 2-$(q^2 + q + 1, q + 1, 1)$.

   (b) *Show that, conversely, any block design of type 2-$(q^2+q+1, q+1, 1)$ is a projective plane of order q.

2. *Show that any block design of type 2-$(q^2, q, 1)$ for $q \geq 2$ is as affine plane of order q (see Exercise 8.1.10 for the definition).

3. Construct a Steiner triple system with 15 elements.

4. *Put $n = 3m$, where m is an odd natural number. We define a set system $(X, \mathcal{M})$ as follows: $X = \{(x, i)\colon i = 1, 2, 3,\ x = 0, 1, \ldots, m - 1\}$, and $\mathcal{M}$ consists of all triples $\{(x, 1), (x, 2), (x, 3)\}$ and all triples of the form $\{(x, i), (y, i), (z, i + 1)\}$, where $x \neq y$, $x + y \equiv 2z\ (\mathrm{mod}\ m)$, and $i = 1, 2, 3$. Prove that $(X, \mathcal{M})$ is a Steiner triple system.

11.2 Fisher's inequality

One of the founders of the theory of block designs was an English statistician, R. A. Fisher. Although examples of block designs have been known for a long time (Steiner triple systems for almost 100 years), Fisher was the first to identify the general definition and its significance in the statistical context. He also found additional necessary conditions restricting the existence of block designs.

11.2.1 Theorem (Fisher's inequality). *Let $(V, \mathcal{B})$ be a block design of type 2-(v, k, λ) with $v > k$. Then $|\mathcal{B}| \geq |V|$. Hence, the design of tests for v treatments requires at least v experiments.*

Note that block designs with $|\mathcal{B}| = |V|$ exist (for example, the finite projective planes), and so, in this sense, Fisher's inequality is optimal. The following example illustrates its strength.

Example. Fisher's inequality implies that there is no block design of type 2-$(16, 6, 1)$. Indeed, by the remark following Theorem 11.1.3, the number of blocks in such a design must be $\frac{16 \cdot 15}{6 \cdot 5} = 8 < 16$. At the same time, the integrality conditions are satisfied for this choice of parameters: we have already verified that the number of blocks is integral, and also the number $r = \frac{15}{5} = 3$ is an integer. The reader may check that block designs of type 2-$(21, 6, 1)$ and of type 2-$(25, 10, 3)$ are excluded by Fisher's inequality as well, although they again satisfy the integrality conditions.

Fisher's inequality can be proved by a remarkable application of elementary linear algebra, as was discovered by an Indian mathematician R. C. Bose. Before we begin with the proof, let us introduce the incidence matrix of the set system $(V, \mathcal{B})$; this is a similar concept to the incidence matrix of a graph (for oriented graphs it was discussed in Section 7.5). Let us denote the elements of the set V by $x_1, x_2, \ldots, x_v$ and the sets of $\mathcal{B}$ by $B_1, B_2, \ldots, B_b$. We define a $v \times b$ matrix $A = (a_{ij})$, with rows corresponding to points of V and columns to sets of $\mathcal{B}$, by the formula

$$a_{ij} = \begin{cases} 1 & \text{if } x_i \in B_j \\ 0 & \text{otherwise.} \end{cases}$$

The matrix A is called the *incidence matrix* of the set system $(V, \mathcal{B})$.

Proof of Fisher's inequality. For a given block design $(V, \mathcal{B})$, consider its incidence matrix $A = (a_{ij})$. The matrix transposed to A, A^T, has size $b \times v$, and hence the matrix product AA^T has size $v \times v$. We show that the matrix $M = AA^T$ has a very simple form.

Let us consider an entry m_{ij} of M. By the definition of matrix multiplication, we have

$$m_{ij} = \sum_{k=1}^{b} a_{ik} a_{jk}$$

(the term in the jth column and kth row of the matrix A^T is a_{jk}). Thus, the entry m_{ij} expresses the number of sets B_k containing both x_i and x_j. By the definition of a block design, m_{ij} can only attain two possible values:

$$m_{ij} = \begin{cases} \lambda & \text{for } i \neq j \\ \lambda \frac{v-1}{k-1} & \text{for } i = j. \end{cases}$$

The number $\lambda \cdot \frac{v-1}{k-1}$ has been denoted by r in the preceding text, and so the matrix M is

$$\begin{pmatrix} r & \lambda & \cdots & \lambda \\ \lambda & r & \cdots & \lambda \\ \vdots & \vdots & \ddots & \vdots \\ \lambda & \lambda & \cdots & r \end{pmatrix}.$$

We want to show that this matrix is nonsingular, i.e. that its determinant is nonzero. Elementary row operations give

$$\det M = \det \begin{pmatrix} r + (v-1)\lambda & r + (v-1)\lambda & \cdots & r + (v-1)\lambda \\ \lambda & r & \cdots & \lambda \\ \vdots & \vdots & \ddots & \vdots \\ \lambda & \lambda & \cdots & r \end{pmatrix}$$

$$= \left(r + (v-1)\lambda \right) \det \begin{pmatrix} 1 & 1 & \cdots & 1 \\ \lambda & r & \cdots & \lambda \\ \vdots & \vdots & \ddots & \vdots \\ \lambda & \lambda & \cdots & r \end{pmatrix}$$

$$= \left(r + (v-1)\lambda \right) \det \begin{pmatrix} 1 & 1 & \cdots & 1 \\ 0 & r-\lambda & \cdots & 0 \\ \vdots & \vdots & \ddots & \vdots \\ 0 & 0 & \cdots & r-\lambda \end{pmatrix}$$

$$= (r + (v-1)\lambda) \cdot (r-\lambda)^{v-1}.$$

We now recall that $r = \lambda \cdot \frac{v-1}{k-1}$. Clearly, $r + (v-1)\lambda \neq 0$, and since $v > k$, we also have $r > \lambda$, and so $\det M \neq 0$. Therefore, the matrix

M has rank[2] v. But if we had $b < v$ then the rank of both the matrices A and A^T would be strictly smaller than v, and consequently also the matrix $M = AA^T$ would have rank $< v$ (here we use a simple property of matrix rank; see Exercise 2). We conclude that $b \geq v$. This finishes the proof of Fisher's inequality. $\qquad\qquad\square$

This application of matrix rank became the basis of many similar (and important) proofs in combinatorics. Another proof of the fact that $r(M) = v$ is indicated in Exercise 4.

Exercises

1. For $\lambda = 1$, prove Fisher's inequality directly (without using linear algebra).

2. Show that if A is an $n \times k$ matrix and B a $k \times m$ matrix then $r(AB) \leq \min(r(A), r(B))$.

3. *Let F be some field and $G \subseteq F$ some subfield of F (you may want to imagine the real numbers for F and the rationals for G). Let A be a matrix with elements from the field G. The rank of A can be considered over the field G (with the linear dependencies in the definition having coefficients in G) or over the field F (linear dependence with coefficients in F). Explain why we obtain the same rank in both cases.

4. A square $n \times n$ real matrix M is called *positive definite* if we have $x^T M x > 0$ for each nonzero (column) vector $x \in \mathbf{R}^n$.

   (a) Why does any positive definite $n \times n$ matrix M have the full rank n?

   (b) Show that the matrix M used in the proof of Fisher's inequality in the text is positive definite (and hence it has rank v, without a calculation of the determinant).

5. (a) Let $C_1, C_2, \ldots, C_m$ be subsets of an n-element set X. Suppose that each C_i has an odd cardinality, and that the cardinality of each intersection $C_i \cap C_j$ (for $i \neq j$) is even. Prove that then $m \leq n$. Look at the matrix $A^T A$, where A is the incidence matrix of the considered set system, but work over the 2-element field (i.e. modulo 2).

---

[2]Let us recall that the *rank* of a matrix M, usually denoted by the symbol $r(M)$, is the maximum possible number of linearly independent rows in M. The rank can also be defined analogously using the columns instead of the rows, and a basic linear algebra theorem tells us that both definitions always yield the same number.

(b) Consider a similar problem as in (a), but this time we require that the sizes of the sets themselves be *even* while all $|C_i \cap C_j|$ be *odd* for $i \neq j$. Prove that $m \leq n$ again holds.

(c) *And this time we require that the sets C_i all be distinct, and their sizes and the sizes of all pairwise intersections be *even*. Show that one can construct such a system with $2^{\lfloor n/2 \rfloor}$ sets.

6. (Generalized Fisher's inequality) Let X be an n-element set, and let q be an integer number, $1 \leq q < n$. Let $C_1, C_2, \ldots, C_m$ be subsets of X, and suppose that all the intersections $C_i \cap C_j$ (for $i \neq j$) have cardinality exactly q.

(a) *Using the method of Exercise 4, prove that $m \leq n$. (Treat the case when $|C_i| = q$ holds for some i separately.)

(b) *Why can the claim in (a) be called a "generalized Fisher's inequality"? Derive Fisher's inequality from it!

11.3 Covering by complete bipartite graphs

The following question has been motivated by a problem in telecommunications:

Problem. The set of edges of a complete graph K_n should be expressed as a disjoint union of edge sets of m complete bipartite graphs. What is the smallest value of $m = m(n)$ for which this is possible?

One possible way of expressing $E(K_n)$ as a disjoint union of edge sets of $n - 1$ complete bipartite graphs uses graphs of type K_{1,n_i} ("stars"). Here is such a disjoint covering for $n = 5$:

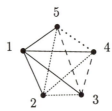

To produce such a disjoint covering for an arbitrary n, suppose that $E(K_{n-1})$ has already been expressed using $n - 2$ stars. In the graph K_n, consider one vertex, and cover all edges containing that vertex by the edge set of a star $K_{1,n-1}$. Then it remains to cover all edges of a graph isomorphic to K_{n-1}, and we can already do this.

It is not at all obvious that we couldn't do better, say by using some complete bipartite graphs both of whose color classes are large.

But Graham and Pollak discovered an ingenious way of showing that no better disjoint covering may exist:

11.3.1 Theorem (Graham–Pollak theorem). *We have* $m(n) \geq n - 1$.

Proof. Suppose that complete bipartite graphs $B_1, B_2, \ldots, B_m$ disjointly cover all the edges of K_n, i.e. we have $V(B_k) \subseteq V(K_n) = \{1, 2, \ldots, n\}$ and $E(K_n) = E(B_1) \dot\cup E(B_2) \dot\cup \cdots \dot\cup E(B_m)$. Let X_k and Y_k be the color classes of B_k; this means that the edges of B_k all go between X_k and Y_k.

To each graph B_k, we assign an $n \times n$ matrix A_k, whose entry in the ith row and jth column is

$$a_{ij}^{(k)} = \begin{cases} 1 & \text{if } i \in X_k \text{ and } j \in Y_k \\ 0 & \text{otherwise.} \end{cases}$$

The definition of A_k resembles the adjacency matrix of the graph B_k, except for the fact that A_k is not symmetric—each edge of the graph B_k only contributes one 1. For example, for the subgraph

$$B_k = $$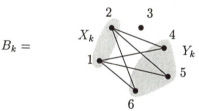

the matrix A_k is

$$\begin{pmatrix} 0 & 0 & 0 & 1 & 1 & 1 \\ 0 & 0 & 0 & 1 & 1 & 1 \\ 0 & 0 & 0 & 0 & 0 & 0 \\ 0 & 0 & 0 & 0 & 0 & 0 \\ 0 & 0 & 0 & 0 & 0 & 0 \\ 0 & 0 & 0 & 0 & 0 & 0 \end{pmatrix}.$$

We claim that each of the matrices A_k has rank 1. This is because all the nonzero rows of A_k equal the same vector, namely the vector with 1s at positions whose indices belong to Y_k and with 0s elsewhere.

Let us now consider the matrix $A = A_1 + A_2 + \cdots + A_m$. Each edge $\{i, j\}$ belongs to exactly one of the graphs B_k, and hence for each $i \neq j$, we have either $a_{ij} = 1$, $a_{ji} = 0$, or $a_{ij} = 0$, $a_{ji} = 1$, where a_{ij} denotes the entry of the matrix A at position (i, j). We also have $a_{ii} = 0$. From this we get $A + A^T = J_n - I_n$, where J_n is the $n \times n$ matrix having 1s everywhere and I_n is the $n \times n$ identity matrix, with 1s on the diagonal and 0s elsewhere. We now want to show that the rank of such a matrix A is at least $n - 1$. Once we know this, we get $n - 1 \le r(A) \le r(A_1) + \cdots + r(A_m) = m$, because for arbitrary two matrices M_1 and M_2 (of the same shape) we have $r(M_1 + M_1) \le r(M_1) + r(M_2)$; this is easy to check from the definition of matrix rank (Exercise 1).

For contradiction, let us suppose that $r(A) \le n - 2$. If we add an extra row consisting of all 1s to the matrix A, the resulting $(n+1) \times n$ matrix still has rank $\le n-1$, and hence there exists a nontrivial linear combination of its columns equal to the zero vector. In other words, there exists a (column) vector $x \in \mathbf{R}^n$, $x \ne (0, 0, \dots, 0)^T$, such that $Ax = 0$ and also $\sum_{i=1}^n x_i = 0$.

From the last mentioned equality, we get $J_n x = 0$. We calculate

$$x^T \left(A + A^T \right) x = x^T (J_n - I_n) x = x^T (J_n x) - x^T (I_n x)$$

$$= 0 - x^T x = -\sum_{i=1}^n x_i^2 < 0.$$

On the other hand, we have

$$x^T \left(A^T + A \right) x = \left(x^T A^T \right) x + x^T (Ax) = 0 \cdot x + x \cdot 0 = 0,$$

and this is a contradiction. Hence $r(A) = n - 1$ and Theorem 11.3.1 is proved. $\qquad\square$

Exercises

1. Let M_1 and M_2 be two $n \times m$ matrices. Check that $r(M_1 + M_2) \le r(M_1) + r(M_2)$, and show with examples that sometimes equality holds and sometimes not.

2. (a) The edges of K_n should be covered by edges of complete bipartite subgraphs of K_n, but we do not insist on their edge sets being disjoint (i.e. one edge may be covered several times). Show that there exists a covering consisting of $\lceil \log_2 n \rceil$ bipartite graphs.

   (b) *Prove that a covering as in (a) really requires at least $\lceil \log_2 n \rceil$ bipartite graphs.

3. *This time, we want to cover all the edges of K_n by edges of complete bipartite subgraphs, in such a way that each edge is covered an *odd*

number of times. Prove that we need at least $\frac{1}{2}(n-1)$ complete bipartite subgraphs. Proceed similarly as in the proof in the text, but work with matrices over the 2-element field. Instead of the matrix A_k, consider the adjacency matrix of the graph B_k, whose rank is 2.

4. *In an ice-hockey tournament, each of n teams played against every other team, with no draws. It turned out that for every nonempty group G of teams, a team exists (it may or may not belong to G) that won an odd number of games with the teams of G. Prove that n is even. Hint: Work with a suitable matrix over the 2-element field.

11.4 Cycle space of a graph

Let $G = (V, E)$ be an undirected graph. Let the symbol $\mathcal{K}_G$ denote the set of all cycles, of all possible lengths, in the graph G (more exactly, the set of all subgraphs that are cycles). It is clear that this set deserves a complicated notation by a calligraphic letter since it is quite large. Check that, for instance, for the complete graph K_n we have

$$|\mathcal{K}_{K_n}| = \sum_{k=3}^{n} \binom{n}{k} \cdot \frac{(k-1)!}{2}, \tag{11.2}$$

and for the complete bipartite graph $K_{n,n}$ we get

$$|\mathcal{K}_{K_{n,n}}| = \sum_{k=2}^{n} \binom{n}{k}^2 \cdot \frac{k!(k-1)!}{2}. \tag{11.3}$$

On the other hand, we may note that $\mathcal{K}_G = \emptyset$ if and only if G is a forest (a disjoint union of trees).

It might seem that the set $\mathcal{K}_G$ has no reasonable structure. In this section, we introduce a suitable generalization of the notion of a cycle (related to the material explained in Section 3.4 on Eulerian graphs) which extends the set of all cycles to a larger set with a very simple structure, namely with the structure of a vector space. Ideas presented in the sequel originally arose in the investigation of electrical circuits.

11.4.1 Definition (Even set of edges). *Let $G = (V, E)$ be an (undirected) graph. A set $E' \subseteq E$ is called* even *if the degrees of all vertices in the graph (V, E') are even.*

For example, the empty set and the edge set of an arbitrary cycle are even sets.

In the sequel, it will be advantageous to identify a cycle with its edge set.

11.4.2 Lemma. *A set E' of edges is even if and only if pairwise disjoint cycles $E_1, E_2, \ldots, E_t$ exist such that $E' = E_1 \dot\cup E_2 \dot\cup \cdots \dot\cup E_t$.*

Proof. If E' is a nonempty even set then the graph (V, E') is not a forest and hence it contains some cycle, E_1. The set $E' \setminus E_1$ is again even, and so we can proceed by induction on the number of edges in E'. $\square$

We now describe the structure of the family of all even sets of edges in a given graph $G = (V, E)$ algebraically. We denote the edges of G by $e_1, e_2, \ldots, e_m$, and to each set $A \subseteq E$, we assign its *characteristic vector* $\mathbf{v}_A = (v_1, v_2, \ldots, v_m)$ defined by

$$v_i = \begin{cases} 1 & \text{if } e_i \in A \\ 0 & \text{otherwise.} \end{cases}$$

The vectors are added and multiplied modulo 2 (which means that $1 + 1 = 0$). Let $A, B \subseteq E$ be even sets. Then the reader should be able to check without difficulties that

$$\mathbf{v}_A + \mathbf{v}_B = \mathbf{v}_C,$$

where $C = (A \cup B) \setminus (A \cap B)$ is the *symmetric difference* of the sets A and B.

Let the symbol $\mathcal{E}$ denote the set of the characteristic vectors of all even sets of edges in G. For an easier formulation, we need to generalize the notion of a spanning tree a little: by a *spanning forest* of an arbitrary, possibly disconnected, graph G we understand any subgraph $(V(G), \bar{E})$ of G containing no cycles and inclusion maximal with respect to this property (i.e. adding any edge of G to $\bar{E}$ creates a cycle). For a connected graph G, this definition coincides with the definition of a spanning tree in Section 4.3. For a disconnected graph we get a forest consisting of spanning trees of all the components.

11.4.3 Theorem (Cycle space theorem).

(1) *For any graph G, the set $\mathcal{E}$ is a vector space over the 2-element field $GF(2)$. The dimension of this vector space is $|E| - |V| + k$, where k is the number of components of the graph G.*

(2) *Fix an arbitrary spanning forest $T = (V, E')$ of the graph G, and for each edge $e \in E \setminus E'$, let C_e denote the (unique) cycle contained in the graph $(V, E' \cup \{e\})$. Then the characteristic vectors of the cycles C_e, $e \in E \setminus E'$, form a basis of $\mathcal{E}$.*

In this theorem, the symbol $GF(2)$ stands for the 2-element field consisting of the numbers 0 and 1, with arithmetic operations performed modulo 2. The letters GF come from "Galois field", which is a traditional term for a finite field. A vector space over $GF(2)$ simply means that if the characteristic vectors of two even sets are added modulo 2 we again obtain a characteristic vector of an even set.

The cycle C_e is called an *elementary cycle* (determined by the edge e, with respect to a given spanning forest).

Proof. First we show that $\mathcal{E}$ is a vector space. To this end, we have to check that by adding two vectors from $\mathcal{E}$ we again get a vector from $\mathcal{E}$, and by multiplying a vector from $\mathcal{E}$ by a number from the field $GF(2)$ we also get a vector from $\mathcal{E}$. The latter is rather trivial, since we can only multiply by 0 (getting $0\mathbf{v}_A = \mathbf{v}_\emptyset$) or by 1 (and then $1\mathbf{v}_A = \mathbf{v}_A$). According to the remark above the theorem, the addition of vectors corresponds to taking the symmetric difference of the respective sets, and so it suffices to show that the symmetric difference of two even sets A and B is again an even set. Let us pick an arbitrary vertex $v \in V$. Suppose that d_A of the edges incident to v belong to A, d_B of them belong to B, and d of them belong to both A and B. Here d_A, d_B are both even numbers. The number of edges incident to v belonging to the symmetric difference $(A \cup B) \setminus (A \cap B)$ is $d_A + d_B - 2d$, and hence the degree of v in the symmetric difference is even. Therefore, the symmetric difference is an even set and so $\mathcal{E}$ is a vector space over $GF(2)$.

Let (V, E') be a spanning forest of the graph G. Then $|E'| = |V| - k$, where k is the number of components of the graph G. It suffices to prove that the characteristic vectors of all elementary cycles constitute a basis of the vector space $\mathcal{E}$.

We first show that the elementary cycles are linearly independent. Consider an edge $e_i \notin E'$. The characteristic vector of the elementary cycle C_{e_i} is the only one among the characteristic vectors of all elementary cycles having 1 at position i. For this reason, a characteristic vector of an elementary cycle cannot be a linear combination of other such vectors.

Next, let us show that the elementary cycles generate $\mathcal{E}$. Choose an even set A and define a set B by giving its characteristic vector:

$$\mathbf{v}_B = \sum_{e \in A \setminus E'} \mathbf{v}_{C_e}.$$

Which edges are contained in B? These are exactly the edges belonging to an odd number of elementary cycles (with respect to the spanning forest (V, E'). Surely B does contain the set $A \backslash E'$ because each of its edges lies in a unique elementary cycle. Let C denote the symmetric difference of the sets A and B. This is an even set, and at the same time, it must be contained in E'. Since E' has no cycles, we get that $C = \emptyset$, which means that $A = B$. We have thus expressed the set A as a linear combination of elementary cycles. □

Example. Let us consider the following graph $G = (V, E)$:

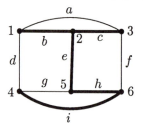

The dimension of its cycle space is $9 - 6 + 1 = 4$. If we choose the spanning tree drawn by thick lines, then the corresponding basis of the cycle space consists of the characteristic vectors of the following elementary cycles:

$$C_a = \{a, b, c\}$$
$$C_d = \{b, e, h, i, d\}$$
$$C_g = \{g, h, i\}$$
$$C_f = \{c, f, h, e\}.$$

For example, for the cycle $C = \{a, f, i, d\}$, we have the expression $\mathbf{v}_C = \mathbf{v}_{C_a} + \mathbf{v}_{C_d} + \mathbf{v}_{C_f}$.

We easily obtain the following consequence of Theorem 11.4.3:

11.4.4 Corollary. *The number of even sets of a graph $G = (V, E)$ with k components is $2^{|E|-|V|+k}$.*

In spite of the large number of even sets, their structure is simple and they can be easily generated from a much smaller basis. The dimension of the cycle space, i.e. the number $|E| - |V| + k$, is called the *cyclomatic number* of the graph $G = (V, E)$. (Also the older name *Betti number* is sometimes used. This is related to a topological view of a graph.)

Exercises

1. Verify the formulas (11.2) and (11.3).

2. Prove Corollary 11.4.4.

3. Determine the cyclomatic number of an $m \times n$ grid graph; for example,

3×4

4. Prove that for any topological planar 2-connected graph (i.e. a planar 2-connected graph with a fixed planar drawing), the border cycles of the bounded faces form a basis of the cycle space $\mathcal{E}$.

11.5 Circulations and cuts: cycle space revisited

In this section, we put the material of the previous section into a somewhat different perspective. We will express things using matrices, and we will work with both undirected graphs and their orientations. In this more complicated setting, a cycle space theorem similar to 11.4.3 will appear in new connections.

Circulations. Let us recall the notion of an orientation from Section 7.5. An *orientation* of a graph $G = (V, E)$ is a directed graph $\vec{G} = (V, \vec{E})$, where the set $\vec{E}$ contains, for each edge $\{x, y\} \in E$, exactly one of the directed edges (x, y) and (y, x). The following picture shows a graph and one of its possible orientations:

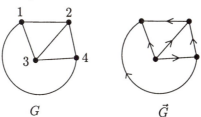

G $\qquad\qquad$ $\vec{G}$

(The number of all possible orientations of a graph G is $2^{|E|}$, but many of them can be isomorphic.)

We assume that some orientation $\vec{G} = (V, \vec{E})$ has been chosen once and for all. A real function $f: \vec{E} \to \mathbf{R}$ is called a *circulation* if we have, for each vertex v of the graph G,

$$\sum_{x \in V: (v,x) \in \vec{E}} f(v, x) = \sum_{x \in V: (x,v) \in \vec{E}} f(x, v). \qquad (11.4)$$

(Let us remark that a circulation may also attain negative values on some edges.) The set of the undirected edges e with $f(\vec{e}) \neq 0$, where $\vec{e}$

is the chosen orientation of e, is called the *carrier* of f. It is easy to see
that the carrier of any nonzero circulation contains a cycle (but here it
is important that we work with an orientation of an undirected graph,
i.e. that edges (x, y) and (y, x) never occur simultaneously).

The notion of a circulation has various intuitive interpretations. For
example, if $\vec{G}$ is a design of an electrical circuit and if $f(e)$ expresses the
current flowing through an edge e, then Eq. (11.4) says that for each
vertex, the amount of current flowing in is the same as the amount flowing out. This is the first Kirchhoff law. This "electrical" interpretation
was one of the first motivations for the theory discussed in this section.

One example of a circulation can be obtained as follows. Let $C =
(v_1, v_2, \ldots, v_{k+1} = v_1)$ be the sequence of vertices of an (undirected)
cycle in the graph G. We define a circulation f in the orientation $\vec{G}$ by
putting

$$
\begin{aligned}
f(v_i, v_{i+1}) &= 1 && \text{if } (v_i, v_{i+1}) \in \vec{E} \\
f(v_{i+1}, v_i) &= -1 && \text{if } (v_{i+1}, v_i) \in \vec{E} \\
f(x, y) &= 0 && \text{if } \{x, y\} \text{ is not an edge of } C.
\end{aligned}
$$

We say that this circulation *belongs* to the cycle C. (Note that it depends
of the direction in which the edges of the undirected cycle C were listed.)
For the graph in the preceding picture and for the cycle with the vertices
1, 2, 3, 4 (in this order), the circulation looks as follows:

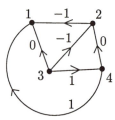

It is easy to see that if f_1 and f_2 are circulations, then the function
$f_1 + f_2$ is a circulation on $\vec{G}$ as well. For a real number c, the function
cf_1 is also a circulation. (In both cases, it suffices to check the condition
(11.4).) Hence the set of all circulations has the structure of a vector
space. This time it is a vector space over the real numbers. Let us denote
this vector space by $\mathcal{C}$ and call it the *circulation space* of G.

It can be shown that for any graph G, the circulation space $\mathcal{C}$ is
generated by circulations belonging to cycles in G (see Exercise 1).
Soon we will obtain a more exact information about the structure of $\mathcal{C}$.

Potentials and cuts. Let $p: V \to \mathbf{R}$ be an arbitrary function (p for
potential). We define a function δp on the set of directed edges of $\vec{G}$ by
the formula

$$\delta p(x, y) = p(x) - p(y) \tag{11.5}$$

for each directed edge $(x, y) \in \vec{E}$.

The function $\delta p \colon \vec{E} \to \mathbf{R}$ is called a *potential difference* (anyone remembering a little of physics can probably see a relation to electrical circuits). Each function $g \colon \vec{E} \to \mathbf{R}$ for which a potential p exists with $g = \delta p$ is also called a *potential difference*.

It is easy to verify that the sum of two potential differences is a potential difference, and similarly for multiplying a potential difference by a real number. Let us denote the vector space of all potential differences by $\mathcal{R}$ and call it the *cut space*.

Why did we choose this name? Let us consider the following situation: Let a potential p attain the values 0 and 1 only. Let us put $A = \{v \in V \colon p(v) = 1\}$, $B = V \setminus A$. Then the potential difference $g = \delta p$ is nonzero only for the directed edges with one vertex in A and the other vertex in B:

$$
\begin{array}{rcll}
g(x,y) & = & 1 & \text{for } x \in A,\ y \in B \\
g(x,y) & = & -1 & \text{for } x \in B,\ y \in A \\
g(x,y) & = & 0 & \text{otherwise.}
\end{array}
$$

It is natural to call the set of all edges going between A and B a *cut*, because the number of components increases by its removal (assuming that there is at least one edge between A and B). In this picture, we have indicated a potential p

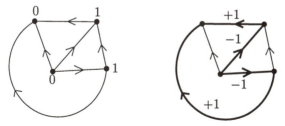

and its potential difference. The edges of the corresponding cut are drawn bold.

We now describe the whole situation using the incidence matrix. To this end, we number the vertices, $V = \{v_1, \ldots, v_n\}$, and the edges, $E = \{e_1, \ldots, e_m\}$. The symbol $\vec{e}_i$ denotes the directed edge corresponding to the edge e_i in the orientation $\vec{G}$. From Section 7.5 we recall the *incidence matrix* of the orientation $\vec{G}$. This matrix, D, has size $n \times m$, and its entry d_{ik} is defined by the following rule:

$$
d_{ik} = \begin{cases}
-1 & \text{if } v_i \text{ is the tail of } \vec{e}_k \\
1 & \text{if } v_i \text{ is the head of } \vec{e}_k \\
0 & \text{otherwise.}
\end{cases}
$$

The following elegant description of the circulation space and of the cut space can be given:

11.5.1 Theorem. *For any graph G, the cut space $\mathcal{R}$ is generated by the rows of the incidence matrix D. The circulation space $\mathcal{C}$ is the orthogonal complement of the cut space $\mathcal{R}$, i.e. $\mathcal{C} = \{x \in \mathbf{R}^m : x^T y = 0 \text{ for all } y \in \mathcal{R}\}$.*

Proof. Let $D = (d_{ij})$ be the $n \times m$ incidence matrix. It is easy to see that each row is a potential difference (the corresponding potential has a 1 sitting at the row's vertex and 0 elsewhere). Next, consider an arbitrary potential difference $g = \delta p$. For a directed edge $\vec{e}_j = (v_r, v_s)$, we have $g(\vec{e}_j) = p(v_r) - p(v_s) = \sum_{i=1}^{n} d_{ij} p(v_i)$; this follows by comparing Eq. (11.5) to the definition of the incidence matrix. Hence, if we regard the function g as a row vector, it is a linear combination of the rows of the incidence matrix. Therefore, the rows generate the cut space $\mathcal{R}$.

Let us now rewrite the condition (11.4) for a circulation f using the incidence matrix. From the equality

$$\sum_{x \in V: (x,v) \in \vec{E}} f(x,v) = \sum_{x \in V: (v,x) \in \vec{E}} f(v,x)$$

we get

$$\sum_{(x,v) \in \vec{E}} f(x,v) - \sum_{(v,x) \in \vec{E}} f(v,x) = 0.$$

For $v = v_i$, this can also be written as

$$\sum_{j=1}^{m} f(\vec{e}_j) d_{ij} = 0.$$

The function f, regarded as an m-element vector, thus has a zero scalar product with the ith row of the matrix D. So f is orthogonal to each row of D; actually, we see that the orthogonality of f to all rows of D is equivalent to f being a circulation. Therefore, the vector spaces $\mathcal{C}$ and $\mathcal{R}$ (regarded as subspaces of the space $\mathbf{R}^m$) are orthogonal complements of each other. $\qquad\square$

What is the dimension of the cut space $\mathcal{R}$? By the previous theorem, this is the dimension of the space generated by the rows of the incidence matrix D. By a basic result of linear algebra we have used several times already, this is the same as the dimension of the space generated by the columns of D, and this dimension we are now going to determine.

Let d_j denote the jth column of D (corresponding to the directed edge $\vec{e}_j$).

Let us consider some set $J \subseteq \{1, 2, \ldots, m\}$ of column indices, and let us ask when the set of columns $\{d_j : j \in J\}$ is linearly dependent. A linear dependence means that there exist numbers c_j, $j \in J$, not all of them zero, such that $\sum_{j \in J} c_j d_j$ is the zero n-element vector. Let us write this out for each component separately. The ith component, corresponding to the vertex v_i, gives

$$\sum_{j \in J} c_j d_{ij} = 0.$$

If we define $c_t = 0$ for $t \notin J$, we get the condition (11.4) from the definition of a circulation, and so $c = (c_1, c_2, \ldots, c_m)$ is a nonzero circulation. Hence the carrier of c contains a cycle (see Exercise 2).

On the other hand, as we already know, for each cycle C there exists a nonzero circulation with carrier C. Altogether we get that the set of columns $\{d_j: j \in J\}$ is linearly independent if and only if the edge set $\{e_j: j \in J\}$ contains no cycle. Hence the rank of the matrix D is $n - k$, where k is the number of components of the graph G. We have proved:

11.5.2 Theorem. *For a graph G with n vertices, m edges, and k components, the cut space $\mathcal{R}$ has dimension $n - k$ and the circulation space $\mathcal{C}$ has dimension $m - n + k$.*

This result can be regarded as a more sophisticated version of Theorem 11.4.3 (over the field of real numbers).

The relation of the even sets of edges and the cuts, or of the circulations and the potential differences, has the nature of a duality. The considerations above can also be extended to other combinatorial objects than graphs. This is considered in the so-called *matroid theory* (see Oxley [24], for instance).

Exercises

1. Define an "elementary circulation with respect to a given spanning forest". Show that such elementary circulations generate the circulation space.

2. (a) Let $f: \vec{E} \to \mathbf{R}$ be a nonzero circulation on an orientation $\vec{G}$ of an undirected graph G. Prove that the carrier of f contains a cycle.

   (b) True or false? The carrier of any circulation is a even set of edges.

11.6   Probabilistic checking

Checking matrix multiplication. Multiplying two square matrices is a very important operation in many applications. A straightforward algorithm multiplying two $n \times n$ matrices according to the definition of matrix multiplication requires about n^3 arithmetic operations. But, quite surprisingly, ingenious algorithms have been discovered that can do such a multiplication in an asymptotically much better time. You can read about such algorithms in Aho, Hopcroft, and Ullman [9], for instance. The current asymptotic record is an algorithm executing only $O(n^{2.376})$ operations. Because of a really huge constant hidden in the $O()$ notation, this algorithm is only theoretically interesting.

But what is a theory today may enter the commercial sector tomorrow, and so you can imagine that a software company sells a program called MATRIX WIZARD and claims it can multiply square matrices at a lightning speed. However, since your matrix multiplication application is quite critical and wrong results could be disastrous, you would like to be sure that MATRIX WIZARD multiplies matrices not only fast but also correctly. You can test it on many examples, but even if it succeeds on many test examples this doesn't necessarily mean it's always correct. It would be most satisfactory to have a simple *checking program* appended to MATRIX WIZARD that would always check whether the resulting matrix, C, really is the product of the input matrices A and B. Of course, a checking program that would actually multiply A and B and compare the result with C makes little sense; you don't know how to multiply matrices as quickly as MATRIX WIZARD does and so all the advantages from its fast multiplication would be lost. But it turns out that if we allow some slight probability of error in the checking, there really is a very simple and efficient checker for matrix multiplication. For simplicity, we assume that the considered matrices consist of rational numbers, although everything works without change for matrices over any field.

11.6.1 Theorem (Freivalds' matrix multiplication checker).
There exists a randomized algorithm which for any three input $n \times n$ matrices A, B, C with rational entries performs $O(n^2)$ arithmetic operations and answers either CORRECT or INCORRECT. If A, B, C are such that $AB = C$ then the algorithm always says CORRECT. If, on the other hand, $AB \neq C$, then the algorithm answers INCORRECT with probability at least $\frac{1}{2}$, for any given A, B, C.

By a randomized algorithm, we mean an algorithm that may make some of its decisions at random. We may imagine that at some steps, it "tosses a coin" and chooses one of two possible continuations according to the result heads/tails.

> In real computers, there is naturally no dwarf tossing coins, mainly because such a dwarf or any other known physical device would slow down the computation considerably, not to mention of other technical problems. Randomness is simulated by complicated but non-random

computations, and doing this well is quite challenging. But in most programming languages there is a more or less satisfactory random number generator at our disposal, so we will ignore this issue here and assume that truly random numbers are available. The reader can find out much more about randomized algorithms, in particular more material related to Freivalds' checker, in the book by Motwani and Raghavan [23]. This book is delightful reading for anyone interested in combinatorics and theoretical computer science.

In our case, if A, B, C are fixed matrices with $AB \neq C$, the algorithm in the theorem may sometimes answer INCORRECT and sometimes CORRECT, depending on the outcome of its random internal decisions. The theorem claims that the (wrong) answer CORRECT has probability no more than $\frac{1}{2}$.

But wait, the reader may cry, does this mean that if the multiplication done by MATRIX WIZARD is incorrect, the checker may claim the result correct with probability $\frac{1}{2}$? Come on, this is much worse than the reliability of a weather forecast! Well, yes, but there's a subtle difference: we can run the checking algorithm several times for the same matrices A, B, C, and a single answer INCORRECT suffices to detect that $AB \neq C$. If we make k runs for the same A, B, C with $AB \neq C$, the probability that we get the wrong answer CORRECT every time, and thus fail to detect the error in the multiplication, is only at most $\left(\frac{1}{2}\right)^k$ because the results of the runs are independent. For instance, for $k = 50$, this probability is below 10^{-15}, which is most likely much smaller than the chance of the computer running the matrix multiplication algorithm being destroyed by ants or nuclear war next Friday. (In contrast, listening to 50 weather forecasts in a row doesn't seem to provide such a near-absolute certainty about tomorrow's weather.) So we can re-formulate the theorem's conclusion as follows: with running time $O(n^2 \log \frac{1}{\delta})$, the probability of a wrong answer can be bounded by δ, for any prescribed parameter $\delta > 0$.

Proof of Theorem 11.6.1. The algorithm is actually very simple. We choose a random vector $x \in \{0, 1\}^n$, all the 2^n possible vectors having the same probability (the underlying probability space is thus $\mathcal{C}_n$ as in Definition 9.2.2). We calculate the vector $y = ABx - Cx$, where x and y are considered as $n \times 1$ matrices. If $y = 0$ we output CORRECT, otherwise we output INCORRECT.

A vector can be multiplied by an $n \times n$ matrix using $O(n^2)$ operations. The right way of computing the product ABx is $A(Bx)$, where we need two matrix–vector multiplications. Hence y can be computed in $O(n^2)$ steps. It remains to prove the claim about probability. Obviously, if $AB = C$ then $y = ABx - Cx = (AB - C)x = 0$; hence in this case the algorithm always says CORRECT. So assume that $AB \neq C$ and write $D = AB - C$. It suffices to prove the following:

Lemma. *Let D be any rational $n \times n$ matrix with at least one nonzero entry. Then the probability of $y = Dx$ being the zero vector, for a random vector $x \in \{0,1\}^n$, is at most $\frac{1}{2}$.*

Proof of the lemma. Suppose that d_{ij} is a nonzero entry of D. We show that in such a case, $y_i = 0$ holds with probability at most $\frac{1}{2}$. We have

$$y_i = d_{i1}x_1 + d_{i2}x_2 + \cdots + d_{in}x_n = d_{ij}x_j + S,$$

where

$$S = \sum_{\substack{k=1,2,\ldots,n \\ k \neq j}} d_{ik}x_k.$$

Imagine that we choose the values of the entries of x according to successive coin tosses and that the toss deciding the value of x_j is made as the last one (since the tosses are independent it doesn't matter). Before this last toss, the quantity S is already fixed, because it doesn't depend on x_j. After the last toss, we either leave S unchanged (if $x_j = 0$) or add the nonzero number d_{ij} to it (if $x_j = 1$). In at least one of these two cases, we must obtain a nonzero number. Consequently, the probability of $y_i = 0$ is at most $\frac{1}{2}$. This proves the lemma and also concludes the proof of Theorem 11.6.1. □

It is very well known that writing error-free software for a complicated task is almost impossible. This is quite alarming since we are being increasingly surrounded by computer-controlled devices, and the possibility of their failure is very real (well-known examples were serious software-related problems in the Apollo or Space Shuttle projects, but today one really doesn't have to go to outer space to encounter a software malfunction). The idea of using an independent checker to verify the correctness of results obtained by complicated software is gaining in popularity. Research in theoretical computer science revealed that efficient probabilistic checking is possible for almost any computation. In principle, the work of a very fast computer can be checked, with a high reliability in the sense of probability, by a much slower computer. Although such results haven't yet been put widely into practice, they certainly raise interesting possibilities. Moreover, the "program-checking" stream of ideas has been merged with ideas from some other lines of research and has led to some of the deepest and most unexpected results in theoretical computer science. These results prove the "hardness of approximation" for certain algorithmic problems. For example, it has been known for a long time that the problem of finding the largest complete subgraph of a given graph is a problem in a certain class of difficult problems called "NP-complete problems". The new results say that if there were an efficient algorithm for approximating the size of the largest complete subgraph within any reasonable factor then there would also be an efficient algorithm for finding that size exactly. It seems somewhat mysterious that the proof of such a result should be related to probabilistic program checking, but that's the way it is. These results are very new and there is no "classical" textbook treatment. A comprehensive survey was written by Arora and Lund [11].

Probabilistic associativity testing. Let S be a set and let $\circ$ be some *binary operation* on S. This means that any two elements $a, b \in S$ are assigned some element $c \in S$, and this c is denoted by $a \circ b$. Formally, $\circ$ is thus a mapping $S \times S \to S$. You know many binary operations with numbers. For example, "+" (addition) is a binary operation on the set of all natural numbers and "−" (subtraction) is a binary operation on the set of all integers. On the other hand, "−" is not a binary operation on **N** and "/" (division) is not a binary operation on the set of all real numbers in the sense of our definition. These everyday operations usually satisfy various laws, like commutativity $(a + b = b + a)$ or associativity $((a + b) + c = a + (b + c))$. Here we will consider a quite arbitrary operation $\circ$ that may be quite lawless and anarchistic; we assume no commutativity, no associativity, nothing like that.

In algebra, a set S with an arbitrary binary operation on it is called a *groupoid*.

It may be instructive to compare the definition of a binary operation on S to the definition of a binary relation on S: a relation assigns to every pair $(a, b) \in S \times S$ a yes/no answer (the elements are either related or not), while an operation assigns to (a, b) a "result" lying again in S. While nonmathematical examples of relations abound, it seems that operations in the considered sense are peculiar to mathematics; at least we can't see any convincing and nontrivial nonmathematical examples. Of course, there are many situations when two objects give rise to a third object of the same kind, but for a binary operation we require

that any two objects can be so combined and that the result is again
from the same a priori chosen set of objects.

We will be interested in the case when S is a finite set, and we
suppose that the operation $\circ$ is specified by a table, such as the
following for the case $S = \{A, B, \Gamma, \Delta\}$:

$\circ$	A	B	Γ	Δ
A	A	A	A	A
B	A	B	Γ	Δ
Γ	A	Γ	A	Γ
Δ	A	Δ	A	B

We would like to check whether the given operation $\circ$ is associative or not. Call a triple $(a, b, c) \in S^3$ an *associative triple* if
$(a \circ b) \circ c = a \circ (b \circ c)$ holds, and a *nonassociative triple* otherwise.
An obvious method is to go through all triples $(a, b, c) \in S^3$ and test
whether they are associative. For each triple (a, b, c), we need two
lookups in the table to find $(a \circ b) \circ c$ and two more lookups to com-
pute $a \circ (b \circ c)$. Hence if $|S| = n$, the number of operations needed
for this straightforward associativity checking grows roughly as n^3.

We will present an ingenious algorithm for associativity testing,
due to Rajagopalan and Schulman, which only needs $O(n^2)$ operations. It is again a randomized algorithm and it doesn't give the
correct answer with 100% certainty.

11.6.2 Theorem. *There is a randomized algorithm with the following properties. It accepts as an input a binary operation $\circ$ on an
n-element set given by a table. The running time is $O(n^2)$, and the
algorithm outputs one of the answers ASSOCIATIVE or NONAS-
SOCIATIVE. In the case when $\circ$ is associative, the algorithm always
declares it associative, and in the case when $\circ$ is not associative, it
detects the nonassociativity with probability at least $\frac{1}{8}$ (for any given
operation $\circ$).*

The probability of an incorrect answer can be made arbitrarily
small by repeating the algorithm enough times, similar to Theorem 11.6.1. That is, if we repeat the algorithm k times for a nonassociative operation it will give the answer NONASSOCIATIVE at
least once with probability at least $1 - \left(\frac{7}{8}\right)^k$.

An obvious randomized algorithm for associativity testing would
be to repeatedly pick a random triple $(a, b, c) \in S^3$ and to test its associativity. But the catch is that the nonassociativity need not man-

ifest itself on many triples. For example, the operation specified in the above table has only two nonassociative triples, namely (Δ, Δ, Γ) and (Δ, Γ, Δ), while there are $4^3 = 64$ triples altogether. In fact, for any $n \geq 3$ there is an example of an operation on an n-element set with just one nonassociative triple (Exercise 4). Hence, even if we test n^2 random triples, the chance of detecting nonassociativity is only $\frac{1}{n}$, and not a constant as for the algorithm in Theorem 11.6.2.

Proof of Theorem 11.6.2. Let S be the given n-element set. First we define a vector space over $GF(2)$ related to S. This will be very similar to the definitions in Section 11.4 on the cycle space of a graph.

Let V denote the set of all n-tuples of 0s and 1s whose components are indexed by the elements of S. For $a \in S$ and $v \in V$, we let $(v)_a \in \{0,1\}$ denote the component of v corresponding to the element a. Moreover, for an element $a \in S$, let $\mathbf{v}_a$ be the characteristic vector of the set $\{a\}$; that is, the vector with the component corresponding to a equal to 1 and with all other components 0.

The set V can be regarded as a vector space over the 2-element field $GF(2)$, which is again similar to the considerations related to the cycle space of a graph. Vectors are added and multiplied by an element of $GF(2)$ componentwise.

Since we will be dealing with objects of several different kinds, we will distinguish them notationally as follows: letters u, v, w stand for elements of V, i.e. n-tuples of 0s and 1s, Greek letters α, β, γ denote elements of $GF(2)$, i.e. 0 or 1, and a, b, c, p, q, r are elements of S.

Next, based on the operation $\circ$ on S, we define a binary operation, denoted also by $\circ$, on V. For $u, v \in S$, we set

$$u \circ v = \sum_{a,b \in S} (u)_a (v)_b \mathbf{v}_{a \circ b}.$$

On the right-hand side, the multiplication $(u)_a (v)_b$ is in $GF(2)$, the result multiplies $\mathbf{v}_{a \circ b}$ as a vector, and the sum is addition of vectors in V. To make this more intuitive, suppose for a moment that $S = \{p, q, r\}$, and write $u = \alpha_p \mathbf{v}_p + \alpha_q \mathbf{v}_q + \alpha_r \mathbf{v}_r$, $v = \beta_p \mathbf{v}_p + \beta_q \mathbf{v}_q + \beta_r \mathbf{v}_r$. Of course, α_p, α_q, $\alpha_r \in \{0,1\}$ are just the components of u, and similarly for v. To find the vector $u \circ v$, we first "multiply out" the parentheses:

$$(\alpha_p \mathbf{v}_p + \alpha_q \mathbf{v}_q + \alpha_r \mathbf{v}_r) \circ (\beta_p \mathbf{v}_p + \beta_q \mathbf{v}_q + \beta_r \mathbf{v}_r)$$

$$= \alpha_p \beta_p (\mathbf{v}_p \circ \mathbf{v}_p) + \alpha_p \beta_q (\mathbf{v}_p \circ \mathbf{v}_q) + \cdots + \alpha_r \beta_r (\mathbf{v}_r \circ \mathbf{v}_r).$$

Then this expression is "simplified" using the definition $\mathbf{v}_a \circ \mathbf{v}_b = \mathbf{v}_{a \circ b}$ for all $a, b \in S$.

We claim that the operation $\circ$ on V is associative if and only if $\circ$ on S was associative. Clearly, if (a, b, c) is a nonassociative triple in S, with $(a \circ b) \circ c = p \neq q = a \circ (b \circ c)$, we have $(\mathbf{v}_a \circ \mathbf{v}_b) \circ \mathbf{v}_c = \mathbf{v}_p \neq \mathbf{v}_q = \mathbf{v}_a \circ (\mathbf{v}_b \circ \mathbf{v}_c)$. Checking the associativity of $\circ$ on V for an associative $\circ$ on S is an easy exercise (Exercise 5).

For the associativity-checking algorithm, we define a function $g: V^3 \to V$ by setting

$$g(u, v, w) = [(u \circ v) \circ w] - [u \circ (v \circ w)].$$

By the above discussion, all values of g are 0 if and only if $\circ$ is associative (on V and also on S). Note that if vectors $u, v \in V$ are given, $u \circ v$ can be calculated using $O(n^2)$ operations using the table specifying $\circ$ on S. Therefore $g(u, v, w)$ can be evaluated in $O(n^2)$ time too.

Now we are ready to formulate the algorithm for Theorem 11.6.2.

11.6.3 Algorithm. Select vectors $u, v, w \in V$ at random and independently of each other (each of the 2^n vectors of V has the same probability of being selected for u etc.). Compute $g(u, v, w)$ and answer ASSOCIATIVE if $g(u, v, w) = 0$ and NONASSOCIATIVE otherwise.

It remains to prove that for a nonassociative operation $\circ$, the answer NONASSOCIATIVE is returned with probability at least $\frac{1}{8}$. This means that at least $\frac{1}{8}$ of the possible triples $(u, v, w) \in V^3$ are nonassociative. To this end, fix some nonassociative triple $(a, b, c) \in S^3$. Call two triples (u_1, v_1, w_2) and (u_2, v_2, w_2) equivalent if u_1 and u_2 agree in all components but possibly the component corresponding to a (i.e. $u_1 - u_2 = \alpha \mathbf{v}_a$, $\alpha \in GF(2)$), v_1 and v_2 coincide everywhere except possibly in the component of b, and w_1, w_2 differ only possibly in the c component. Each class of this equivalence has exactly 8 elements. We show that each class contains at least one nonassociative triple:

Lemma. *Let (a, b, c) be a fixed nonassociative triple. For all u, v, $w \in V$, there exist $\alpha, \beta, \gamma \in GF(2)$ such that*

$$g(u + \alpha \mathbf{v}_a, v + \beta \mathbf{v}_b, w + \gamma \mathbf{v}_c) \neq 0.$$

Proof of the lemma. We actually show that the sum

$$\sigma = \sum_{\alpha,\beta,\gamma \in GF(2)} g(u + \alpha \mathbf{v}_a, v + \beta \mathbf{v}_b, w + \gamma \mathbf{v}_c)$$

is nonzero. By the definition of the operation $\circ$ on V, we obtain, for any $u, v, w \in V$,

$$g(u, v, w) = \sum_{p,q,r \in S} (u)_p (v)_q (w)_r \mathbf{v}_{g(p,q,r)} .$$

Substituting this into the sum σ and exchanging the summation order, we get

$$\sigma = \sum_{p,q,r \in S} \left[\sum_{\alpha,\beta,\gamma \in GF(2)} (u + \alpha \mathbf{v}_a)_p (v + \beta \mathbf{v}_b)_q (w + \gamma \mathbf{v}_c) \right] \mathbf{v}_{g(p,q,r)}.$$

The sum in the brackets can be rewritten, using distributivity in $GF(2)$, as follows:

$$\sum_{\alpha,\beta,\gamma \in GF(2)} (u + \alpha \mathbf{v}_a)_p (v + \beta \mathbf{v}_b)_q (w + \gamma \mathbf{v}_c)$$

$$= [(u)_p + (u + \mathbf{v}_a)_p] \, [(v)_q + (v + \mathbf{v}_b)_q] \, [(w)_r + (w + \mathbf{v}_c)_r]$$

$$= (2u + \mathbf{v}_a)_p (2v + \mathbf{v}_b)_q (2w + \mathbf{v}_c)_r = (\mathbf{v}_a)_p (\mathbf{v}_b)_q (\mathbf{v}_c)_r$$

because $2 = 1 + 1 = 0$ in $GF(2)$. Consequently, the only possibly nonzero term is obtained for $p = a$, $q = b$, and $c = r$, and so $\sigma = \mathbf{v}_{g(a,b,c)} \neq 0$. This proves the lemma and also finishes the proof of Theorem 11.6.2. $\qquad\square$

> *Remarks.* The algorithm and its analysis could also be formulated without introducing the vector space V. We could just talk about random subsets of S etc., but this seems to obscure the whole thing instead of making it more accessible. The approach can also be generalized to verify other identities for one or several binary or k-ary operations on a given finite set, although interesting open problems remain. More details can be found in the original paper Rajagopalan and Schulman [39].

Exercises

1. Suppose that we chose the components of the vector x randomly from the set $\{0, 1, \ldots, m\}$ instead of from $\{0, 1\}$ in Freivalds' checker (Theorem 11.6.1). Show that the probability of a wrong answer is at most $\frac{1}{m}$ in this case.

2. *Suppose that we have a program for multiplying two polynomials $p(x)$, $q(x)$ with integer coefficients very quickly. Can you design a method for a fast probabilistic checking of the correctness of the result, analogous to Freivalds' matrix multiplication checker? That is, given three polynomials $p(x)$, $q(x)$, and $r(x)$, the algorithm should check whether $r(x) = p(x)q(x)$. Use the fact that a nonzero polynomial of degree d has at most d roots.

3. How many binary operations on an n-element S are there?

4. *For each $n \geq 3$, find an example of an n-element set S with a nonassociative binary operation that possesses only one nonassociative triple.

5. Prove that if $\circ$ is an associative operation on S then also the operation $\circ$ on V defined in the proof of Theorem 11.6.2 is associative.

6. ^{CS} Implement the randomized algorithm for associativity testing. Compare its speed with the straightforward $O(n^3)$ algorithm for various inputs.

7. Let S be a finite set with a binary operation $\circ$. For a set $G \subseteq S$, define $G^{(1)} = G$ and $G^{(k+1)} = G^{(k)} \cup \{a \circ b : a, b \in G^{(k)}\}$, and put $\langle G \rangle = \bigcup_{k=1}^{\infty} G^{(k)}$. A set $G \subseteq S$ is called a *generator set* if $\langle G \rangle = S$.

   (a) Show that if G is some generator set and $(a \circ g) \circ c = a \circ (g \circ c)$ holds for any $a, b \in S$ and any $g \in G$ then $\circ$ must be associative.

   (b) *Call the operation $\circ$ *cancellative* if for every $a, b \in S$, there exists a unique $x \in S$ with $a \circ x = b$ and a unique $y \in S$ with $y \circ a = b$. Prove that if $\circ$ is cancellative, $A \subset S$ is a nonempty set, and $b \in S \setminus \langle A \rangle$ then $|\langle A \cup \{b\} \rangle| \geq 2|\langle A \rangle|$.

   (c) Using (b), prove that if $|S| = n$ and $\circ$ is cancellative then a generator set of size at most $\log_2 n + 1$ exists and ^{CS,*}it can be found in $O(n^2 \log n)$ time.

   Hence, for a cancellative operation $\circ$, (a) and (c) together give an associativity testing algorithm with no randomness involved and with $O(n^2 \log n)$ running time.

8. (Final exercise)

   (a) Having read the book, can you now give solutions to all the problems raised in Section 1.1?

   (b) Explain a solution of at least one of them to an educated nonspecialist (to one who has only learned some high-school mathematics, say)! This exercise may be one of the most important ones in this book for the student's future career.

Appendix: Prerequisites from algebra

This is a microcourse of matrices, vector spaces, fields, and a few other things from algebra that are used at some places in the book. Unlike other chapters, this part is not meant to be a textbook introduction to the subject. It is mainly intended for the reader who has some knowledge of the subject but may have forgotten the exact definitions or may know them in a slightly different form.

Matrices. A matrix is a rectangular table of numbers. The entries may be real numbers, complex numbers, or sometimes elements of other algebraic structures. An $m \times n$ matrix has m rows and n columns. If a matrix is called A, then its entry in the ith row and jth column is usually denoted by a_{ij}. So, for example, a 3×4 matrix A has the general form

$$\begin{pmatrix} a_{11} & a_{12} & a_{13} & a_{14} \\ a_{21} & a_{22} & a_{23} & a_{24} \\ a_{31} & a_{32} & a_{33} & a_{34} \end{pmatrix}.$$

Note that a matrix is denoted by writing large parentheses to enclose the table of elements.

A matrix is multiplied by a number α by multiplying each entry by α. Two $m \times n$ matrices A and B are added by adding the corresponding entries. That is, if we denote $C = A + B$, we have $c_{ij} = a_{ij} + b_{ij}$ for $i = 1, 2, \ldots, m$ and $j = 1, 2, \ldots, n$.

Matrix multiplication is more complicated. A product AB, where A and B are matrices, is defined only if the number of columns of A is the same as the number of rows of B. If A is an $m \times n$ matrix and B is an $n \times p$ matrix, then the product $C = AB$ is an $m \times p$ matrix given by

$$c_{ij} = a_{i1}b_{1j} + a_{i2}b_{2j} + \cdots + a_{in}b_{nj}.$$

Pictorially,

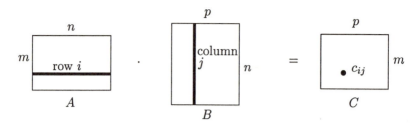

If A is an $m \times n$ matrix and $x = (x_1, x_2, \ldots, x_n)$ is a vector, we may regard x as an $n \times 1$ matrix (so we think of the vector as being written in a column) and consider the product Ax as a product of two matrices.

If A is an $m \times n$ matrix then A^T denotes the $n \times m$ matrix having the element a_{ji} in the ith row and jth column. The matrix A^T is called the *transposed* matrix A. For example, if $x = (x_1, x_2, \ldots, x_n)$ is regarded as a column vector, i.e. as an $n \times 1$ matrix, then x^T is a $1 \times n$ matrix or a row vector. (In the literature, vectors are sometimes regarded as row vectors even without the transposition symbol.) If x and y are n-component (column) vectors, the product $x^T y$ is a 1×1 matrix, i.e. a single number, called the *scalar product* of x and y. Explicitly, we have $x^T y = \sum_{i=1}^{n} x_i y_i$. The vectors x and y are called *orthogonal* if $x^T y = 0$.

For transposing the matrix product, we have the formula $(AB)^T = B^T A^T$.

Here is some more terminology for matrices. A *square matrix* is an $n \times n$ matrix, i.e. one with the same number of rows and columns. The *diagonal* (sometimes called the *main diagonal*) of an $n \times n$ matrix A consists of the elements $a_{11}, a_{22}, a_{33}, \ldots, a_{nn}$; see a schematic figure:

An *upper triangular matrix* is one having 0s everywhere below the diagonal, i.e. with $a_{ij} = 0$ whenever $i > j$. A *diagonal matrix* may only have nonzero entries on the diagonal, i.e. $a_{ij} = 0$ for $i \neq j$. The $n \times n$ *identity matrix* I_n has 1s on the diagonal and 0s elsewhere.

Determinants. Every square matrix A is assigned a number $\det(A)$ called the *determinant* of A. The determinant of A is defined by the formula

$$\det(A) = \sum_{\pi \in S_n} \mathrm{sgn}(\pi) \prod_{i=1}^{n} a_{i,\pi(i)},$$

where the sum is over all permutations π of the set $\{1, 2, \ldots, n\}$, and where $\mathrm{sgn}(\pi)$ denotes the *sign* of a permutation π. The sign of any permutation is either $+1$ or -1, and it can be compactly defined as the sign of the expression

$$\prod_{1 \le i < j \le n} (\pi(j) - \pi(i)).$$

For example, the determinant of a 2×2 matrix A equals $a_{11}a_{22} - a_{12}a_{21}$. This is also almost the only case when the determinant is computed from the definition. For larger matrices, the determinant is calculated by various other methods. Such methods are usually based on the rules for determinants listed below:

1. The determinant of an upper triangular matrix (in particular, of a diagonal matrix) equals the product of the entries on the diagonal.
2. ("Elementary row operation I") If all entries in some row of a matrix A are multiplied by some number α then the determinant is multiplied by α. In particular, if A has a row of all zeros then $\det(A) = 0$. Similarly for a column.
3. ("Elementary row operation II") The determinant is not changed by adding a multiple of one row to another row. That is, if $i \ne k$ are row indices and if the entry a_{ij} is replaced by $a_{ij} + \alpha a_{kj}$ for $j = 1, 2, \ldots, n$, where α is a fixed number, then the determinant remains the same. Similarly for columns.
4. If A, B are $n \times n$ matrices then $\det(AB) = \det(A)\det(B)$.
5. ("Expanding the determinant according to a row") For any row index i, we have the formula

$$\det(A) = \sum_{j=1}^{n} (-1)^{i+j} \det(A_{ij}),$$

where A_{ij} denotes the matrix arising from A by deleting the ith row and the jth column.

A square matrix A is called *nonsingular* if $\det(A) \ne 0$. A *submatrix* of a matrix A is any matrix that can be obtained from A by deleting some rows and/or columns of A. The *rank* of a matrix A, denoted by $r(A)$, is the maximum number k such that a nonsingular $k \times k$ submatrix of the matrix A exists (below, an alternative definition of rank will be given). For the product and sum of matrices, we have the inequalities $r(AB) \leq \min(r(A), r(B))$ and $r(A + B) \leq r(A) + r(B)$.

Groups and fields. Let X be a set. A *binary operation* on X is a mapping of $X \times X$ into X; that is, any two elements $a, b \in X$ are assigned some element of X (see also Section 11.6). For example, addition and multiplication are binary operations on the set of all natural numbers (while subtraction is not). Binary operations are usually written using symbols like $*$, $\circ$, $+$, $\cdot$, etc., so for example if $*$ is a binary operation, we write $a * b$ for the element assigned by $*$ to the pair (a, b).

A binary operation $*$ is called *associative* if we have $a * (b * c) = (a * b) * c$ for all $a, b, c \in X$. A binary operation $*$ is *commutative* if $a * b = b * a$ for all $a, b \in X$. Practically all binary operations one encounters in everyday mathematics are associative, but there are common and important operations that are not commutative, such as the multiplication of $n \times n$ matrices or the composition of mappings on some set.

A *group* is a set G with an associative binary operation $*$ satisfying the following axioms (there are many equivalent versions of the axioms):

(i) There exists an element $e \in G$, called the *unit element* or the *neutral element*, such that $a * e = e * a = a$ holds for all $a \in G$.

(ii) For every $a \in G$, there exists an element $b \in G$ with $a * b = b * a = e$, where e is the unit element. Such a b is called the *inverse* of a and it is usually denoted by a^{-1}.

Some important examples of groups are the integers with the operation of addition, the set of all positive reals with the operation of multiplication, the set of all regular $n \times n$ real matrices with multiplication, the set of all permutations on a given set X with the operation of composition, and the set of all rotations around the origin in the 3-dimensional space with the operation of composition (performing one rotation and then another). On the other hand, the natural numbers with addition or the real numbers with multiplication do not form a group.

A *field* is a set K with two binary operations $+$ and $\cdot$ (here these symbols need not denote the usual operations with numbers), such that the following hold:

1. The set K with the operation $+$ is a commutative group with the unit element denoted by 0.

2. The set $K \setminus \{0\}$ with the operation $\cdot$ is a commutative group with the unit element denoted by 1.

3. *Distributivity* holds: $a \cdot (b+c) = (a \cdot b) + (a \cdot c)$ for any $a, b, c \in K$.

The multiplication $a \cdot b$ is often written just ab. The most important fields are the rational numbers, the real numbers, and the complex numbers (with addition and multiplication as operations). On the other hand, the integers with addition and multiplication do not form a field.

If K is a field and L is a subset of K that contains 0 and 1 of K, is closed under both the operations of K (i.e $a + b \in L$ and $a \cdot b \in L$ holds for any $a, b \in L$), and is a field with these operations then L is called a *subfield* of K. For example, the field of the rationals is a subfield of the reals.

In combinatorics, we are often interested in finite fields. A finite field with q elements, if one exists, is commonly denoted by $GF(q)$. The simplest finite field is $GF(2)$, which is just the set $\{0, 1\}$ with the usual multiplication of integers and with addition given by $0 + 0 = 1 + 1 = 0$, $0 + 1 = 1 + 0 = 1$. A finite field $GF(q)$ is known to exist if and only if q is a power of a prime number. If q is a prime, $GF(q)$ can be easily described: it is the set $\{0, 1, 2, \ldots, q - 1\}$, where the addition and multiplication are performed as for integers, but after adding or multiplying two numbers we take the remainder of the result after dividing it by q (we say that we do *arithmetic modulo q*). The above definition of $GF(2)$ is a special case of this construction. On the other hand, if $q = p^k$ for a prime p and $k > 1$, then the set $\{0, 1, \ldots, q - 1\}$ with arithmetic modulo q is *not* a field, and the construction of $GF(q)$ is different and more complicated.

Vector spaces. Let K be a field (most of the time one can think of the real numbers). A *vector space* over K is a commutative group V with an operation $+$ and neutral element 0 plus a mapping (operation) assigning to every pair (α, v) with $\alpha \in K$ and $v \in V$ an element of V. That is, any element of V can be "multiplied" by any element of K. This multiplication is usually written without any multiplication sign, so we write just αv for v multiplied by α. We require the following conditions, for any $u, v \in V$ and any $\alpha, \beta \in K$: $\alpha(u + v) = \alpha u + \alpha v$, $(\alpha + \beta)v = \alpha v + \beta v$ (these are two kinds of distributivity), $\alpha(\beta v) = (\alpha \beta)v$ (a kind of associativity), and $1v = v$.

From this one can prove various other properties, such as $0v = 0$ (the left 0 is in the field K and the right 0 in the vector space V). The elements of V are called *vectors*.

The most usual and most important example of a vector space is one consisting of all n-tuples of real numbers (for some given integer n), with componentwise addition and multiplication by a real number. This is a vector space over the field of real numbers, but similarly one can consider the vector space of all n-tuples of elements of some field K, which gives a vector space over K. This vector space is usually denoted by K^n.

A set $A \subset V$ is called *linearly dependent* if there exist elements $v_1, v_2, \ldots, v_n \in V$, $n \geq 1$, and nonzero $\alpha_1, \alpha_2, \ldots, \alpha_n \in K$ with $\alpha_1 v_1 + \alpha_2 v_2 + \cdots + \alpha_n v_n = 0$. If A is not linearly dependent it is called *linearly independent*.

The maximum possible cardinality of a linearly independent set in V is called the *dimension* of V, and any linearly independent set in V of that cardinality is called a *basis* of V. The vector space K^n has, as everyone would expect, dimension n.

Every vector space has a basis, and every inclusion maximal linearly independent set is a basis. In combinatorics, we almost always deal with vector spaces of a finite dimension. If $(e_1, e_2, \ldots, e_n)$ is a basis of an n-dimensional vector space V, then any vector $v \in V$ can be uniquely represented as $v = \sum_{i=1}^{n} \alpha_i e_i$ for some $\alpha_1, \alpha_2, \ldots, \alpha_n \in K$. Here $\alpha_1, \ldots, \alpha_n$ are called the *coordinates* of v with respect to the basis $(e_1, \ldots, e_n)$. Every vector $v \in V$ is thus assigned a vector from K^n, namely the n-tuple of its coordinates. In this way, V is put into a bijective correspondence with the vector space K^n (there are many such correspondences, depending on the particular choice of a basis). All n-dimensional vector spaces over a given field are "the same", i.e. they are all isomorphic in a suitably defined sense. Hence K^n can be viewed as *the* n-dimensional vector space over K, and vectors can usually be thought of as n-tuples of numbers. On the other hand, sometimes it is convenient to work with other versions of n-dimensional vector spaces over K as well.

A *subspace* of a vector space V is a subset $W \subseteq V$ that is closed on addition and on multiplication by elements of K, i.e. such that $u + v \in W$ and $\alpha u \in W$ holds for any $\alpha \in K$ and $u, v \in W$. If $X \subseteq V$ is a set of vectors, then the *subspace generated by* X or the *linear span* of X is the smallest subspace of V containing X. Explicitly, it is the set $\{\sum_{i=1}^{n} \alpha_i v_i: n \in \mathbf{N}, \alpha_1, \ldots, \alpha_n \in K, v_1, \ldots, v_n \in X\}$.

The rank of an $m \times n$ matrix A, with entries from a field K, equals the dimension of the subspace generated by the rows of A (considered as n-element vectors) in the space K^n. It also equals the dimension of the subspace generated by the column vectors of A in K^m. In fact, this definition of rank can be regarded as a "more basic" one than the definition via determinants given above.

If V and W are vector spaces over a field K, we define a *linear mapping* of V into W as a mapping $f: V \to W$ satisfying $f(\alpha u) = \alpha f(u)$ and $f(u + v) = f(u) + f(v)$ for any $\alpha \in K$ and $u, v \in V$. The *kernel* of f is the set $\ker(f) = \{v \in V: f(v) = 0\}$. The kernel of f is a subspace of V and the set $f(V)$ is a subspace of W. The dimension of $f(V)$ is called the *rank* of f. For any linear mapping $f: V \to W$, we have $\dim \ker(f) + \dim f(V) = \dim V$.

Let $(e_1, e_2, \ldots, e_n)$ be a basis of V and $(f_1, f_2, \ldots, f_m)$ a basis of W. Then linear mappings $f: V \to W$ are in a bijective correspondence with the $m \times n$ matrices with entries from K. As we know, any vector $v \in V$ is uniquely written $v = \sum_{i=1}^{n} \alpha_i e_i$. Any linear mapping $f: V \to W$ can be expressed as follows:

$$f(v) = f\left(\sum_{i=1}^{n} \alpha_i v_i\right) = \sum_{j=1}^{m}\left(\sum_{i=1}^{n} a_{ji}\alpha_i\right)f_j,$$

where a_{ij} are the elements of the $m \times n$ matrix A corresponding to f. In other words, if α is the column vector of coordinates of v with respect to the basis $(e_1, \ldots, e_n)$ and β denotes the column vector of coordinates of $f(v)$ with respect to the basis $(f_1, \ldots, f_m)$ then we have $\beta = A\alpha$.

Thus, linear mappings can be regarded as abstract versions of matrices. Multiplication of matrices corresponds to composition of linear mappings. The rank of a linear mapping coincides with the rank of its matrix (with respect to any choice of bases).

In conclusion, let us remark that many notions of linear algebra have nice geometric interpretations in the usual Euclidean space which may be helpful for understanding both geometry and linear algebra. Especially for applications in discrete mathematics, one should learn linear algebra together with these geometric connections.

Bibliography

Basic textbooks

[1] N. Biggs: *Discrete Mathematics*, Revised edition, Clarendon Press, Oxford, 1989.

[2] B. Bollobás: *Graph Theory. An Introductory Course*, corrected 3rd printing, Graduate Texts in Mathematics 63, Springer-Verlag, Berlin etc., 1990.

[3] P. Cameron: *Combinatorics: Topics, Techniques, Algorithms*, Cambridge University Press, Cambridge, 1994.

[4] R. Diestel: *Graph Theory*, Graduate Texts in Mathematics 173, Springer-Verlag, Berlin etc., 1996.

[5] R. Graham, D. Knuth, and O. Patashnik: *Concrete Mathematics: A Foundation for Computer Science*, Addison-Wesley, Reading, Massachusetts, 1989.

[6] J. H. van Lint and R. M. Wilson: *A Course in Combinatorics*, Cambridge University Press, Cambridge, 1992.

[7] L. Lovász: *Combinatorial Problems and Exercises*, 2nd edition, Akadémiai Kiadó, Budapest, and North-Holland, Amsterdam, 1993.

[8] I. Stewart and D. Tall: *The Foundations of Mathematics*, Oxford University Press, Oxford, 1977.

Further reading on related topics

[9] A. Aho, J. Hopcroft, and J. Ullman: *The Design and Analysis of Computer Algorithms*, Addison-Wesley, Reading, Massachusetts, 1983.

[10] N. Alon and J. Spencer: *The Probabilistic Method*, John Wiley, New York, 1992.

[11] S. Arora and C. Lund: Hardness of Approximation, in *Approximation Algorithms for NP-hard Problems*, edited by D. Hochbaum, PWS Publishing, Florence, Kentucky, 1997.

[12] L. Babai and P. Frankl: *Linear algebra methods in combinatorics* (Preliminary version 2), Department of Computer Science, The University of Chicago, 1992.

[13] E. R. Berlekamp, J. H. Conway, and R. K. Guy: *Winning Ways for Your Mathematical Plays.* Vol. 1: Games in general. Vol. 2: Games in particular. Academic Press, London, 1982.

[14] N. Biggs: *Algebraic Graph Theory*, 2nd edition, Cambridge University Press, Cambridge, 1993.

[15] J. M. Borwein and P. B. Borwein: *Pi and the AGM*, John Wiley, New York, 1987.

[16] G. Bredon: *Topology and Geometry*, Graduate Texts in Mathematics 139, Springer-Verlag, Berlin etc., 1993.

[17] D. Cox, J. Little, and D. O'Shea: *Ideals, Varieties, and Algorithms. An Introduction to Computational Algebraic Geometry and Commutative Algebra*, 2nd edition, Springer-Verlag, Berlin etc., 1996.

[18] G. R. Grimmett and D. R. Stirzaker: *Probability and Random Processes*, 2nd edition, Oxford University Press, Oxford, 1992.

[19] F. Harary and E. M. Palmer: *Graphical Enumeration*, Academic Press, New York and London, 1973.

[20] N. Koblitz: *A Course in Number Theory and Cryptography*, 2nd edition, Graduate Texts in Mathematics 114, Springer-Verlag, Berlin etc., 1994.

[21] J. Littlewood: *A Mathematician's Miscellany*, London, Methuen, 1953.

[22] B. Mohar and C. Thomassen: *Graphs on Surfaces*, Johns Hopkins University Press, Baltimore, Maryland, 1997.

[23] R. Motwani and P. Raghavan: *Randomized Algorithms*, Cambridge University Press, New York, 1995.

[24] J. G. Oxley: *Matroid Theory*, Oxford University Press, Oxford, 1992.

[25] J. Pach and P. K. Agarwal: *Combinatorial Geometry*, John Wiley, New York, 1995.

[26] A. de Saint-Exupéry: *Le petit prince*, English translation *The Little Prince*, Harcourt Brace, San Diego, California, 1993.

[27] J. Stillwell: *Classical Topology and Combinatorial Group Theory*, Graduate Texts in Mathematics 72, Springer-Verlag, Berlin etc., 1980.

[28] W. T. Trotter: *Combinatorics and Partially Ordered Sets: Dimension Theory*, Johns Hopkins Series in the Mathematical Sciences, Johns Hopkins University Press, Baltimore, Maryland, 1992.

[29] G. Ziegler: *Lectures on Polytopes*, Graduate Texts in Mathematics 152, Springer-Verlag, Berlin etc., 1994.

Other references

[30] G. E. Andrews: *The Theory of Partitions* (Encyclopedia of mathematics and its applications, vol. 2), Addison-Wesley, Reading, Massachusetts, 1976.

[31] D. Bayer and P. Diaconis: Trailing the dovetail shuffle to its lair, *Annals of Applied Probability* 2(1992), No. 2, 294–313.

[32] Th. Beth, D. Jungnickel, and H. Lenz: *Design Theory*, B.I. Wissenschaft Verlag, Mannheim, Wien, Zürich, 1985.

[33] K. Chandrasekhar: *Introduction to Analytic Number Theory*, Springer-Verlag, Berlin etc., 1968.

[34] P. Flajolet, B. Salvy, and P. Zimmerman: Automatic average-case analysis of algorithms, *Theoretical Computer Science* 79(1991), 37–109.

[35] R. L. Graham, M. Grötschel, and L. Lovász (editors): *Handbook of Combinatorics*, Vol. 1–2, North-Holland, Amsterdam, 1995.

[36] G. Kant: Drawing planar graphs using the canonical ordering, *Algorithmica* 16(1996), 4–32.

[37] D. Karger, P. Klein, and R. Tarjan: A randomized linear-time algorithm to find minimum spanning trees, *Journal of the ACM* 42(1995) 321–328.

[38] D. Knuth: *The Art of Computer Programming*, Vol. I: Fundamental Algorithms, Addison-Wesley, Reading, Massachusetts, 1968.

[39] S. Rajagopalan and L. J. Schulman: Verifying identities (extended abstract), *Proc. 37th Annual IEEE Symposium on Foundations of Computer Science*, pages 612–616, IEEE, Los Alamitos, California, 1996.

[40] N. Robertson, D. P. Sanders, P. D. Seymour, and R. Thomas: The Four Color Theorem, *Journal of Combinatorial Theory* Ser. B 70(1997), 2–44.

[41] N. Robertson, P. D. Seymour, and R. Thomas: Permanents, Pfaffian orientations, and even directed circuits, Preprint, Geor-

gia Institute of Technology, Atlanta, 1997 (available electronically at `www.math.gatech.edu/~thomas`).

[42] M. Sharir and P. K. Agarwal: *Davenport-Schinzel Sequences and Their Geometric Applications*, Cambridge University Press, Cambridge, 1995.

[43] I. Stewart: Mathematical recreations, *Scientific American* 272(1995), No. 1, 76–79.

[44] C. Thomassen: The Jordan-Schoenflies theorem and the classification of surfaces, *American Mathematical Monthly* 99(1992), No. 2, 116–130.

"Modul X" (drawing by Jiří Načeradský and Jaroslav Nešetřil).

Hints to selected exercises

1.2.2. Denote the right-hand side by m. Hence m is the unique integer satisfying $m^2 \leq \lfloor x \rfloor < (m+1)^2$. This holds if and only if $m^2 \leq x < (m+1)^2$, and hence $m = \lfloor \sqrt{x} \rfloor$.

1.2.3(d). See Section 10.4, Exercise 4.

1.2.7. Write each number in the form $2^k(2m+1)$, $k, m \geq 0$ integers. Since m can attain at most 500 distinct values, the considered set contains two numbers of the form $2^k(2m+1)$ and $2^{k'}(2m+1)$, $k < k'$.

1.3.3(a). Adding the next line to n lines already present slices exactly $n+1$ of the existing regions into two.

1.3.3(b). Use (a) in an induction step. The result is $(n^3 + 5n + 6)/6$.

1.3.5. Prove that the numerator decreases in each step.

1.3.6. Cut the chessboard into four $2^{n-1} \times 2^{n-1}$ parts. Place one L-shape so that it has one square in each part, except for the one with a missing square.

1.3.7(b). The strategy is to invert the previous moves of the first player. In proving that it works, proceed by contradiction (if there was a repetition caused by the second player, find an earlier repetition caused by the first player).

1.3.8. Induction on n. Let R be the lowest row initially containing a black square, and let C be the leftmost such column. By the inductive hypothesis, after moment $n-1$ all squares above R are white, and also all squares to the right of C. The only possible remaining black square at the intersection of R and C disappears at the moment n.

1.3.9. 4. By induction on m, describe the distribution of particles at time $2^m - 1$.

1.3.10. Define $x_1 = \min(M)$, $x_{i+1} = \min(M \setminus \{x_1, \ldots, x_i\})$. Should M be infinite, the set $\{x_1, x_2, \ldots\}$ would have no largest element.

1.3.12. For instance, first show that if $1 < n_i \leq n_j$, then the sum of squares doesn't decrease by replacing n_i by $n_i - 1$ and n_j by $n_j + 1$. Then prove that by repeating this operation, any initial values of the n_i can be converted to the situation described in the incomplete solution.

1.3.13. One possibility is by induction on n. Suppose the claim holds for all $n < n_0$, and consider some M for $n = n_0$. Set $M' = M \setminus \{[1, n]\}$. If no interval of M' contains 1 then $|M'| \leq n - 2$ by the inductive hypothesis.

Otherwise let $q < n_0$ be largest with $[1, q] \in M'$, let $M_1 = \{I \in M': I \subseteq [1, q]\}$, and $M_2 = M' \setminus M_1$. Use the inductive hypothesis for M_1 and M_2.

1.3.14. $k\lfloor n/2 \rfloor$. To prove $m \leq k\lfloor n/2 \rfloor$, use the fact that each I_j contains at least one even number.

1.4.7(c). It suffices to prove that there is no bijection from the set $\{1, 2, \ldots, n\}$ onto a proper subset $A \subset \{1, 2, \ldots, n\}$. Proceed by induction on n. The case $n = 1$ is clear. Suppose there were such a bijection $f: \{1, 2, \ldots, n\} \to A$, $n > 1$. If $f(n) = n$ or $n \notin A$ then f restricted to $\{1, 2, \ldots, n-1\}$ gives a bijection of $\{1, 2, \ldots, n-1\}$ to its proper subset. If $f(n) = i \neq n$ and $f(j) = n$ for some $j < n$ then define $g(j) = i$, $g(k) = f(k)$ for $k \neq j, n$. This g is again a bijection of $\{1, 2, \ldots, n-1\}$ on its proper subset.

1.5.3(a). There are only finitely many relations on X.

1.5.3(c). Take $(\mathbf{N}, <)$.

1.5.4(b). The pairs related by $R \circ S$ correspond to the nonzero entries of the matrix $A_R A_S$.

1.6.6(b). If the considered congruence $\sim$ is not the diagonal, let $q = \min\{|x - y|: x \neq y, x \sim y\}$. Prove that $\equiv_q$ coincides with $\sim$.

1.6.6(c). Yes in (a), no in (b) (make all even numbers equivalent, while each odd number is only equivalent to itself).

1.7.4. Define a relation $S = \{(x, x): x \in X\} \cup R \cup R \circ R \cup R \circ R \circ R \cup \ldots$, prove that this is an ordering.

1.7.6. For instance, the relation of immediate predecessor is empty for the set of all rational numbers with the usual ordering.

1.7.8. By induction on $|X|$. Remove some minimal element of X and place it as the first element in the linear ordering by $\leq$.

1.7.9(b). A finite linearly ordered set has a (unique) largest element. Map it to the largest element of the other set, and proceed by induction.

1.7.9(c). The natural ordering, and the opposite ordering ($1 \succ 2 \succ 3 \succ \ldots$).

1.7.9(d). Uncountably many: partition $\mathbf{N}$ into infinite subsets $S_1, S_2, \ldots$, place S_2 after S_1, S_3 after S_2, $\ldots$; each S_i is ordered either as the natural ordering of $\mathbf{N}$ or as its reverse.

1.7.10. Choose $A = X$, and to every $x \in X$ assign the set $M_x = \{y \in X: y \leq x\}$.

1.7.11(d). Show that for any $A \subseteq X$, we have $\inf A = \sup\{x \in X: x \leq a$ for all $a \in A\}$.

2.1.2. Encode a pair (A, B) by a mapping $f: \{1, 2, \ldots, n\} \to \{0, 1, 2\}$: an element $x \in A$ receives value 2, $x \in B \setminus A$ value 1, and $x \notin B$ value 0.

2.1.5(b). Choose the first column as any nonzero vector over the q-element field. The second column must not be a multiple of the first, which excludes

q vectors. The third column must not be a linear combination of the first 2 columns, and this excludes q^2 possibilities (no two distinct linear combinations of the first 2 columns give the same vector—check!). In general, there are $q^n - q^i$ possibilities for the ith column.

2.1.6. Pair up each divisor d with n/d. If n is not a perfect square then this partitions all divisors into disjoint pairs.

2.2.1. $(n-1)!$.

2.2.3(b). What is the order of a permutation with a single cycle of length k? What if there are two cycles of length k_1 and k_2?

2.2.5(a). Show that if a permutation with k increasing segments is written backwards, we get a permutation with $n + 1 - k$ increasing segments.

2.2.5(b). Consider a permutation of the set $\{1, 2, \ldots, n-1\}$ with k segments (in a one-line notation), and insert the number n into it on some of the possible $n + 1$ positions. For how many positions will the resulting permutation have $k + 1$ segments, and for how many k segments?

2.2.5(c). A general formula, derived by Euler, is $f(n, k) = \sum_{j=0}^{k} (-1)^j \cdot (k-j)^n \binom{n+1}{j}$.

2.2.5(d). Divide the permutations into classes according to the set of numbers occurring on the first $k + 1$ positions. All orderings of the first $k + 1$ numbers have the same probability. For how many of the $(k + 1)!$ possible orderings of the first $k + 1$ numbers do we get an initial increasing segment of length exactly k? The resulting probability is $k/(k + 1)!$.

2.2.6(c). If the numbers are initially ordered according to a permutation π, then exchanging two neighbors removes at most 1 inversion of the current permutation.

2.2.7(b). How many among the numbers $1, 2, \ldots, n$ are divisible by p^j, for $j = 1, 2, \ldots$? The resulting formula is $\lfloor n/p \rfloor + \lfloor n/p^2 \rfloor + \lfloor n/p^3 \rfloor + \cdots$.

2.3.3(b). The right-hand side is the number of $(r+2)$-tuples of nonnegative integers satisfying the equation $X_1 + X_2 + \cdots + X_{r+2} = n - r$. Divide these $(r+2)$-tuples into $n-r$ groups according to the value of X_{r+2}, and calculate the number of solutions in each group separately. This yields the left-hand side.

2.3.4. Transform the kth term into $\binom{n}{m}\binom{n-m}{n-k}$ and take $\binom{n}{m}$ in front of the sum. The result is $\binom{n}{m}2^{n-m}$.

2.3.5(a). Use $\frac{1}{k}\binom{k}{m} = \frac{1}{m}\binom{k-1}{m-1}$ and formula (2.9).

2.3.5(b). This time $k\binom{k}{m} = (k+1)\binom{k}{m} - \binom{k}{m} = (m+1)\binom{k+1}{m+1} - \binom{k}{m}$.

2.3.6. Let M be an m-element set and N an n-element set. Both sides count the number of ordered pairs (X, Y) with $X \subseteq M$, $Y \subseteq N \cup X$, and $|Y| = m$. For the left-hand side, we first select X and then Y, and for the right-hand side, we first pick $Y \cap N$, then $Y \cap M$, and finally X.

2.3.7. Set $k_i = f(i+1) - f(i)$, $k_0 = f(1)$, $k_n = n - f(n)$. The desired number is the number of nonnegative integer solutions to the equation $k_0 + k_1 + \cdots + k_n = n$.

2.3.9. Consider such a subset $\{a_1, \ldots, a_k\}$, $a_1 < a_2 < \cdots < a_k$. Assign the k-tuple $\{a_1, a_2 - 1, a_3 - 2, \ldots, a_k - k + 1\}$ to it. This given a bijection between all subsets of the required kind and $\binom{\{1,2,\ldots,n-k+1\}}{k}$.

2.3.10(b). The result is $\binom{n+1}{2}^2$.

2.3.10(c). Calculate the area of the big square in two ways. Its side is $2(1 + 2 + 3 + \cdots + n)$. The k-th "layer" (counted from the center) consists of $4k$ squares of size $k \times k$.

2.3.15(b). In the formula (2.1) for the binomial coefficient, exactly one of the factors in the numerator is divisible by p. So $\binom{n}{p}$ is divisible by p if and only if this factor is divisible by p^2 too.

2.3.16(a). Substitute the following values into the binomial theorem: $x = 1$, $x = -1$, $x = i$, and $x = -i$ ($i = \sqrt{-1}$ stands for the imaginary unit). Add the resulting equations together. Rewrite the nth powers on the right-hand side using de Moivre's formula. The required number is given by the remarkable expression $2^{n-2} + 2^{n/2-1} \cos \frac{\pi n}{4}$. Verify its correctness for small values of n.

2.3.16(b). Proceed as in (a) but substitute the 3 (complex) roots of the equation $x^3 = 1$ for x.

2.3.19. First arrange the elves, which determines 8 possible positions for the goblins, pick 5 of these positions, and finally arrange the goblins.

2.3.23. Those with A_1 as a vertex: $\binom{n-k-1}{k-1}$. All: this number times $\frac{n}{k}$.

2.3.24(b). Show that there are always precisely 2 triangles sharing 2 sides with the n-gon. Starting from a triangle T_0 sharing 2 sides with the n-gon, we can pick a triangle T_1 neighboring T_0, then T_2 neighboring T_1, etc. For T_0, we have n possibilities, and in each subsequent steps there are 2 possibilities. Each of the considered triangulations is obtained twice by this process, since there are 2 choices for the T_0 to start with. The result is $n2^{n-5}$.

2.3.26(a). The left-hand side is the number of distinct orderings of n objects, with k_i objects of the ith kind. The right-hand side: first we choose the kind of object at the first position and then we arrange the remaining objects.

2.3.27(a). $\binom{r_1+r_2+\cdots+r_k}{r_1,r_2,\ldots,r_k}$.

2.3.27(b). $(n-1)!/\prod_{i=1}^{k}(r_i + 1)$, where $n = |X| = 1 + \sum_{i=1}^{k}(r_i + 1)$.

2.4.2. For example, $f(n) = n^{\lfloor n/2 \rfloor}$, $g(n) = n^{\lfloor (n+1)/2 \rfloor - 1/2}$.

2.4.4. $|f|$ is bounded by a constant, $|g|$ is bounded from below by a positive constant, $|h|$ is bounded by some polynomial function.

2.4.5. First, rewrite all the functions in the form $e^{f(n)}$.

2.5.2(b). Write $1 + \frac{1}{n} = \frac{n+1}{n} = 1/(1 - \frac{1}{n+1}) \geq 1/e^{-1/(n+1)}$.

2.5.3. The line $y = x + 1$ is a tangent to the graph of the function e^x at the point $(0, 1)$. The function e^x is convex, since its second derivative is nonnegative, and so it cannot intersect its tangent anywhere else than at the point of tangency.

2.5.7(b). Put $A = (x_1 + x_2 + \cdots + x_{n-1})/(n - 1)$. By $AG(n)$ applied to numbers $x_1, \ldots, x_{n-1}$, and A, we have $A^n = ((x_1 + x_2 + \cdots + x_{n-1} + A)/n)^n \geq x_1 x_2 \ldots x_{n-1} A$, and from this we get $A \geq (x_1 \ldots x_{n-1})^{1/(n-1)}$.

2.5.10. The function $\ln x$ is concave, and so each triangle with vertices at points $(i, \ln i)$, $(i + 1, \ln i)$, and $(i + 1, \ln(i + 1))$ lies completely below its graph (draw a picture). The area of this triangle is $\frac{1}{2}(\ln(i + 1) - \ln i)$. Hence $\ln n! \leq \ln n + \int_1^n \ln x \, dx - \frac{1}{2} \sum_{i=1}^{n-1} (\ln(i+1) - \ln i)$. A further formula manipulation leads to the required result.

2.5.11. Induction on n.

2.5.12. Lower bound: by induction; prove $2(\sqrt{n + 1} - \sqrt{n}) \leq 1/\sqrt{n}$ by transforming the left-hand side to a fraction with denominator $\sqrt{n + 1} + \sqrt{n}$. Upper bound: similarly.

2.6.2(a). All such primes divide $\binom{2m}{m}$.

2.6.2(b). If P denotes the product of primes as in (a), we have $\log_2 P \leq 2m$, and on the other hand, $P \geq m^{\pi(2m) - \pi(m)}$. This gives $\pi(2m) \leq \pi(m) + O(m/\ln m)$.

2.6.2(c). The largest power of p dividing $n!$ has exponent $\lfloor n/p \rfloor + \lfloor n/p^2 \rfloor + \cdots$. Use the equation $\binom{2m}{m} = (2m)!/(m!)^2$ and express the difference of the maximum powers of p in the numerator and in the denominator.

2.6.2(d). By (c), we have $2^{2n}/(n + 1) \leq \binom{2n}{n} \leq (2n)^{\pi(2n)}$.

2.7.2. For instance, one can proceed as in the second proof of the inclusion–exclusion principle, i.e. consider the contribution of a single element.

2.8.2(b). Choose the points that are fixed. The rest is a permutation without a fixed point as in the hatcheck lady problem.

2.8.4. Partition all permutations into classes depending on the number of fixed points. Express the number of permutations in each class using the function $D(k)$.

2.8.5(a). Consider a distribution of hats to n gentlemen without fixed points. Gentleman No. 1 exchanges the hat he received for his hat with some gentleman No. i and leaves. If gentleman No. i received his own hat in this exchange, he leaves too and we have a situation with $n-2$ gentlemen. Otherwise we are left with $n - 1$ gentlemen.

2.8.7(c). Let the mapping go onto the set $\{1, 2, \ldots, m\}$. Let A_i be the set of mappings that do not map anything to the element i. Use inclusion–exclusion to determine $|A_1 \cup \cdots \cup A_n|$, i.e. the number of bad mappings. For $m = n$ we must get $n!$.

2.8.7(d). An onto mapping $f: N \to M$, where $|M| = m$, defines an equivalence relation on N with m equivalence classes. Show that each equivalence with m classes on N corresponds to exactly $m!$ mappings onto.

2.8.8(a). You can use the hint to (d) of the preceding exercise and the result of (c) in that exercise.

2.8.8(b). We sum the result of (a) over all k. We obtain the following result: $\sum_{k=0}^{n} \frac{1}{k!} \sum_{j=0}^{k} (-1)^j \binom{k}{j} (k-j)^n$.

2.8.8(c). In the result of (b), write i for $k - j$, let k formally run to ∞ (this is possible since the inner sum, expressing the number of equivalences with k classes on an n-element set, is 0 for $k > n$), and change the order of summation.

2.8.9. A number q is relatively prime to m if and only if its remainder after division by n is relatively prime to m. Prove that for any pair (a, b), $0 \leq a < n, 0 \leq b < m$, there exists exactly one $q < mn$ giving the remainder a after division by n and remainder b after division by m.

2.8.11(b). Rewrite each term in the product as $1 + p_i + p_i^2 + \cdots + p_i^{\alpha_i}$.

2.8.11(c). Let $n = 2^q \prod_{i=1}^{r} p_i^{\alpha_i}$, p_i odd primes. By (b), we must have $2n = t \prod (p_i^{\alpha_i} + p_i^{\alpha_i - 1} + \cdots + 1)$, $t = 2^{q+1} - 1$. After dividing this by the expression $t \prod p_i^{\alpha_i}$ we have $1 + 1/t = \prod (1 + 1/p_i + \cdots + 1/p_i^{\alpha_i})$. At the same time, some p_i divides t. In order that the right-hand side is not larger than the left-hand side, we must have $r = 1$, $t = p_1$.

2.8.12. For every prime $p_i \leq N$, let A_i be the set of pairs (m, n) such that $p_i | n$ and $p_i | m$. Use inclusion–exclusion to determine the number of bad pairs.

2.8.13(a). Define A_i as the set of all graphs in which the vertex i has degree 0, and calculate $|A_1 \cup \cdots \cup A_n|$.

2.8.14. Define A_i as the set of all ways of seating in which the ith couple is adjacent.

3.1.4(b). A simple counterexample arises as follows: take triangles with vertex sets $\{1, 2, 3\}$ and $\{4, 5, 6\}$, and add the edges $\{1, 4\}$, $\{2, 5\}$, and $\{3, 6\}$.

3.1.5. $(2n - 1)(2n - 3)(2n - 5) \cdots 5 \cdot 3$.

3.1.7. Here is a way of constructing $2^{n^2/2 - O(n \log n)}$ nonisomorphic graphs (this is almost as many as we got by the counting argument in the text). Let n be large enough, and let m be the smallest integer with $2^m \geq n$. Denote the n vertices by $a, b, c, d, u_0, u_1, \ldots, u_{m-1}$, and $v_0, v_1, \ldots, v_{n-m-5}$; write $U = \{u_0, \ldots, u_{m-1}\}$, $V = \{v_0, \ldots, v_{n-m-5}\}$. Connect a to b, and b to c, to d, and to all vertices of U. Choose some asymmetric graph on U, say the path $u_0, u_1, \ldots, u_{m-1}$ plus the edge $\{u_1, u_3\}$. Connect the vertices c and d to every vertex of V (so a is the only vertex of degree 1). Choose an arbitrary graph on V. Finally, connect each vertex $v_i \in V$ to the vertices $u_{j_1}, \ldots, u_{j_k}$, where $0 \leq u_1 < u_2 < \cdots < u_k \leq m - 1$ are the (uniquely determined) numbers with $2^{j_1} + 2^{j_2} + \cdots + 2^{j_k} = i$ (this corresponds to writing i in binary). It is easy to check that for distinct graphs on V we obtain nonisomorphic graphs on n vertices. Hence we have at least $2^{\binom{n-m-5}{2}}$ nonisomorphic graphs on n vertices.

3.2.2. The most edges are clearly obtained if the components are complete graphs. If their numbers of vertices are $n_1, \ldots, n_k$, $\sum n_i = k$, we have $\binom{n_1}{2} + \cdots + \binom{n_k}{2}$ edges and we need to maximize this expression as a function of $n_1, \ldots, n_k$. Show that if, for example, $n_1 \geq n_2 > 1$, the value doesn't decrease by increasing n_1 by 1 and decreasing n_2 by 1. Hence the maximum is attained for $n_1 = n - k + 1$ and $n_i = 1$ for $i = 2, 3, \ldots, n$.

3.2.4. If there is no odd cycle, assume that the graph is connected, and label the vertices by $+1$ and -1 by the following process: label an arbitrarily chosen vertex by $+1$, and whenever v is a vertex adjacent to a vertex labeled x label v by $-x$. Show that the whole graph gets labeled in this way and no two vertices with the same label are adjacent.

3.2.5. If v is a vertex with at least 2 neighbors it has to be connected to all vertices of its component. Each component has to be K_3 or $K_{1,n}$.

3.2.6. One component: any connected graph on ≤ 4 vertices, a star $K_{1,n}$, a star $K_{1,n}$ with one extra edge (forming a triangle), two stars $K_{1,n}$ and $K_{1,m}$ whose centers are connected by an edge. (By the center of a star we mean the vertex connected to all others.)

3.2.8. Graph diameter is usually defined as $\max\{d_G(u,v) \colon u, v \in V(G)\}$, and radius as $\min_{v \in V(G)} \max_{u \in V(G)} d_G(u,v)$.

3.2.11. For a given vertex v_i, the entry at position (i, i) in A_G^4 is the number of walks of length 4 starting and ending in v_i. Each 4-cycle in G appears $8\times$ as such a walk. But such a walk may also pass the same edge $4\times$ (and each edge is counted twice for such walks), or pass 2 edges with a common vertex $2\times$ (and each pair of edges with a common vertex appears $4\times$).

3.3.1. For instance, the first graph has no cycle of length 4, the second one 2 such cycles, and the third one 5 of them.

3.3.8(a). Never. Its score would have to be $(0, 1, \ldots, n-1)$, but if there is a vertex adjacent to all other vertices no vertex may have degree 0.

3.3.8(b). For all $n \geq 2$. By induction, we construct graphs G_n with score containing the numbers $0, 1, \ldots, n-2$, with $\lfloor (n-1)/2 \rfloor$ repeated. G_2 are 2 isolated vertices, G_{n+1} arises by adding an isolated vertex to the complement of G_n (the *complement* of a graph $G = (V, E)$ is the graph $(V, \binom{V}{2} \setminus E)$).

3.3.10. Consider the coloring with the smallest possible number of monochromatic edges. If there were a monochromatic path of length 2, one could switch the color of its middle vertex.

3.3.11. Consider a path in G of maximum possible length. Its end-vertex is connected to at least 2 other vertices of this path; this gives the desired subgraph.

3.3.12. We must have $k \le n-1$, and kn must be odd. This is also sufficient; for instance, set $V = \{0, \ldots, n-1\}$, $E = \{\{i, j\}: (i-j) \bmod n \in S\}$, where $S = \{1, -1, 2, -2, \ldots, \frac{k}{2}, -\frac{k}{2}\}$ for k even, $S = \{1, -1, \ldots, \frac{k-1}{2}, -\frac{k-1}{2}, \frac{n}{2}\}$ for k odd, n even.

3.3.16. It suffices to show that if $\{u, v\} \in E(G)$ then $\deg_G(u) = \deg_G(v)$. Let U be the set of all neighbors of u except for v, and let V be the set of all neighbors of v except for u. Each vertex in U has 4 neighbors in V, and similarly each vertex of V has 4 neighbors in U. Hence $|U| = |V|$ (by double-counting the edges among U and V).

3.3.17. By contradiction. Suppose that any two vertices have an odd number of common neighbors. For an arbitrary vertex v, look at the subgraph induced by the neighbors of v. All degrees in this subgraph are odd, and hence the degree of v is even. Let us count walks of length 2 beginning at v. Their total number is even (because every vertex has an even degree). At the same time, an odd number of these walks come back to v, but any other vertex is reached by an odd number of walks, and hence the number of vertices distinct from v must be even—a contradiction.

3.4.5(a). For a fixed vertex set, proceed by induction by the number of edges of G. Show that if $E(G) \ne \emptyset$, G contains a cycle; remove it.

3.4.7. The number of odd terms in the sequence is even.

3.4.10(a). Consider the longest path in G and the sets of neighbors of its end-vertices.

3.4.10(b). In general, no (consider a suitable complete bipartite graph).

3.6.7. Remove every knight with an ally sitting to his right. The remaining knights are from alternate castles as we go around the table, so their number is even, and the number of removed knights is the same.

3.6.8. By induction on the number of vertices. For an arbitrary vertex v, let T_1 be the tournament induced by the vertices with arrows going to v and T_2 the tournament induced by the vertices with arrows going from v. Take a path in T_1, continue to v, and then by a path in T_2.

3.6.9. Induction on the number of vertices n. In a tournament T on n vertices, delete a vertex u, and let v be a vertex as required in the problem for the remaining tournament. If v cannot be reached from u by a length 2 directed path in T then $(v, u) \in E(T)$ and also all vertices with an arrow going to v have an arrow to u, and hence u can be reached by at most 2 arrows from everywhere.

3.7.2(b). In K_{n+1}, subdivide each edge.

3.7.2(c). Let T be a tree (see Chapter 4) in which all vertices except for leaves have degree n, and having a vertex at distance n from all leaves (a complete n-ary tree). Make a copy T' of T, and identify each leaf of T' with its pre-image in T.

3.7.3(a). Yes.

3.7.3(b). Yes.

3.7.4. For instance, by induction, using a synthesis of the graph by successive "gluing of ears".

3.7.5. It is k. Prove that any two vertices of the cube can be connected by k vertex-disjoint paths.

3.7.6. Let A be a set of k vertices. Double-count the edges connecting A to the other vertices, assuming that $G - A$ has more than k components.

3.7.7. For $n = 2k - 1$, G must be a K_{2k-1}. For $n \geq 2k$, if there is a vertex of degree $\leq 2k - 3$, delete it and use induction. Otherwise, suppose that G minus some $\leq k-1$ vertices can be partitioned into two nonempty parts G_1 and G_2 with no edges connecting them. Show that the inductive hypothesis holds for G_1 or for G_2, for otherwise G would have too few edges.

4.1.6. We have $2n - 2 = 2|E(T)| = \sum_{v \in V(T)} \deg_T(v) = \sum_{i=1}^{n-1} i p_i$. The required inequality follows by a simple manipulation.

4.1.8(c). Induction on n. Let $n \geq 3$, and let $(d_1, \ldots, d_n)$ satisfy the condition. Then there is some $d_i = 1$ and also some $d_j > 1$ (for a simple notation, assume $d_n = 1$, $d_{n-1} > 1$). By the inductive hypothesis, we construct a tree with score $(d_1, \ldots, d_{n-2}, d_{n-1} - 1)$ and we connect a new end-vertex to the vertex number $n - 1$.

4.2.5. Divide the sequences into two groups, those beginning with 0 and those beginning with 1, and sort both groups recursively. In an actual implementation of the algorithm, the sequences themselves are not actually moved; one only works with suitable pointers to them.

4.2.6. The code of a tree on n vertices has length $2n$. Hence there are at most 4^n distinct codes.

4.3.1. Prove that the mark of any single vertex is changed at most $(\log_2 n)$ times.

4.3.2. Let the graph G be given by lists S_v of edges containing each vertex $v \in V(G)$. Maintain a doubly linked list N of all edges going between the set V_i and its complement, and, moreover, for each edge belonging to N keep a pointer to its occurrence in N. If the newly added vertex y_i has degree d then all the lists and pointers can be updated in $O(d)$ time.

4.3.3(a). Induction on the number of vertices. Show that for every edge $\{v, v'\} \in E(T)$, an ordering $v_1, v_2, \ldots, v_n$ of the vertices exists such that $v_1 = v$, $v_n = v'$, and the distance of v_i and v_{i+1} is at most 3, $i = 1, 2, \ldots$, $n - 1$.

4.4.1. Use Kruskal's algorithm for the weight function $-w$.

4.4.4. The computation of Kruskal's algorithm is determined uniquely and it gives some minimum spanning tree. The proof of correctness shows that any other spanning tree has a strictly larger weight.

4.4.5. In each group of edges of the same weight, put the edges of T first. Let T' be the tree computed by Kruskal's algorithm, let $e'_1, \ldots, e'_{n-1}$ be its edges in the order of their selection, and let $e_1, \ldots, e_{n-1}$ be the edges of T numbered according to the chosen ordering. For contradiction, let k be the smallest index with $e_k \neq e'_k$. We get $w(e'_k) < w(e_k)$, but from the proof in the text we have $w(e'_i) \leq w(e_i)$ for all i, and so T is not minimum.

4.4.6. Follows from Exercise 5.

4.4.8(a). If v has degree ≥ 7 then there exist edges $\{v, u_1\}$ and $\{v, u_2\}$ with angle < 60 degrees. Show that one of them can be replaced by the edge $\{u_1, u_2\}$.

4.4.8(b). Prove and use the fact that if $ABCD$ is a convex quadrangle then $|AB| + |CD| \leq |AC| + |BD|$ and $|BC| + |AD| \leq |AC| + |BD|$. Show that two crossing edges may be replaced by two noncrossing edges.

4.4.9. Cover the unit square by an $\sqrt{n} \times \sqrt{n}$ chessboard. Order the squares of the chessboard into a sequence $s_1, s_2, \ldots, s_n$ so that consecutive squares are adjacent. Construct a path on the given points by first going through all points in s_1, then all points in s_2, etc.

4.4.10(a). Let E_M be a maximum matching and E_G the matching found by the greedy algorithm. To each edge $e \in E_M$ assign the first edge of E_G intersecting it. Each edge $\breve{e} \in E_G$ is thereby assigned to at most two edges $e_1, e_2 \in E_M$ with $w(e_1), w(e_2) \leq w(\breve{e})$.

4.4.11(b). Let $e_1, \ldots, e_k$ be the edges selected by the greedy algorithm and $\breve{e}_1, \ldots, \breve{e}_t$ edges of some smallest edge cover. Let $k_1 = |\{i: e_i \cap (e_1 \cup \cdots \cup e_{i-1}) = \emptyset\}|$, and similarly $t_1 = |\{i: \breve{e}_i \cap (\breve{e}_1 \cup \cdots \cup \breve{e}_{i-1}) = \emptyset\}|$. We have $|V| = k + k_1 = t + t_1$. Key observation: in the steps of the greedy algorithm for $i > k_1$, at least one point of each edge $\breve{e}_j$ contributing to t_1 must have been covered, and hence $k_1 \geq \frac{1}{2} t_1$. From this we get $k = t + t_1 - k_1 \leq t + \frac{1}{2} t_1 \leq \frac{3}{2} t$.

4.4.12. Let $V = \{1, 2, \ldots, 2^{k+2} - 2\} \cup \{a_1, a_2, b_1, b_2, \ldots, b_k\}$. The vertex a_1 is connected to $1, 3, 5, \ldots$; a_2 to $2, 4, 6, \ldots$; b_i to $2^i - 1, 2^i, \ldots, 2^{i+1} - 2$; a_1, a_2 are, moreover, connected to all the b_i. The greedy algorithm selects all the b_i while the optimum dominating set is $\{a_1, a_2\}$.

4.5.5(a). Prove that after the ith phase, every component has at least 2^i vertices.

4.5.5(b). For each component, maintain a list of its outgoing edges. In each phase, each list can be traversed and the edge of minimum weight in it can be found.

5.1.1(a). Fig. 5.1.

5.1.1(b). Graphs G_1, G_2 as in (a) can be glued together. For instance, remove an edge $\{a, b\}$ adjacent to the outer face of G_1, an edge $\{c, d\}$ adjacent to the outer face of G_2, and add edges $\{a, c\}$ and $\{b, d\}$.

5.1.3. By induction on the number of edges. Consider a vertex v of degree at least 4, and look at two adjacent edges incident to v (in the circular order around v). If possible, choose these edges so that they go into distinct components of $G - v$. Replace these two edges by a single edge connecting the other ends. (It may be advantageous to allow for graphs with multiple edges.)

5.2.6. On each 5-tuple of vertices we have a copy of K_5, and some two of its edges must cross. We thus obtain at least $\binom{n}{5}$ pairs of crossing edges. One pair of crossing edges has been counted in $n - 4$ copies of K_5.

5.2.7(a). Choose a disk D such that $k \cap D$ is a segment. From any point of $\mathbf{R}^2 \setminus k$ we can reach D by a polygonal arc (going along k). But $D \setminus k$ has at most two equivalence classes since any two points on the same side can be connected by a segment.

5.3.5. First, one has to handle the case when G is not 2-connected. For a 2-connected G, prove that one can add a diagonal into a face bounded by at least 6 edges. But one has to be careful to create no triangle! Probably it is necessary to discuss several possible cases.

5.3.7(a). Let us say that a dot with $k \leq 3$ outgoing edges contributes $3 - k$ degrees of freedom. Initially, there are $3n$ degrees of freedom, finally at least 1, and each move reduces the number of degrees of freedom by 1.

5.3.7(c). Show that in the final position, there is exactly one "unused" arm of a cross protruding into each face. The total number of unused arms remains constant during the game, and therefore there are $4n$ faces in the end. Each move either connects two components into one or increments the number of faces by 1.

5.3.8(a). By induction on n, prove $e = v + f - 1$; when inserting a new line distinguish newly created intersections and already existing ones.

5.3.8(c). Derive $e \geq 3f - n$ from (b). If d_i is the number of lines passing through the ith intersection, we calculate $\sum d_i = e - n$ (since each edge contributes by 2 ends, except for the $2n$ semiinfinite rays). By substituting for f from (a) we finally get $\sum d_i \leq 3v - 3$.

5.3.9. Assume connectedness. Let a red–blue corner be a pair (f, v), where f is a face and v is a vertex adjacent to f such that if we go along the boundary of f clockwise (counterclockwise around the unbounded face), v is preceded by a red edge and followed by a blue edge. By double-counting and by Euler's formula, show that there is a vertex with at most one red–blue corner.

5.4.2. Proceed by induction on the number of vertices. For a given graph, remove a smallest-degree vertex, color the rest, and put the vertex back.

5.4.3. A subgraph of an outerplanar graph is outerplanar. Hence by Exercise 2 it is enough to show that an outerplanar graph always has a vertex of degree ≤ 2. This can be extracted from Euler's formula using the fact that some face has at least n edges adjacent to it.

5.4.4. Show that $\delta(G) \leq 3$ and use Exercise 2.

5.4.6(b). One can even take a tree.

5.4.6(c). Add sufficiently many new leaves to adjust the degrees.

5.4.10(a). By induction on the number of edges. Consider a face F and the set of edges E_F lying on its boundary. Show that the degrees of all vertices in the graph (V, E_F) are even. Delete the edges from E_F, color by induction, put E_F back, and recolor the face F by the other color.

5.4.10(b). Color the faces by 2 colors. The number of edges can be counted as the sum of the circumferences of the faces of one color, and also as the sum of circumferences of the faces of the other color. But one way of counting gives a number divisible by 3, the other a number not divisible by 3.

5.4.12(b). For one class, take k vertices $v_1, \ldots, v_k$, where v_i has the list $L(v_i) = \{k(i-1)+1, k(i-1)+2, \ldots, ki\}$. For each set S containing exactly one element from each of the $L(v_i)$, put a vertex v_S with the list $L(v_S) = S$ in the other class.

5.4.13. In the inductive step, distinguish two cases. If there is an edge $\{v_i, v_j\}$ with v_i and v_j lying on the cycle C but not consecutively, split G along that edge. Otherwise, remove v_k and adjust the lists of its neighbors suitably.

6.1.1. Define a graph on the faces exactly as in the proof of Sperner's lemma.

6.1.2. Imagine two tourists on the same day: one ascending, one descending.

6.1.4. Yes only for (b), (e). In (a), use a rotation around the center; in (d), let $f(x)$ be the point of the sphere antipodal to x.

6.1.5(a). Imitate the proof of the planar version. Connect by an edge tetrahedra sharing a face labeled 1, 2, 3, and use the planar version to show that the degree of the outer face is odd.

6.1.6(b). Ensure that a is much farther from c than b from d.

6.1.6(c). By contradiction. If Betty has a winning strategy, Alice will make her first move arbitrarily and find herself essentially in Betty's situation (one marked node from the first move can only help). Then she will use Betty's winning strategy.

6.1.7(b). Show that if Alice starts then Betty has a very simple strategy to force a draw.

6.2.3. An independent set system $\mathcal{M}$ satisfies $\sum_{M \in \mathcal{M}} \binom{n}{|M|}^{-1} \leq 1$, and the system $\binom{X}{\lfloor n/2 \rfloor}$ gives equality. Hence all the sets in a system $\mathcal{M}$ of largest size have size $\lfloor n/2 \rfloor$ or $\lceil n/2 \rceil$ (thus, for an even n we are done). Show that if $\mathcal{M}$ has t sets of size $\lfloor n/2 \rfloor$ and $0 < t < \binom{n}{\lfloor n/2 \rfloor}$ then $\mathcal{M}$ has $< \binom{n}{\lfloor n/2 \rfloor} - t$ sets of size $\lceil n/2 \rceil$.

6.2.4. For an automorphism h, define a mapping $f : X \to X$, where $f(x)$ is the y such that $h(\{x\}) = \{y\}$. Show that $h = f^{\#}$ (hence all automorphisms come from permutations and there are $n!$ of them).

6.2.6(a). Apply Sperner's theorem on the set system $\{\{i\colon \varepsilon_i = 1\}\colon \sum \varepsilon_i a_i \in (-1,1)\}$.

6.2.7. Let $n = p_1 p_2 \ldots p_n$ be the decomposition of n into prime factors. To a divisor $d = p_{i_1} p_{i_2} \ldots p_{i_k}$ assign the set $M_d = \{i_1, \ldots, i_k\} \subseteq \{1, 2, \ldots, n\}$; then $d_1 | d_2 \Leftrightarrow M_{d_1} \subseteq M_{d_2}$. Use Sperner's theorem.

6.2.8(a). By induction on c. Take the set of all minimal elements as the first antichain, and show that by removing these elements, c drops by 1.

6.2.8(b). An elementary proof can be found in Van Lint and Wilson [6] (Theorem 6.1).

6.2.9. Define $i \preceq j$ if and only if $i \leq j$ and $a_i \leq a_j$. A chain in the poset $(\{1, \ldots, n\}, \preceq)$ corresponds to an increasing subsequence, an antichain to a decreasing subsequence.

6.3.1. Imitate the proof for $K_{2,2}$, the only difference being that one vertex v contributes at most $t - 1$ elements to M.

6.3.6. Instead of $f(x) = x(x-1)/2$ consider the function $f(x)$ which equals 0 for $x \leq 2$ and $x(x-1)(x-2)/6$ for $x > 2$. Show it is convex, and use this similarly as for excluded $K_{2,2}$.

6.3.7(a). The graph with vertex set $P \cup L$ and edges corresponding to the pairs being counted contains no $K_{2,2}$.

6.3.7(b). This time, the graph $K_{2,3}$ is excluded.

7.1.1. Proceed as we did for the lower bound for the number of nonisomorphic graphs in Section 3.1. Use Theorem 7.1.1 and the estimate for $n!$ from Theorem 2.5.5.

7.1.2. Count spanning trees containing a given edge of K_n. By symmetry, this number is the same for all edges. Use Cayley's formula.

7.1.3. From each spanning tree of K_n, we can remove one edge in $n - 1$ ways; this yields a pair of trees. Conversely, from a spanning tree on some k-point subset of the vertices plus a spanning tree on the remaining $n - k$ vertices, a spanning tree of the whole K_n can be manufactured by adding one edge, in $k(n - k)$ ways.

7.1.4. Given a spanning tree T of G, consider the subgraph of G^* consisting of the edges not crossed by the edges of T, and prove that this is a spanning tree of G^*. See Lovász [7] for details.

7.2.1(a). It's 0 for n odd. For n even we have $\frac{n}{2} + 1$ leaves and $\frac{n}{2} - 1$ vertices of degree 3 (by induction). We thus sum the expression $(n - 2)!/2^{n/2-1}$ over all choices of a vector $(d_1, \ldots, d_n)$ with $d_i \in \{1, 3\}$ and $\sum d_i = 2n - 2$. By the substitution $k_i = \frac{1}{2}(d_i - 1)$ we see that the number of summands equals the number of $(\frac{n}{2} - 1)$-element subsets of an n-element set, and therefore the answer is $(n - 2)!\binom{n}{n/2-1}2^{-n/2+1}$.

7.2.1(b). Perhaps the easiest solution uses generating functions. The answer is the coefficient of x^{2n-2} in the expression $(n - 2)!(x + x^2 + \frac{1}{2}x^3)^n$, which can be found using the multinomial theorem (part (a) can be solved similarly as well).

7.2.2(a). Both sides of the equality count the number of pairs spanning trees (T, T^*) where $\deg_T(n) = k$, $\deg_{T^*}(n) = k + 1$, and T^* arises from T by the following operation: pick an edge $\{i, j\} \in E(T)$ with $i \neq n \neq j$, delete it, and add either the edge $\{i, n\}$ or the edge $\{j, n\}$, depending on which of these edges connects the two components of $T - \{i, j\}$. From one T we can get $n - 1 - k$ different T^*, and one T^* can be obtained from $k(n - 1)$ different T.

7.2.2(c). $\sum_k N_k$ happens to be the expansion of $((n-1)+1)^{n-2}$ according to the binomial theorem.

7.3.3. Imagine that you send two pedestrians from the vertex i in the directed graph of f. One of them traverses one arrow per minute, the other one maintains the speed of one arrow in two minutes. Think about the times the pedestrians will meet at the same vertex.

7.4.1. The inequality $\deg_T(i) \geq m_i + 1$ is easy to see. At the same time, we have $\sum_i m_i = n - 2$ and $\sum_i (\deg_T(i) - 1) = 2(n - 1) - n = n - 2$, and so equality must hold for all i.

7.5.1(a). All row sums of Q are 0.

7.5.1(b). Since $T(G) > 0$, we have $\det Q_{11} \neq 0$ by Theorem 7.5.1.

7.5.1(c). The sum of the rows of Q corresponding to the vertices of a component of G is 0, and hence the vector space generated by the rows of Q has dimension $< n - 1$.

7.5.1(d). The kernel of the linear mapping $x \mapsto Qx$ contains $(1, 1, \ldots, 1)$ because the sum of rows is 0, and the kernel is one-dimensional by (b).

7.5.1(e). The product of the ith row of Q and the ith column of Q^* is the expansion of $\det Q$ according to the ith row. Use $\det Q = 0$. Other entries in the product QQ^* are expansions of determinants for matrices in which one row is repeated twice; hence also 0.

7.5.2(b). 384.

7.5.4. The result is $n^{m-1} m^{n-1}$.

7.5.5. For one implication, compute the determinant of the incidence matrix of an odd-length cycle. The required property of M for bipartite graphs can be proved by induction, similarly as in the proof of Lemma 7.5.3.

8.1.3(c). Possibilities: $\mathcal{L} = \{X\}$, $\mathcal{L} = \{X, \{a\}\}$ for some $a \in X$, and $\mathcal{L} = \{X \setminus \{a\}\} \cup \{\{a, x\}: x \in X \setminus \{a\}\}$ for some $a \in X$. In a proof, Exercise 4 may be useful.

8.1.4. Show that 2 points of $L_1 \setminus L_2$ and 2 points of $L_2 \setminus L_1$ together form a configuration F as in axiom (P0).

8.1.6. At most 4 sets can contain any given point.

8.1.7. Yes. Draw 8 lines in the plane in general position (no 2 parallel, no 3 intersecting at a common point). Let the intersections represent stops and the lines bus routes.

8.1.8(a). One pair is in at most one set. One set covers $\binom{n+1}{2}$ pairs. The total number of covered pairs comes out the same as the number of all pairs, and hence all pairs must be covered.

8.1.8(b). If there were more such sets they have more than $n^2 + n + 1$ points altogether.

8.1.8(c). Double-count the pairs (x, L), $x \in L$, and use (b).

8.1.8(d). Count that there are $n^2 + n + 1$ distinct lines intersecting a given line.

8.1.9. Fix a point $a \in A$. The $n + 1$ lines containing it cover all points of X, and each of them has at most one point of A besides a.

8.1.10(a). Delete the points of one, arbitrarily chosen, line.

8.1.10(b). Transitivity: suppose $A_1 \parallel A_2$, $A_2 \parallel A_3$, and $x \in A_1 \cap A_3$. Then A_1 and A_3 are two lines passing through x and parallel to A_2, contradicting the third axiom.

8.1.10(c). For showing that two lines A, A' have the same cardinality, construct a bijection using lines parallel to some line $\overline{xy}$ with $x \in A$ and $y \in A'$.

8.1.10(d). For each equivalence class of $\parallel$, add one new point "at infinity", add it to the lines from that class, and also make all the new points into a line. Check the axioms.

8.3.4(a). To given t orthogonal Latin squares, add one square with all the entries of the ith row equal to i, $i = 1, 2, \ldots, n$, and one square having all the entries j in the jth column, $j = 1, 2, \ldots, n$.

8.3.4(b). In order that a liberated square be orthogonal to another, it has to contain each $i \in \{1, 2, \ldots, n\}$ exactly $n\times$. Permute entries of the given $t + 2$ orthogonal liberated squares (the same permutation in each square) in such a way that the first square has all numbers i in the ith row, $i = 1, \ldots, n$. Then permute entries inside each row (again, in the same way for all squares) so that the second square has all the j in the jth column. Check that the remaining t squares are Latin.

8.3.6. $n! \times$ (the number of permutations with no fixed point).

8.4.3(a). Let the vertex classes be A and B; double-counting gives $\sum \binom{d_i}{2} = \binom{n}{2}$, where the d_i are the degrees of the vertices in A, and we have $m = |E(G)| = \sum_i d_i$. This time one has to calculate exactly: Cauchy–Schwarz gives $m^2 \leq n \sum d_i^2 = 2n \sum \binom{d_i}{2} + nm \leq n^2(n-1) + nm$. Solve the resulting quadratic inequality for m.

9.1.1(a). A function of n variables defines two functions of $n - 1$ variables, one for $x_n = 0$ and one for $x_n = 1$. One can proceed by induction on n.

9.1.3. Each weighing has 3 possible outcomes, and hence 3 weighings can only distinguish one among 3^3 possibilities.

9.1.4. Code all possible ways of the cars passing through the depicted system of rails, e.g. "a car from rail A goes to rail I, car from A to C, car from I to C, car from C to II, ...". Show that at most C^n possible codes exist (for a large enough constant C); hence there are at most this many possible car orders on rail B, and this is smaller than $n!$ for n large.

9.2.2(a). $P(A_1 \cup \cdots \cup A_n) = \sum_{\emptyset \neq I \subseteq \{1,...,n\}} (-1)^{|I|-1} P\left(\bigcap_{i \in I} A_i\right)$.

9.2.2(b). All the three proofs of inclusion–exclusion in Chapter 2 generalize easily.

9.2.4. For a disconnected graph, there exists a nonempty proper subset A of vertices such that there are no edges between A and its complement. Calculate the probability of this event for a fixed A, and then sum up over all possible choices of A.

9.2.8. If we replace some of the A_i by their complements we still have a collection of independent events, and hence the intersection of all these events has a nonzero probability. In this way, one can produce 2^n disjoint events with nonzero probability.

9.2.9. It's $\frac{1}{3}$. (Let the children be A and B. There are 3 equally probable possibilities: A=boy, B=boy; A=boy, B=girl; A=girl, B=boy.)

9.3.2. Show $0 \leq \mathbf{E}\left[(f - \mathbf{E}[f])^2\right] = \mathbf{E}\left[f^2\right] - \mathbf{E}[f]^2$.

9.3.3. Use indicators. Let A_i be the event "$\pi(i) = i$". Result: $\mathbf{E}[f] = 1$.

9.3.4(a). Let A_i be the event "1 is contained in a cycle of length i". Show that $P(A_i) = \frac{1}{n}$, and express the desired expectation as $\sum_{i=1}^{n} i \cdot \mathbf{E}[I_{A_i}]$.

9.3.4(b). Let A_{ij} be the event "j lies in a cycle of length i". By part (a), we have $P(A_{ij}) = \frac{1}{n}$. Check that the resulting expectation equals $\sum_{i,j=1}^{n} \frac{1}{i} \mathbf{E}[I_{A_{ij}}]$.

9.3.6. It's better to count boundaries among runs. The probability that a given position between two tosses is a boundary is $\frac{1}{2}$.

9.3.7. We have $\mu = \sum_{\omega \in \Omega} P(\{\omega\}) X(\omega) \geq \sum_{\omega \in \Omega: X(\omega) \geq t\mu} P(\{\omega\}) t\mu = t\mu P(\{\omega \in \Omega: X(\omega) \geq t\mu\})$.

9.4.4. A K_r-free graph with at least $r - 1$ vertices and with the maximum possible number of edges has to contain a K_{r-1}. Tear off a K_{r-1} and count the number of deleted edges.

9.4.6(b). Let L be the set of numbers smaller than the dividing element and R the set of the larger numbers. Show that in both the considered algorithms, the number $\min(|L|, |R|)$ has the same probability distribution. This number fully determines the sizes of the two pieces in the recursion.

9.4.7. Let R be the r circles that remain black, c the circle that was colored and emptied again, and $S(R)$ the set of empty circles seen by the observer for a given R. Let p be the probability of $c \in S(R)$. Since c is a random element of $R \cup \{c\}$ we get $p \le 2/(r + 1)$. On the other hand, since we could first pick R and then c at random among the remaining $n - r$ empty circles, p is the average, over all choices of R, of $|S(R)|/(n - r)$. Hence $p = \mathbf{E}\left[|S(R)|\right]/(n - r)$.

9.4.9. There are at most 2 peaks at level 0. Compute the expected number of peaks at level 0 for a random sample R of the lines in two ways as in the proof in the text.

10.1.4. In order to get the term $x_1^{k_1} \ldots x_m^{k_m}$ in multiplying out n parentheses, we have to choose k_1 parentheses from which we take $x_1, \ldots, k_m$ parentheses from which we take x_m. The number of such choices is just the multinomial coefficient.

10.1.6. Begin with the equality $(1 - x)^n (1 + x)^n = (1 - x^2)^n$.

10.2.5. $a(x) = (1 - x)^{-1}x^6(1 - x^3)^{-2} = x^6(1 + x + x^2)/(1 - x)^3$. Use the generalized binomial theorem.

10.2.8(a). The generating function is $1/((1 - x)(1 - x^2)(1 - x^5))$.

10.2.9(d). Result: $(-1)^m \binom{n-1}{m}$.

10.2.10. For $1 \le k \le n - r + 1$, there are $\binom{n-k}{r-1}$ subsets with smallest element k, so we need to sum $\sum_k k\binom{n-k}{r-1}$. This can be done (for instance) by considering the equality $[x^{r-1}/(1-x)^r][x/(1-x)^2] = x^r/(1-x)^{r+2}$ and comparing the coefficients of x^n. Alternatively, if we denote the expectation by $f(n,r)$, we have the recurrence $f(n,r) = p + (1 - p)(1 + f(n - 1, r))$ with $p = \frac{r}{n}$ (since either R contains 1, which happens with probability p, or it can be regarded as a random r-element subset of $\{2, 3, \ldots, n\}$). Then $f(n, r) = \frac{n+1}{r+1}$ follows by induction.

10.2.12(a). For the continuity, we need to prove that for $x_0 \in [0, \rho)$, we have $\lim_{x \to x_0} |a(x) - a(x_0)| = 0$. Use the estimate $|x_0^i - x^i| \le |x_0 - x|(x_0^{i-1} + x_0^{i-2}x + \cdots + x^{i-1}) \le |x - x_0|im^{i-1}$, where $m = \max(x, x_0)$.

10.2.12(c). For instance, $a_i = 1/i^2$.

10.2.13(a). First, show (by induction) that each derivative of the function f at a point $x \ne 0$ has the form $R(x)e^{-1/x^2}$, where $R(x)$ is a ratio of two polynomials. Then prove that each derivative is 0 at 0.

10.2.13(b). If it were a power series, its coefficients could be expressed using the derivatives, and hence they would have to be all 0s.

10.2.14(a). For the ring arrangement: $-\ln(1 - x)$.

10.2.14(c). $D(x) = 1/(1 - A(x))$.

10.2.14(d). $E(x) = e^{A(x)}$.

10.2.14(e). The exponential generating function is $e^{x^2/2}$.

10.2.15. Use part (d) with the subgroups arranged into unorganized crowds with at least one member, i.e. $A(x) = e^x - 1$. The result is $e^{e^x - 1}$.

10.2.16. The exponential generating function is $(x^2/2! + x^3/3! + x^4/4!)^5$. The answer is $12! \times$ (the coefficient of x^{12}).

10.2.17(a). Such arrangements are in a bijective correspondence with permutations, since they encode the cycles of the permutation.

10.2.17(b). The result is $e^{-\ln(1-x)-x} = (1+x+x^2+\cdots)(1-x/1!+x^2/2!-x^3/3!+\cdots)$.

10.3.1. Let a_n, b_n be the number of such sequences ending with 0, 1 respectively. Derive recurrence relations for a_n and b_n. Result: the Fibonacci numbers.

10.3.2. Induction on n. Call a representation of n as a sum of distinct Fibonacci numbers *reduced* if it uses no two consecutive Fibonacci numbers. Any representation can be converted to a reduced one by repeatedly replacing the largest two consecutive Fibonacci numbers by their sum. To go from n to $n + 1$ consider a reduced representation of n, add a 1 to it, and make it reduced again. Let us remark that a reduced representation is unique.

10.3.5. Consider the sequence $b_n = \log_2 a_n$.

10.3.7. A suitable expression is $\sum_{k=0}^{\infty} x^k (1 + 2x)^k = 1/(1 - x - 2x^2) = \frac{1}{3}[1/(1 + x) + 2/(1 - 2x)]$, and the formula is $\frac{2}{3}4^n + \frac{1}{3}$.

10.3.8. $(n + 2)/2^{n+1}$.

10.3.9. It satisfies the recurrence $a_{n+2} = 2a_{n+1} + a_n$ with $a_0 = a_1 = 1$.

10.3.10. Show that $x_n = (6+\sqrt{37})^n - (6-\sqrt{37})^n$ satisfies $x_{n+2} = 12x_{n+1} + x_n$ with initial conditions $x_0 = 2$, $x_1 = 12$ and hence is integral for all n. Further use $\sqrt{37} - 6 < 0.1$.

10.3.11. The equation $\sqrt{x + 1} - \sqrt{x} = y$ solves to $x = ((1 - y^2)/2y)^2$ $(0 < y < 1)$. Hence it suffices to show that this x is an integer for $y = (\sqrt{2}-1)^n$. We get $x = a_n^2$ where $a_n = \frac{1}{2}[(\sqrt{2}+1)^n - (\sqrt{2}-1)^n]$. Find a recurrence for a_n and show by induction that for n even, a_n is integral, and for n odd, a_n is an integer multiple of $\sqrt{2}$.

10.3.12. Let u_n denote the number of such sequences beginning with a or b, and let v_n be the number of such sequences beginning with c or d. One gets the recurrences $u_n = u_{n-1} + 2v_{n-1}$ and $v_n = 2u_{n-1} + 2v_{n-1}$. Write down the corresponding equations for the generating functions $u(x)$ and $v(x)$, solve them, and calculate the coefficients. The result is $u_n + v_n = \frac{\sqrt{17}+1}{4\sqrt{17}}(\frac{1}{2}(3 + \sqrt{17}))^{n+1} + \frac{\sqrt{17}-1}{4\sqrt{17}}(\frac{1}{2}(3 - \sqrt{17}))^{n+1}$.

10.3.13(a). The number is F_{n+1}.

10.3.13(b). Recurrence: $a_{n+3} = a_{n+2} + 4a_{n+1} + 2a_n$, $a_0 = 1$, $a_1 = 1$, $a_2 = 5$; $a_n \sim \frac{1}{\sqrt{3}}(1 + \sqrt{3})^n$.

10.3.13(c). Let a_n be the required number, and let b_n be the number of ways to fill an $n \times 2$ rectangle (the longer side is vertical) with the top left corner square removed. Recurrences: $a_n = a_{n-1} + 2b_{n-1} + a_{n-2}$, $b_n = a_{n-2} + b_{n-1}$, $a_0 = a_1 = 1$, $b_0 = b_1 = 0$. Generating function equations: $b(x) = xb(x) + x^2a(x)$, $a(x) = 1 + x(a(x) + 2b(x)) + x^2a(x)$. From this we get $a(x) = (1 - x)/(1 - 2x - x^3)$ and $a_n \sim (0.4607\ldots)(2.0556\ldots)^n$.

10.3.14(a). For the language with no consecutive letters a: write $A = \{a\}$, $B = \{b\}$. Then $B^*.(A.B.B^*)^* \cup B^*.(A.B.B^*)^*.A$ expresses the desired language as a very regular one.

10.3.14(b). If $a(x)$ is the generating function of L_1 and $b(x)$ of L_2, then $a(x)b(x)$ is the generating function of $L_1.L_2$, $1/(1 - a(x))$ is the generating function of L_1^*, and $a(x)+b(x)$ is the generating function of $L_1 \cup L_2$ provided that $L_1 \cap L_2 = \emptyset$.

10.3.14(c). One can go via deterministic finite automata; see a textbook on automata and formal grammars.

10.3.16(b). Show that the sequences whose jth term is 1 and the other terms among $y_0, \ldots, y_{k-1}$ are 0, for $j = 0, 1, \ldots, k - 1$, form a basis.

10.3.16(c). To show linear independence, it suffices to check that the vectors formed by the first k terms of the considered sequences are linearly independent. To this end, one can use the criterion of linear independence via determinants. This leads to the so-called Vandermonde determinant (usually discussed in linear algebra courses).

10.3.16(e). Perhaps the simplest method is using the order of growth of the sequences. If the sequences were linearly dependent, the fastest-growing one could be expressed as a linear combination of the more slowly growing ones, which is impossible. A problem in this approach appears only if there are several complex roots with the same absolute value. This case can be handled separately, for instance using the Vandermonde determinant as in the hint to (c). Alternatively, one can consider the determinant for the first k terms as in (c) and prove that it is nonzero.

10.4.2. Encode a binary tree with n vertices by a string of $O(n)$ letters and digits, say, thereby showing that $b_n \leq C^n$ for some constant C.

10.4.3(a). $\binom{2n}{n}$.

10.4.3(b). A path that never goes below the diagonal encodes a binary tree in the following manner. Divide the path into two parts at the point where it reaches the diagonal for the first time (the second part may be empty). Remove the first and last edges from the first part and leave the second part intact. Both parts then encode the left and right subtrees recursively (an empty path encodes an empty tree).

10.4.3(c). For the description of the solution, we introduce coordinates: A is $(0,0)$, B is (n,n). Extend the chessboard by one column on the right. Show that the paths that *do* reach below the diagonal are in a bijective correspondence with the shortest paths from A to the point $B_1 = (n + 1, n - 1)$. Follow the given AB-path until the end of its first edge lying below the diagonal, and flip the part of the path from that point on around the line $y = x - 1$. This yields an AB_1-path. Check that this defines a bijection.

10.4.4. Find a bijective correspondence with paths on the chessboard not reaching below the diagonal from the preceding exercise.

10.4.8(b). $c_{2n+1} = b_n$. A bijection between the planted trees considered in the exercise and the binary trees considered in the text is obtained by deleting all the leaves of a given planted tree.

10.4.9(a). Result: $t_n = b_{n-1}$.

10.4.10. An old tree either is the root itself or arises by connecting some k planted trees with at least 2 vertices each to the root, whence $s(x) = x + x/(1 - t(x) + x)$.

10.6.1(a). Let a_i be the number of trajectories starting at 1 and first entering 0 after i moves. The required probability is $a(\frac{1}{2})$. Derive the relation $a(x) = x + xa(x)^2$. The value of a_i can also be calculated explicitly, for instance using the Catalan numbers.

10.6.2(b). The catch is that S_1 is infinite (although 1 will be reached almost surely, the expected time needed to reach it is infinite!).

10.7.1(a). Such an ordered partition can be encoded by dividing the numbers $1, 2, \ldots, n$ into k segments of consecutive numbers. Define a $(k-1)$-element subset of $\{1, 2, \ldots, n-1\}$ by taking all numbers that are last in their segment (except for n).

10.7.1(b). A direct solution is as in (a) (all subsets of $\{1, 2, \ldots, n-1\}$). The generating function in (a) is $x^k/(1-x)^k$, and the sum over all $k \geq 1$ is $x/(1-2x)$.

10.7.3(a). $(1+x)(1+x^2) \ldots (1+x^n)$.

10.7.3(c). Find a bijection between all partitions of n with k distinct summands and all partitions of $n - \binom{k}{2}$ with k summands (not necessarily distinct). Then use the lower bound method demonstrated in the text.

10.7.4(c). The generating functions are $(1+x)(1+x^2)(1+x^3) \ldots$ and $1/((1-x)(1-x^3)(1-x^5) \ldots)$. Multiply both the numerator and denominator of the first expression by $(1-x)(1-x^2)(1-x^3) \ldots$. This leads to the expression $((1-x^2)(1-x^4)(1-x^6) \ldots)/((1-x)(1-x^2)(1-x^3) \ldots)$, and the factors with even powers of x cancel out. For a direct argument see e.g. Van Lint and Wilson [6].

10.7.6(a). $p_0 + p_1 + \cdots + p_m$.

10.7.6(b). There is a bijection mapping such trees to partitions of $m - 1$: the sizes of the components after deleting the root determine the partition of $m - 1$.

10.7.6(c). $\prod_{i=1}^{m} 1/(1 - x^i)^{p_i}$.

10.7.6(d). Proceeding as in the proof of Theorem 10.7.2, one gets $\ln r_n \leq -n \ln x + \sum_{j=1}^{\infty} P_n(x^j)/j$. In the text, it was shown that $\ln P_n(x) \leq Cx/(1-x)$. Choose $x = 1 - \frac{c}{\ln n}$ for a small enough $c > 0$. Then $-n \ln x$ is about $\frac{n}{c \ln n}$, and it still requires some ingenuity to show that the sum $\sum_{j=1}^{\infty} e^{Cx^j/(1-x^j)}/j$ has about the same order as its first term (which is about $n^{c \cdot C}$).

10.7.6(e). Consider trees of a special form, where the root is attached to k subtrees with q leaves each. Their number is at least $p_q^k/k!$. Use the estimate for p_n derived in the text, and adjust k and q suitably.

11.1.1(b). This is essentially a problem dual to Exercise 8.1.8.

11.2.2. Each row of the matrix AB is a linear combination of rows of the matrix B (with coefficients given by the corresponding row of A), and hence $r(AB) \leq r(B)$.

11.2.3. The nonsingularity of a square matrix is equivalent to its determinant being nonzero, and the definition of determinant is independent of the field. The rank of a nonsquare matrix can be expressed as the size of the largest nonsingular square submatrix.

11.2.4(a). If it had rank $< n$ then the system of linear equations $Mx = 0$ has a nonzero solution, and such a solution gives $x^T M x = 0$.

11.2.4(b). M is a sum of a diagonal matrix D with positive elements on the diagonal and of a matrix L with all entries $\lambda > 0$. For each nonzero $x \in \mathbf{R}^v$ we have $x^T D x > 0$ and $x^T L x \geq 0$.

11.2.6(a). If A is the incidence matrix of the set system, them $A^T A$ is a sum of a matrix Q with all entries equal to q and of a diagonal matrix D whose diagonal entries are $|C_i| - q > 0$ (assuming $|C_i| > q$). Hence $A^T A$ is positive definite and consequently nonsingular.

11.2.6(b). In the situation of Fisher's inequality, consider the set system dual to $(V, \mathcal{B})$ and apply (a) to it.

11.3.2(b). By induction on k, prove that if E is a union of edge sets of k bipartite graphs on the vertex set $\{1, 2, \ldots, n\}$, then there exists a set of at least $\lceil n/2^k \rceil$ vertices with no edge of E on it.

11.3.4. Let $A = (a_{ij})$ be the $n \times n$ matrix with $a_{ij} = 0$ or 1 depending on whether team i won with team j. The condition translates to $Ax \neq 0$ for all nonzero vectors x (arithmetic over $GF(2)$); hence A is nonsingular. For n odd, the nonzero terms in the expansion of $\det A$ as a sum over all permutations can be paired up and hence $\det A = 0$.

11.5.2(b). False.

11.6.2. If $r(x)$ has degree d, calculate $r(z) - p(z)q(z)$ for z chosen uniformly at random in the set $\{1, 2, \ldots, 2d\}$, say. If the result is 0 then $r(x) - p(x)q(x)$ is identically 0 with probability at least $\frac{1}{2}$.

11.6.4. Suppose that $0 \in S$. Set $a \circ b = 0$ for all $a, b \in S$ with a single exception: $x \circ y = x$ for some $x, y \in S \setminus \{0\}$.

11.6.7(a). Show that if all triples (a, b, c) with $b \in G^{(k)}$ are associative then also all triples with $b \in G^{(k+1)}$ are associative.

11.6.7(c). Start with $G_1 = \{g_1\}$ for an arbitrary $g_1 \in S$, maintain $\langle G_k \rangle$, and set $G_{k+1} = G_k \cup \{g_{k+1}\}$ for some $g_{k+1} \in S \setminus \langle G_k \rangle$. By (b), the size of $\langle G_k \rangle$ doubles in each step. Showing that $\langle G_k \rangle$ can be maintained in the claimed time still requires some ingenuity.

Index

$\binom{V}{2}$, 98

$\binom{X}{k}$, 56(2.3.1)

$\binom{n}{k}$, 55

$\binom{n}{k_1,\ldots,k_m}$, 61

$\prec$, 41

$\preceq$, 41

$\sum$, 8, 12

$\prod$, 9

$a \mid b$, 42(1.7.3)

(a, b), 8

$[a, b]$, 8

$\lfloor x \rfloor$, 8

$\lceil x \rceil$, 8

$\{x, y\}$, 10

(x, y), 10

$\emptyset$, 11

2^X, 12

$\subseteq$, 13

$\subset$, 13

$|X|$, 12, 32(Ex. 7)

X^2, 15

$X \dot\cup Y$, 13

$X \times Y$, 14

$\{\ldots\}$, 10

$R[x]$, 38

$R \circ S$, 35

xRy, 33

$f(X)$, 27

$f(x)$, 26

f^{-1}, 30

$f : X \to Y$, 26

$f : X \hookrightarrow Y$, 29

$f : x \mapsto y$, 26

$f \sim g$, 72

A^T, 364

$G + \bar{e}$, 133(3.7.2)

$G \% e$, 133(3.7.2)

$G - e$, 132(3.7.2)

$G - v$, 133(3.7.2)

$G.e$, 197(3.7.2)

$G \cong H$, 101(3.1.2)

$\overline{ab}$, 241

$\Omega(.)$, 72

$\Theta(.)$, 72

$\alpha(n)$, 154

$\alpha(G)$, 286(9.4.2)

$\chi(.)$, 192(5.4.2)

δp, 350

$\delta(.)$, 199(Ex. 2)

π, 79

    computation, 80(Ex. 8)

$\pi(n)$, 85(2.6.3)

A_G, 109(3.2.3)

acyclic relation, 45(Ex. 4)

adjacency matrix, 34, 109(3.2.3)

adjacent vertices, 98

affine plane, 249(Ex. 10), 338,
        338(Ex. 2)

algebra, linear (application),
        231–238, 251–253,
        333–362

algorithm

    Borůvka's, 163(4.5.3)

    greedy, 157(4.4.2), 159,
        161(Ex. 10),
        161(Ex. 11), 161(Ex. 12)

    Jarník's, 162(4.5.1)

    Kruskal's, 157(4.4.2)

    Prim's, see Jarník's algorithm

    QUICKSORT, 290–292

    randomized, 354

    sorting, 55(Ex. 6), 290–292

antichain, 211, 217(Ex. 5)
antisymmetric relation, 40
arc, 167
arc-connected set, 169
arithmetic mean, 75
associative (operation), 13, 366
asymmetric
    graph, 104(Ex. 2)
    tree, 150(Ex. 1)
asymptotic analysis, 69
automorphism
    of a graph, 104(Ex. 2)
    of a poset, 215, 217(Ex. 5)

band, Möbius, 170
basis, 368
Bell number, 95(Ex. 8),
        309(Ex. 15)
Bernoulli's inequality, 80
Bertrand postulate, 85
Betti number, see cyclomatic
        number
bijection, 28(1.4.3)
binary operation, 357, 366
binary tree, 317–319
Binet–Cauchy theorem,
        234(7.5.4)
binomial coefficient, 55–66,
        296–297, 298(Ex. 6)
    estimate, 81–86
    generalized, 301(10.2.3)
binomial theorem, 59(2.3.3)
    combinatorial meaning, 296
    generalized, 301(10.2.3), 319
bipartite graph, 101, 111(Ex. 4),
        239(Ex. 5)
    complete, 101
block design, 333–341
Bonferroni inequality, 90
Boolean function, 263,
        268(Ex. 1)
Borůvka's algorithm, 163(4.5.3)
Borsuk–Ulam theorem, 207
bottle, Klein, 170
bounded face, 169

bridge, 123
Brouwer's theorem, 205(6.1.3),
        210(Ex. 5)

$\mathcal{C}$, 350
$C(G)$, 148
C_n, 100
$\mathcal{C}_n$, 271(9.2.2)
carrier, 350
Cartesian product, 14
Catalan number, 319–320
Cauchy–Schwarz
    inequality, 220(6.3.2),
        221(Ex. 4)
Cayley's formula, 223(7.1.1)
center of a graph, 148
centroid, 150(Ex. 7)
chain, 45(Ex. 5), 212
    symmetric, 213
characteristic function, 50
characteristic polynomial, 313
chromatic number, 192(5.4.2),
        199(Ex. 2)
    list, 201(Ex. 12)
circuit, see cycle
circulation, 349
    space, 350
closure, transitive, 39(Ex. 3)
code
    of a tree, 145
    Prüfer, 229
coefficient
    binomial, 55–66, 296–297,
        298(Ex. 6)
        generalized, 301(10.2.3)
    multinomial, 61
coloring
    of a graph, 192(5.4.2)
    of a map, 191–201
commutative (operation), 13,
        366
compactness, 207
complement, 111(Ex. 1)
complete
    bipartite graph, 101

graph, 100
complexity (of algorithm), 152
component, 107
composition
    of functions, 27
    of relations, 35
configuration, tactical, 336
connected
    graph, 107
    set, 169
connectedness
    strong, 128(3.6.2)
    weak, 128(3.6.2)
connectivity, 132
contraction, 197
convex
    body, 182
    function, 221(Ex. 5)
cover, edge, 161(Ex. 11)
critical 2-connected graph,
        137(Ex. 2)
cube (graph), 131, 239(Ex. 2)
curve, Jordan, 175
cut, 351
    space, 351
cycle, 100
    elementary, 347
    Hamiltonian, 122(Ex. 8),
        155(Ex. 3)
    in a graph, 107
    of a permutation, 53, 284
    space, 346
cyclomatic number, 348

$d_G(.,.)$, 108
De Bruijn graph, 130
de Moivre's theorem, 23(Ex. 4),
        330
de Morgan laws, 14
decreasing subsequence,
        218(Ex. 9)
$\deg_G(.)$, 112
$\deg_G^+(.)$, 128
$\deg_G^-(.)$, 128
degree (of a vertex), 112

degree sequence, see score
dependence, linear, 368
depth-first search, 108
derangement, 92
design, block, 333–341
determinant, 364
    expansion, 365
diagonal, 364
    matrix, 364
    relation, 40
diagram
    Ferrers, 327
    Hasse, 44
diameter, 111(Ex. 8)
difference, symmetric, 346
digraph, see directed graph
Dilworth's theorem, 218(Ex. 8)
dimension, 368
directed
    cycle, 128
    edge, 127(3.6.1)
    graph, 127(3.6.1)
    tour, 128
distance (in a graph), 108
distributive (operation), 13, 367
dominating set, 161(Ex. 12)
double-counting, 56, 202–222,
        249(Ex. 8), 336
drawing of a graph, 168(5.1.1)
dual
    graph, 193(5.4.3)
        spanning trees, 224(Ex. 4)
    projective plane, 246
duality, 246, 253

$\mathcal{E}$, 346
$\mathbf{E}$, 280(9.3.6)
e, 76
$E(G)$, 98
ear decomposition, 136
edge, 97(3.1.1)
    connectivity, 132
    contraction, 197
    cover, 161(Ex. 11)
    directed, 127(3.6.1)

multiple, 120
  subdivision, 133(3.7.2)
  weight, 156
element
  largest, 45(Ex. 7)
  maximal, 45(Ex. 7)
  minimal, 45(Ex. 7)
  smallest, 45(Ex. 7)
elementary
  cycle, 347
  event, 269
  row operation, 365
empty
  product, 12
  set, 11
  sum, 12
end-vertex, 141
equivalence, 37(1.6.2)
  maintaining, 153(4.3.4),
      155(Ex. 1)
  number of, 95(Ex. 8)
Erdős–Szekeres lemma,
      218(Ex. 9)
estimate
  of binomial coefficient, 81–86
  of factorial, 73–79, 80(Ex. 9)
Euler
  formula, 181(5.3.1)
    for trees, 140(4.1.2)
  function, 93–94, 95(Ex. 9)
  number, 76
Eulerian
  graph, 117–132, 174(Ex. 3)
  tour, 117
even set, 345(11.4.1)
  number of, 348(11.4.4)
event, 269
  elementary, 269
events, independent, 275–276
excentricity, 148
ex$_G$(.), 148
expansion of determinant, 365
expectation, 279–293, 322
  definition, 280(9.3.6)
  linearity, 282(9.3.9)

exponential generating function,
      308(Ex. 14)
extension, linear, 46(Ex. 8),
      65(Ex. 27)
extremal graph theory, 286

face (of a planar graph), 169
face (of a polytope), 183
factorial, 54
  divisibility, 55(Ex. 7)
  estimate, 73–79, 80(Ex. 9)
family of sets, 11
Fano plane, 242, 258(Ex. 1)
father (in rooted tree), 145
Ferrers diagram, 327
Fibonacci number, 309–312
field, 366
finite probability space,
      269(9.2.1)
finite projective plane, 240–261
  definition, 240(8.1.1),
      249(Ex. 4), 249(Ex. 8),
      338(Ex. 1)
  existence, 250–251
  order, 245(8.1.4)
Fisher's inequality, 339(11.2.1),
      342(Ex. 6)
fixed point, 92, 284(Ex. 3)
  theorem, 204(6.1.2), 205(6.1.3)
forest, 151
  spanning, 346
formula
  Cayley's, 223(7.1.1)
  Euler's, 181(5.3.1)
    for trees, 140(4.1.2)
  Heawood's, 199
  Leibniz, 63(Ex. 13)
  logical, 264, 268(Ex. 1)
  Stirling's, 79
fractions, partial, 310
Freivalds' checker, 354(11.6.1)
function, 26(1.4.1)
  bijective, 28(1.4.3)
  Boolean, 263, 268(Ex. 1)
  characteristic, 50

convex, 221(Ex. 5)
Euler, 93–94, 95(Ex. 9)
generating, 294–332
   exponential, 308(Ex. 14)
   of a sequence, 300(10.2.2)
   operations, 301–304
graph, 228, 228(Ex. 1),
     229(Ex. 3)
identity, 31(Ex. 4)
injective, 29
monotone, 62(Ex. 7)
number of, 48(2.1.1)
one-to-one, 28(1.4.3)
   number of, 51(2.1.4)
onto, 28(1.4.3)
   number of, 95(Ex. 7)
period, 228(Ex. 3)
surjective, 29

$GF(q)$, 367
$GF(2)$, 346
$\mathcal{G}_n$, 273(9.2.4)
$\gcd(m, n)$, 93
generalized binomial theorem,
     301(10.2.3), 319
generating function, 294–332
   exponential, 308(Ex. 14)
   of a sequence, 300(10.2.2)
   operations, 301–304
genus, 173(5.1.3)
geometric mean, 75
golden section, 311, 325
Graham–Pollak theorem,
     343(11.3.1)
graph, 97(3.1.1)
   asymmetric, 104(Ex. 2)
   bipartite, 101, 111(Ex. 4),
     239(Ex. 5)
     complete, 101
   chromatic number, 192(5.4.2),
     199(Ex. 2)
   coloring, 192(5.4.2)
   complete, 100
   connected, 107
   De Bruijn, 130

diameter, 111(Ex. 8)
directed, 127(3.6.1)
drawing, 98, 168(5.1.1)
dual, 193(5.4.3)
   spanning trees, 224(Ex. 4)
k-edge-connected, 132
Eulerian, 117–132, 174(Ex. 3)
Heawood, 117(Ex. 15), 247
isomorphism, 101(3.1.2)
Kneser, 104(Ex. 1)
line, 122(Ex. 9)
list chromatic number,
     201(Ex. 12)
metric, 108
number of, 96(Ex. 13), 102
of a function, 228, 228(Ex. 1),
     229(Ex. 3)
of incidence, 246, 260
orientation, 232
oriented, 127
outerplanar, 199(Ex. 3)
Petersen, 104(Ex. 1)
planar, 167–201
   maximal, 185(5.3.3)
   number of edges, 185(5.3.3)
   score, 188(5.3.4)
radius, 111(Ex. 8)
random, 273(9.2.4),
     278(Ex. 4), 278(Ex. 3)
randomly Eulerian, 126(Ex. 2)
regular, 117(Ex. 12)
strongly connected, 128(3.6.2)
topological, 168(5.1.1)
tough, 137(Ex. 6)
triangle-free, 292(Ex. 1)
2-connected, 132–137
   critical, 137(Ex. 2)
k-vertex-connected, 132
weakly connected, 128(3.6.2)
with loops, 120
with multiple edges, 120
without K_k, 286
without $K_{2,2}$, 219, 260
without $K_{2,t}$, 221(Ex. 1)
without $K_{3,3}$, 222(Ex. 6)

greedy algorithm, 157(4.4.2),
    159, 161(Ex. 10),
    161(Ex. 11), 161(Ex. 12)
Grötsch theorem, 199(Ex. 4)
group, 366

Hamiltonian
  cycle, 122(Ex. 8), 155(Ex. 3)
  path, 131(Ex. 8)
handshake lemma, 113
  applications, 202–209
harmonic
  mean, 80(Ex. 6)
  number, 67(2.4.1), 81(Ex. 13)
Hasse diagram, 44
hatcheck lady problem,
    91(2.8.1), 309(Ex. 17)
  recurrence, 95(Ex. 4),
    95(Ex. 5)
Heawood
  formula, 199
  graph, 117(Ex. 15), 247
hydrocarbons, number of,
    321(Ex. 12)
hypergraph, 334
hypothesis, inductive, 17

I_n, 364
identity
  function, 31(Ex. 4)
  matrix, 364
image, 26
incidence
  graph, 246, 260
  matrix, 232, 239(Ex. 5), 339,
    351
inclusion–exclusion, 86–91,
    278(Ex. 2)
  applications, 91–96
increasing
  segment of a permutation,
    54(Ex. 5)
  subsequence, 218(Ex. 9)
indegree, 128
independence, linear, 368

independent
  events, 275–276
  set, 286
  set system, 211
indicator, 281(9.3.7)
induced subgraph, 106(3.2.1)
induction, 16
inductive
  hypothesis, 17
  step, 17
inequality
  Bernoulli's, 80
  Bonferroni, 90
  Cauchy–Schwarz, 220(6.3.2),
    221(Ex. 4)
  Fisher's, 339(11.2.1),
    342(Ex. 6)
  Jensen's, 221(Ex. 5)
  LYM, 213
  Markov, 284(Ex. 7)
inf A, 46(Ex. 11)
infimum, 46(Ex. 11)
injection, 29
inner face, 169
integers, 7
integrality conditions,
    336(11.1.3)
intersection of level k, 287–290,
    293(Ex. 9)
inversion (of a permutation),
    55(Ex. 6)
isomorphism
  of graphs, 101(3.1.2)
  of posets, 46(Ex. 9), 215
  of trees, 144–150
isthmus, see bridge

J_n, 344
Jarník's algorithm, 162(4.5.1)
Jensen's inequality, 221(Ex. 5)
Jordan curve, 175
  theorem, 175(5.2.1)
Jordan–Schönflies theorem, 176

$\mathcal{K}_G$, 345

K_n, 100
$K_{n,m}$, 101
kernel, 369
Klein bottle, 170
Kneser graph, 104(Ex. 1)
Kruskal's algorithm, 157(4.4.2)
Kuratowski's theorem, 179(5.2.4)

Laplace matrix, 231, 238(Ex. 1)
largest element, 45(Ex. 7)
Latin
 rectangle, 258(Ex. 6)
 square, 255–258
 squares, orthogonal, 255
leaf, see end-vertex
left maximum, 279(9.3.4), 283
Leibniz formula, 63(Ex. 13)
lemma
 Erdős–Szekeres, 218(Ex. 9)
 Sperner's, 203(6.1.1),
 210(Ex. 5)
lexicographic ordering, 148
line
 at infinity, 241
 of a projective plane,
 241(8.1.1)
line graph, 122(Ex. 9)
linear
 algebra (application),
 231–238, 251–253,
 317(Ex. 16), 333–362
 extension, 46(Ex. 8),
 65(Ex. 27)
 mapping, 369
 ordering, 41
 span, 368
linearity of expectation,
 282(9.3.9)
linearly dependent (set), 368
list chromatic number,
 201(Ex. 12)
Littlewood–Offord problem,
 217(Ex. 6)
logical formula, 264, 268(Ex. 1)
loop, 120

LYM inequality, 213

Mader's theorem, 137(Ex. 7)
maintaining an equivalence,
 153(4.3.4), 155(Ex. 1)
map, see function
map (coloring), 191–201
mapping, see function
 linear, 369
Markov inequality, 284(Ex. 7)
matching, 161(Ex. 10)
mathematical induction, 16
matrix, 363
 diagonal, 364
 identity, 364
 incidence, 232, 239(Ex. 5),
 339, 351
 Laplace, 231, 238(Ex. 1)
 multiplication, 110(3.2.4), 363
 checking, 354(11.6.1)
 nonsingular, 365
 permutation, 112(Ex. 12)
 positive definite, 341(Ex. 4)
 rank, 341, 365
 totally unimodular, 239(Ex. 5)
 transposed, 364
matroid, 160, 353
maximal element, 45(Ex. 7)
maximum spanning tree,
 160(Ex. 1)
maximum, left, 279(9.3.4), 283
mean
 arithmetic, 75
 geometric, 75
 harmonic, 80(Ex. 6)
Menger's theorem, 134
metric, 108, 111(Ex. 7)
metric space, 108
minimal element, 45(Ex. 7)
minimum spanning tree, 155–166
minor, 180, 200(Ex. 11)
Möbius band, 170
monotone function (number of),
 62(Ex. 7)
multigraph, 120

multinomial
  coefficient, 61
  theorem, 61(2.3.5), 226,
    298(Ex. 4)
multiple edges, 120

N, 7
natural numbers, 7
neighbor, 98
network, 156
node (of a graph), see vertex
nonsingular matrix, 365
number
  Bell, 95(Ex. 8), 309(Ex. 15)
  Betti, see cyclomatic number
  Catalan, 319–320
  chromatic, 192(5.4.2)
    list, 201(Ex. 12)
  cyclomatic, 348
  Euler, 76
  Fibonacci, 309–312
  harmonic, 67(2.4.1),
    81(Ex. 13)
  integer, 7
  natural, 7
  perfect, 96(Ex. 11)
  rational, 7
  real, 7
number of
  alkane radicals, 321(Ex. 12)
  arrangements, 61
  ball distributions, 57,
    64(Ex. 18)
  binary rooted trees,
    321(Ex. 11)
  binary trees, 317–319
  divisors, 96(Ex. 11)
  edges of a planar graph,
    185(5.3.3)
  equivalences, 95(Ex. 8)
  even sets, 348(11.4.4)
  functions, 48(2.1.1)
  functions onto, 95(Ex. 7)
  graphs, 96(Ex. 13), 102
    nonisomorphic, 103

Latin rectangles, 258(Ex. 6)
  monotone functions, 62(Ex. 7)
  one-to-one functions, 51(2.1.4)
  ordered k-tuples, 63(Ex. 17)
  partitions of n, 326–332
  planted trees, 320(Ex. 8),
    321(Ex. 9)
  solutions, 57, 295
  spanning trees, see number of
    trees
    for general graph, 232(7.5.1)
  subsets, 49(2.1.2), 56(2.3.2),
    63(Ex. 16)
    odd-size, 50(2.1.3), 59
  trees, 223–239
    nonisomorphic, 150(Ex. 6),
      224(Ex. 1), 332(Ex. 6)
    with given score, 224(7.2.1)
  triangulations (of a polygon),
    65(Ex. 24), 320(Ex. 6)
  unordered k-tuples, 63(Ex. 17)

$O(.)$, 69(2.4.2)
$o(.)$, 72
one-to-one function, 28(1.4.3)
  number of, 51(2.1.4)
operation, binary, 357, 366
order
  of a Latin square, 255
  of a permutation, 54(Ex. 3)
  of a projective plane,
    245(8.1.4)
ordered
  pair, 10
  set, 40–47
ordering, 40(1.7.1)
  lexicographic, 148
  linear, 41
  partial, 41
orientation, 131(Ex. 4), 232
oriented graph, 127
orthogonal
  Latin squares, 255
  vectors, 364
outdegree, 128

outer face, 169
outerplanar graph, 199(Ex. 3)

$P(.)$, 269(9.2.1)
p_n, 326
P_n, 100
$\mathcal{P}(X)$, 12
pair
  ordered, 10
  unordered, 10
partial fractions, 310
partial ordering, 41
partition
  of n, 326–332
    ordered, 326, 331(Ex. 1)
Pascal triangle, 58
path, 100, 106
  Hamiltonian, 131(Ex. 8)
  uniqueness, 139(4.1.2)
perfect number, 96(Ex. 11)
period of a function, 228(Ex. 3)
permutation, 52–55
  cycle, 53, 284
  fixed point, 92, 284(Ex. 3),
    309(Ex. 17)
  increasing segment, 54(Ex. 5)
  inversion, 55(Ex. 6)
  left maximum, 279(9.3.4), 283
  matrix, 112(Ex. 12)
  order, 54(Ex. 3)
  random, 55, 92, 271(9.2.3),
    276, 284, 286
  sign, 365
Petersen graph, 104(Ex. 1)
planar drawing, 168(5.1.1)
planar graph, 167–201
  maximal, 185(5.3.3)
  number of edges, 185(5.3.3)
  score, 188(5.3.4)
plane
  affine, 249(Ex. 10), 338,
    338(Ex. 2)
  Fano, 242
  graph, 168
  projective, see projective plane

planted tree, 145
Platonic solids, 182
point
  at infinity, 241, 254
  of a projective plane,
    241(8.1.1)
polynomial, characteristic, 313
polytope, regular, 182
poset, 41, 211
  automorphism, 215, 217(Ex. 5)
  isomorphism, 46(Ex. 9), 215
positive definite matrix,
    341(Ex. 4)
postulate, Bertrand, 85
potential, 131(Ex. 5), 350
  difference, 351
power series, 298–300,
    308(Ex. 13)
power set, 12
Prim's algorithm, see Jarník's
    algorithm
prime number theorem, 85(2.6.3)
principle, inclusion–exclusion,
    86–91, 278(Ex. 2)
  applications, 91–96
probability, 55(Ex. 5), 75, 91,
    96(Ex. 12), 262–293,
    322–326
  space, finite, 269(9.2.1)
  space, infinite, 270–271
problem
  four-color, 191
  hatcheck lady, 91(2.8.1),
    309(Ex. 17)
    recurrence, 95(Ex. 4),
      95(Ex. 5)
  Littlewood–Offord, 217(Ex. 6)
  maximum spanning tree,
    160(Ex. 1)
  minimum spanning tree,
    157(4.4.1)
  Sylvester's, 190(Ex. 8)
product, 9
  Cartesian, 14
  empty, 12

scalar, 364
projection, stereographic, 173
projective plane
    construction, 251–253,
        256–257
    duality, 246, 253
    finite, 240–261
        definition, 240(8.1.1),
            249(Ex. 4), 249(Ex. 8),
            338(Ex. 1)
        existence, 250–251
        order, 245(8.1.4)
    real, 241, 251
property B, 259
Prüfer code, 229

QUICKSORT, 290–292

R, 7
$\mathcal{R}$, 351
$r(A)$, 365
radius, 111(Ex. 8)
random
    graph, 273(9.2.4), 278(Ex. 4),
        278(Ex. 3)
    permutation, 55, 92,
        271(9.2.3), 276, 284,
        286
    variable, 279(9.3.1)
    walk, 323–326
randomized algorithm, 354
randomly Eulerian graph,
    126(Ex. 2)
rank (of a matrix), 341, 365
rationals, 7
real projective plane, 241, 251
reals, 7
rectangle, Latin, 258(Ex. 6)
recurrence, 312–315
recurrent relation, see recurrence
reflexive relation, 37(1.6.1)
region (of a Jordan curve),
    175(5.2.1)
regular
    graph, 117(Ex. 12)

polytope, 182
relation, 32(1.5.1)
    acyclic, 45(Ex. 4)
    antisymmetric, 40
    composition, 35
    reflexive, 37(1.6.1)
    symmetric, 37(1.6.1)
    transitive, 37(1.6.1), 55(Ex. 6)
root of a tree, 145
row operation, elementary, 365

S_n, 92
$\mathcal{S}_n$, 271(9.2.3)
scalar product, 364
score
    of a graph, 112–116
    of a planar graph, 188(5.3.4)
    of a tree, 144(Ex. 8)
search, depth-first, 108
section, golden, 311, 325
series, power, 298–300,
        308(Ex. 13)
set
    connected, 169
    dominating, 161(Ex. 12)
    empty, 11
    independent, 286
    ordered, 40–47
    partially ordered, 41
set system, 11
    independent, 211
    2-colorable, 259, 265–268,
        268(Ex. 2)
$\mathrm{sgn}(\pi)$, 365
sign of a permutation, 365
smallest element, 45(Ex. 7)
solid, Platonic, 182
son (in rooted tree), 145
sorting algorithm, 55(Ex. 6),
        290–292
space
    metric, 108
    of circulations, 350
    of cuts, 351
    of cycles, 346(11.4.3)

probability
  finite, 269(9.2.1)
  infinite, 270–271
  vector, 367
span, linear, 368
spanning forest, 346
spanning tree, 151–155
  algorithm, 151(4.3.2),
      154(4.3.5)
  maximum, 160(Ex. 1)
  minimum, 155–166
Sperner's lemma, 203(6.1.1),
      210(Ex. 5)
Sperner's theorem, 211(6.2.1)
sphere with handles, 170
square matrix, 364
square, Latin, 255–258
Steiner system, 335(11.1.2),
      337(11.1.4), 338(Ex. 4)
Steiner tree, 156
Steinitz theorem, 185
step, inductive, 17
stereographic projection, 173
Stirling's formula, 79
strongly connected graph,
      128(3.6.2)
subdivision (of a graph),
      133(3.7.2)
subfield, 367
subgraph, 106(3.2.1)
  induced, 106(3.2.1)
submatrix, 365
subsequence
  decreasing, 218(Ex. 9)
  increasing, 218(Ex. 9)
subsets
  number of, 49(2.1.2),
      56(2.3.2), 63(Ex. 16)
subspace, 368
sum, 8, 12
  empty, 12
sup *A*, 46(Ex. 11)
supremum, 46(Ex. 11)
surjection, 29
Sylvester's problem, 190(Ex. 8)

sym(.), 128
symmetric
  chain, 213
  difference, 346
  relation, 37(1.6.1)
symmetrization, 128
system of sets, 11
system, Steiner, 335(11.1.2),
      337(11.1.4), 338(Ex. 4)

T(.), 223
tactical configuration, 336
theorem
  Binet–Cauchy, 234(7.5.4)
  binomial, 59(2.3.3)
    combinatorial meaning, 296
    generalized, 301(10.2.3), 319
  Borsuk–Ulam, 207
  Brouwer's, 205(6.1.3),
      210(Ex. 5)
  de Moivre's, 23(Ex. 4), 330
  Dilworth's, 218(Ex. 8)
  fixed point, 204(6.1.2),
      205(6.1.3)
  Grötsch, 199(Ex. 4)
  Graham–Pollak, 343(11.3.1)
  Jordan curve, 175(5.2.1)
  Jordan–Schönflies, 176
  Kuratowski's, 179(5.2.4)
  Mader's, 137(Ex. 7)
  Menger's, 134
  multinomial, 61(2.3.5), 226,
      298(Ex. 4)
  on score, 114(3.3.3)
  prime number, 85(2.6.3)
  Sperner's, 211(6.2.1)
  Steinitz, 185
  Turán's, 286(9.4.2), 293(Ex. 4)
  Wilson's, 338(11.1.5)
time complexity, 152
topological graph, 168(5.1.1)
torus, 169
totally unimodular matrix,
      239(Ex. 5)
tough graph, 137(Ex. 6)

tour, 118
  directed, 128
  Eulerian, 117
tournament, 131(Ex. 8), 277
transitive
  closure, 39(Ex. 3)
  relation, 37(1.6.1), 55(Ex. 6)
transposed matrix, 364
tree, 139(4.1.1)
  asymmetric, 150(Ex. 1)
  binary, 317–319
  code, 145
  planted, 145
  rooted, 145
  spanning, 151–155
    algorithm, 151(4.3.2)
    minimum, 155–166
  Steiner, 156
trees, number of, 150(Ex. 6),
        223–239, 317–319,
        320(Ex. 8), 321(Ex. 9),
        321(Ex. 11), 332(Ex. 6)
triangle-free graph, 286
triangular matrix, 364
triangulation, 185, 185(5.3.3)
  of a polygon, 65(Ex. 24),
        320(Ex. 6)
Turán's theorem, 286(9.4.2),
        293(Ex. 4)

2-coloring, 259, 265–268,
        268(Ex. 2)
2-connected graph, 132–137
  critical, 137(Ex. 2)

unbounded face, 169
uniform (set system), 334
UNION–FIND, 153(4.3.4),
        155(Ex. 1)
unordered pair, 10
upper triangular matrix, 364

$V(G)$, 98
variable, random, 279(9.3.1)
variance, 323(Ex. 1)
variations, 51
vector space, 367
vectors, orthogonal, 364
vertex, 97(3.1.1)
  connectivity, 132

walk, 107
  random, 323–326
weakly connected graph,
        128(3.6.2)
well-ordering, 17
Wilson's theorem, 338(11.1.5)

$\mathbf{Z}$, 7